Cyril Tintarev

Concentration Compactness

De Gruyter Series in Nonlinear Analysis and Applications

Volume 33

Cyril Tintarev

Concentration Compactness

Functional-Analytic Theory of Concentration Phenomena

DE GRUYTER

Mathematics Subject Classification 2010
Primary: 46B50, 35J60, 46N20; Secondary: 46E35, 35A15

Author
Cyril Tintarev
Technion
Department of Mathematics
32000 Haifa
Israel
tammouz@gmail.com

ISBN 978-3-11-053034-6
e-ISBN (PDF) 978-3-11-053243-2
e-ISBN (EPUB) 978-3-11-053058-2
ISSN 0941-813X

Library of Congress Control Number: 2019952753

Bibliographic information published by the Deutsche Nationalbibliothek
The Deutsche Nationalbibliothek lists this publication in the Deutsche Nationalbibliografie; detailed bibliographic data are available on the Internet at http://dnb.dnb.de.

Typesetting: VTeX UAB, Lithuania
Printing and binding: CPI books GmbH, Leck

www.degruyter.com

To Sonia

Preface

The subject of this book is convergence of sequences in Banach spaces without a given compact embedding, or more specifically, structural representation of such sequences, known in applications as concentration compactness, addressed on the functional-analytic level.

Concentration compactness became a standard tool of analysis of partial differential equations since the publication of celebrated papers [83, 84] by P.-L. Lions, followed by the profile decomposition approach introduced by Struwe [119], generalized to general sequences in Sobolev spaces by Solimini [112], and further generalized to sequences in Hilbert and Banach spaces, respectively, in [104] and [113].

This book is a sequel to an earlier monograph [127], whose purpose was to give a functional-analytic theory of concentration compactness in general Hilbert spaces, and to illustrate this abstract approach by applications to calculus of variations, mostly in the settings of Lions. In the present book, the focus is shifted from sampling the known applications to a broader presentation of the method, based on the current state of art. The book extends analysis of concentration from Hilbert to Banach spaces, and presents realizations of concentration compactness in a variety of functional spaces, while [127] dealt only with Sobolev spaces. Now into consideration come Besov and Triebel–Lizorkin spaces, embeddings into spaces of continuous functions, embeddings associated with the Moser–Trudinger inequality, Strichartz embedding for the nonlinear Schrödinger equation, and the affine Sobolev inequality. The book also extends the notion of profile decomposition to functional spaces that do not have a nontrivial group of invariance.

Central to this book is the notion of *cocompact* embedding, which in [127] appears only implicitly. Cocompactness of an embedding of two Banach spaces is a property similar to but weaker than compactness, and it plays central role in having well-structured profile decompositions for bounded sequences – sum of asymptotically decoupled "blowups."

Chapter 1 gives a brief introduction to the basic notions of the theory and examples of an "orderly loss" of compactness (profile decomposition) in presence of cocompact embeddings. Chapter 2 contains technical preliminaries concerning Delta-convergence, a less-known cousin of weak convergence, involved in the profile decomposition for Banach spaces, which are considered in Chapter 4 together with its realization in Sobolev and other scale-invariant function spaces. Chapter 3 sums up known results on cocompactness relative to the rescaling group (actions of translations and dilations), in Besov and Triebel–Lizorkin spaces (with Sobolev and fractional Sobolev spaces as a particular case), as well as cocompactness of an embedding of the Moser–Trudinger-type relative to a different group of logarithmic dilations.

Chapters 5 through 9 can be read independently one of the other. Chapter 5 presents further cocompact embeddings and profile decompositions. Chapter 6 dis-

https://doi.org/10.1515/9783110532432-201

cusses defect of compactness for sequences restricted to different subspaces. Chapters 7 and 8 deal with profile decompositions that do not follow from the general framework of Chapter 4 – for nonreflexive spaces and for Sobolev spaces without invariance. Chapter 9 presents a small selection of applications of concentration methods to semilinear elliptic equations.

Corrections, supporting materials, etc. related to this book, will appear on the author's personal website, http://sites/google.com/site/tintarev.

The book was written in difficult circumstances, as since 2016 the author was subjected by his former employer to a complete travel ban (including host- and self-funded travel), together with further restrictions, which brought the author to leave his job at Uppsala University. The author expresses his warm gratitude to Academic Rights Watch and his colleagues and collaborators at Technion, University of Toulouse – La Capitole, Tata Institute for fundamental research, University of Bari and Politecnic University of Bari, for their unwavering support of his academic rights. He thanks Torbjörn Ohlsson, attorney at law, who negotiated author's continued access to the library resources of his former employer.

The work on this book was completed during the author's stay as Lady Davis Visiting Professor at Technion – Israel Institute of Technology.

Haifa, December 2019

Contents

1 Profile decomposition: a structured defect of compactness

Let E be a reflexive Banach space continuously embedded into another Banach space F. If $(u_k)_{k\in\mathbb{N}}$ is a sequence in E, then by the Banach–Alaoglu theorem, it has a (renamed) weakly convergent subsequence, $u_k \rightharpoonup u \in E$. If the embedding $E \hookrightarrow F$ is compact, this gives $u_k \to u$ in F. Otherwise, one regards the sequence $(u_k - u)_{k\in\mathbb{N}}$, taken up to a remainder vanishing in F, as a defect of compactness of the sequence $(u_k)_{k\in\mathbb{N}}$.

This book studies how the defect of compactness is structured. The famous series of four papers by Pierre-Louis Lions [82, 83] described defect of compactness for sequences of functions in Sobolev spaces in terms of concentration phenomena. This book studies *profile decompositions* which are a more detailed structure of the defect of compactness. They not only elaborate concentration in functional spaces, but occur in general Banach spaces as well. The more traditional approach to concentration, based on Lions' version, is outlined in the Appendix, Section 10.4.

In this chapter, we provide definitions, elementary examples, and some quantitative ramifications for this structure.

1.1 Cocompact embeddings: definition and examples

Cocompactness is a property of embedding of two Banach spaces which is similar to (but is generally weaker) than compactness. Cocompactness is defined via the notion of $\mathcal{G}$-weak convergence.

Definition 1.1.1 ($\mathcal{G}$-weak convergence). Let E be a Banach space and let $\mathcal{G}$ be a set of homeomorphisms $E \to E$. One says that a sequence $(u_n)_{n\in\mathbb{N}}$ in E is $\mathcal{G}$-weakly convergent to a point $u \in E$ relative to the set $\mathcal{G}$, if for any sequence $(g_n)_{n\in\mathbb{N}}$ in $\mathcal{G}$, $g_n(u_n - u)$ is weakly convergent to zero in E. In this case, we use the notation $u_n \overset{\mathcal{G}}{\rightharpoonup} u$.

Obviously, if $\mathcal{G} = \{\mathrm{id}\}$, then $\mathcal{G}$-weak convergence coincides with weak convergence. This is also the case if the set $\mathcal{G}$ is small enough, for example, if $\mathcal{G} = \{u \mapsto u \circ \eta\}_{\eta\in O(N)}$ on $L^2(\mathbb{R}^N)$, which is a particular case of the following.

Proposition 1.1.2. *Let $\mathcal{G}$ be a set of bounded linear operators in a reflexive Banach space E such that its set of adjoints $\mathcal{G}^* = \{g^* : g \in \mathcal{G}\}$ is sequentially compact with respect to the strong [i. e., pointwise] operator convergence, that is, any sequence $(g_k)_{k\in\mathbb{N}}$ in $\mathcal{G}$ has a subsequence (g_{k_j}) and there exists $g \in \mathcal{G}$ for which $g_{k_j}^* v \to g^* v$ for every $v \in E^*$. Then every weakly convergent sequence in E is $\mathcal{G}$-weakly convergent.*

Proof. Assume that $u_n \rightharpoonup u$, but for some sequence $(g_n)_{n\in\mathbb{N}}$ in $\mathcal{G}$ the sequence $g_n(u_n - u)$ is not convergent weakly to zero. By the uniform boundedness principle the set

https://doi.org/10.1515/9783110532432-001

$\mathcal{G}^*$ and the weakly convergent sequence $(u_n - u)_{n\in\mathbb{N}}$ are bounded, and since norms of an operator and of its adjoint coincide, the set $\mathcal{G}$ is bounded as well. Therefore, $(g_n(u_n - u))_{n\in\mathbb{N}}$ is a bounded sequence, and there exists $v \in E^*$ such that, on a renamed subsequence, $\langle v, g_n(u_n - u)\rangle \to 1$, and thus, $\langle g_n^* v, (u_n - u)\rangle \to 1$. By the compactness assumption of the proposition, a renamed subsequence of $(g_n^* v)_{n\in\mathbb{N}}$ converges in E^* to some point $w \in E^*$. This $\langle w, (u_n - u)\rangle \to 1$, which contradicts the assumption $u_n \rightharpoonup u$. □

Corollary 1.1.3. *Let E be a uniformly convex Banach space and let a set $\mathcal{G}$ of linear isometries on E be sequentially compact with respect to the strong operator convergence. Then if a sequence $(u_n)_{n\in\mathbb{N}}$ in E is weakly convergent, it is $\mathcal{G}$-weakly convergent.*

Proof. The assertion will follow from Proposition 1.1.2 once we show that the set of adjoints $\mathcal{G}^*$ is compact with respect to strong convergence. Indeed, if a sequence (g_k^*) in $\mathcal{G}^*$ converges strongly to g^*, then it converges weakly in E, and from the definition of weak convergence one has $g_k x \rightharpoonup gx$ for any $x \in E$. At the same time, since g_k^* are isometries, so are g^*, g_k, and g. In particular, $\lim \|g_k x\| = \|x\| = \|gx\|$. Then, by Proposition 10.1.5, $g_k x \to gx$ for any $x \in E$. □

Definition 1.1.4 (Cocompact embedding). One says that a continuous embedding $E \hookrightarrow F$ of two Banach spaces is cocompact relative to a set $\mathcal{G}$ of homeomorphisms $E \to E$ if for any sequence $(u_n)_{n\in\mathbb{N}}$ in E,

$$u_n \overset{\mathcal{G}}{\rightharpoonup} 0 \text{ in } E \Longrightarrow \|u_n\|_F \to 0.$$

Remark 1.1.5. Obviously, if $\mathcal{G} \subset \mathcal{G}'$, then a $\mathcal{G}$-cocompact embedding is also $\mathcal{G}'$-cocompact.

Example 1.1.6 (Stephane Jaffard, [68]). The embedding $\ell^p(\mathbb{Z}) \hookrightarrow \ell^r(\mathbb{Z})$, $1 \le p < r \le \infty$, is not compact, since any sequence of the form $u_n = u(\cdot + n)$, $n \in \mathbb{N}$, $u \in \ell^p$, converges to zero weakly in ℓ^p, while the ℓ^p-norm on the sequence is constant. On the other hand, this embedding is cocompact relative to the group $\mathcal{G}_{\mathbb{Z}} = \{u \mapsto u(\cdot - j)\}_{j\in\mathbb{Z}}$. Indeed, consider a sequence $(u_n)_{n\in\mathbb{N}}$ in ℓ^p that converges to zero $\mathcal{G}_{\mathbb{Z}}$-weakly, that is, such that $u_n(\cdot + j_n) \rightharpoonup 0$ in ℓ^p for any sequence $j_n \in \mathbb{Z}$. Then $u_n(j_n) \to 0$ in $\mathbb{R}$ for any sequence (j_n) in $\mathbb{Z}$, which implies $u_n \to 0$ in ℓ^∞. Since $\|u\|_r^r \le \|u\|_\infty^{r-p}\|u\|_p^p$, one has $u_n \to 0$ in ℓ^r for all $r > p$. We conclude that the embedding $\ell^p(\mathbb{Z}) \hookrightarrow \ell^r(\mathbb{Z})$, $1 \le p < r \le \infty$, is $\mathcal{G}_{\mathbb{Z}}$-cocompact. Furthermore, the same argument shows that ℓ^∞ is $\mathcal{G}_{\mathbb{Z}}$-cocompactly embedded into itself.

Example 1.1.7 (Cocompactness in the Strauss estimate). Let $\dot{H}^{1,2}(\mathbb{R}^N)$, $N > 2$, be the space of measurable functions whose weak derivative lies in $L^2(\mathbb{R}^N)$ and let $\dot{H}^{1,2}_{\mathrm{rad}}(\mathbb{R}^N)$ be its subspace of all radial functions. Let $C_{\mathrm{rad}}(\mathbb{R}^N, r^{\frac{N-2}{2}})$ be the space of radial continuous functions on $\mathbb{R}^N \setminus \{0\}$ with the norm $\|u\| = \sup_{r>0} r^{\frac{N-2}{2}} |u(r)|$. Then the continuous embedding $\dot{H}^{1,2}_{\mathrm{rad}}(\mathbb{R}^N) \hookrightarrow C_{\mathrm{rad}}(\mathbb{R}^N, r^{\frac{N-2}{2}})$ (see [115]) is cocompact relative to the group

$$\mathcal{G} = \{g_t : u \mapsto t^{\frac{N-2}{2}} u(t\cdot)\}_{t>0}. \tag{1.1}$$

Indeed, let $u_n \overset{\mathcal{G}}{\rightharpoonup} 0$ and assume that $r_n > 0$, $n \in \mathbb{N}$, are such that

$$r_n^{\frac{N-2}{2}} |u(r_n)| \geq \frac{1}{2} \sup_{r>0} r^{\frac{N-2}{2}} |u_n(r)|.$$

This can be rewritten as

$$\sup_{r>0} r^{\frac{N-2}{2}} |u_n(r)| \leq 2 |g_{r_n} u_n(1)|.$$

Since the map $u \mapsto u(1)$ is a continuous linear functional on $\dot{H}^{1,2}_{\mathrm{rad}}(\mathbb{R}^N)$, the right- hand side in the inequality above is going to zero, which implies that u_n vanishes in the norm of $C(\mathbb{R}^N, r^{\frac{N-2}{2}})$, and thus embedding $\dot{H}^{1,2}_{\mathrm{rad}}(\mathbb{R}^N) \hookrightarrow C(\mathbb{R}^N, r^{\frac{N-2}{2}})$ is cocompact.

Let us illustrate how cocompactness of an embedding allows to prove existence of minimizers in isoperimetric problems.

Example 1.1.8 (Minimizer in the Strauss estimate). Consider a minimizing problem for the embedding of Example 1.1.7, that is,

$$c_N = \inf_{\sup_{r>0} r^{\frac{N-2}{2}} |u(r)|=1} \|\nabla u\|_2^2. \tag{1.2}$$

Using the scaling operators (1.1), we may rewrite this as

$$c_N = \inf_{\sup_{r>0} |[g_r u](1)|=1} \|\nabla u\|_2^2. \tag{1.3}$$

Let $(u_n)_{n\in\mathbb{N}}$ be a minimizing sequence for (1.3), namely, $\|\nabla u_n\|_2^2 \to c_N$, $|[g_t u_n](1)| \leq 1$ for all $t > 0$ and $[g_{t_n} u_n](1) \to 1$ with some sequence $(t_n)_{n\in\mathbb{N}}$ of positive numbers. Let $w_n = g_{t_n} u_n / [g_{t_n} u_n](1)$. Then we have $\|\nabla w_n\|_2^2 \to c_N$, $|[g_t w_n](1)| \leq 1 + o(1)$ for all $t > 0$ and $w_n(1) = 1$. Then there is a $w \in \dot{H}^{1,2}_{\mathrm{rad}}(\mathbb{R}^N)$ such that, on a renamed weakly convergent subsequence, $w_n \rightharpoonup w$. Then $w(1) = \lim w_n(1) = 1$, while $|[g_t w](1)| \leq 1$ for all $t > 0$. By weak semicontinuity of the norm, $\|\nabla w\|_2^2 \leq \liminf \|\nabla w_n\|_2^2 = c_N$. Thus w is a minimizer for (1.3) (and then for (1.2) as well), $\|\nabla w_n\|_2 \to \|\nabla w\|_2$, and, consequently, $w_n \to w$ and $g_{t_n} u_n \to w$ in the norm of $\dot{H}^{1,2}_{\mathrm{rad}}(\mathbb{R}^N)$. Furthermore, by the scaling invariance of the gradient norm we have

$$\begin{aligned} c_N &= \inf_{\sup_{r>0} |[g_r u](1)|=1} \|\nabla u\|_2^2 \\ &= \inf_{u(1)=1} \|\nabla u\|_2^2 \leq \|\nabla w\|_2^2 = c_N, \end{aligned}$$

which gives

$$c_N = \inf_{u(1)=1} \|\nabla u\|_2^2. \tag{1.4}$$

The infimum in (1.4) is necessarily attained on a continuous function which is harmonic on open intervals $(0,1)$ and $(1,\infty)$, which by the requirement of being an element of $\dot{H}^{1,2}(\mathbb{R}^N)$ defines it uniquely as

$$\psi_N(r) = \begin{cases} 1, & r \le 1; \\ r^{2-N}, & r \ge 1. \end{cases} \tag{1.5}$$

Since $\psi_N(r)r^{\frac{N-2}{2}} \le 1$ for any $r > 0$, by (1.4) it is also a minimizer for (1.3) (as well as for (1.2)), and by an elementary evaluation of $\|\nabla\psi_N\|_2^2$ we have $c_N = (N-2)\omega_N$ where ω_N denotes the measure of the $N-1$-dimensional unit sphere.

We conclude that any minimizing sequence for (1.2) admits a renamed subsequence and a sequence of positive numbers (t_n) such that $g_{t_n}u_n \to \psi_N$ in the norm of $\dot{H}^{1,2}_{\mathrm{rad}}(\mathbb{R}^N)$.

Furthermore, if w is any minimizer for (1.2), the constant minimizing sequence $(w)_{n\in\mathbb{N}}$ admits a sequence of positive numbers $(t_n)_{n\in\mathbb{N}}$ such that $g_{t_n}w \to \psi_N$ in the $\dot{H}^{1,2}(\mathbb{R}^N)$-norm to ψ_N. Then necessarily $t_n \to t$ with some $t > 0$ and $w = g_{\frac{1}{t}}\psi_N$.

The next example is cocompactness of the embedding $H^{m,p}(\mathbb{R}^N) \hookrightarrow L^q(\mathbb{R}^N)$ relative to the group of lattice shifts on $\mathbb{R}^N$:

$$\mathcal{G}_{\mathbb{Z}^N} \overset{\text{def}}{=} \{u \mapsto u(\cdot - y)\}_{y\in\mathbb{Z}^N}. \tag{1.6}$$

Theorem 1.1.9. *Let $m \in \mathbb{N}$, $1 < p < \infty$ and let*

$$p_m^* = \begin{cases} \frac{pN}{N-mp}, & N > pm, \\ \infty \quad, & N \le pm. \end{cases}$$

For any $q \in (p, p_m^*)$, the embedding $H^{m,p}(\mathbb{R}^N) \hookrightarrow L^q(\mathbb{R}^N)$ is cocompact relative to the group $\mathcal{G}_{\mathbb{Z}^N}$.

Proof. Indeed, let $(u_k)_{k\in\mathbb{N}}$ be a sequence in $H^{m,p}(\mathbb{R}^N)$, such that $u_k(\cdot - y_k) \rightharpoonup 0$ for any sequence (y_k) in $\mathbb{Z}^N$. By continuity of the embedding $H^{m,p}((0,1)^N) \hookrightarrow L^q((0,1)^N)$, we have for every $y \in \mathbb{Z}^N$,

$$\int_{(0,1)^N+y} |u_k|^q \le C \int_{(0,1)^N+y} (|\nabla^m u_k|^p + |u_k|^p)\Bigg(\int_{(0,1)^N+y} |u_k|^q\Bigg)^{1-p/q}.$$

Adding the above inequalities over $y \in \mathbb{Z}^N$ and taking into account that (u_k) is bounded in $H^{m,p}(\mathbb{R}^N)$, we have

$$\int_{\mathbb{R}^N} |u_k|^q \le C \sup_{y\in\mathbb{Z}^N}\Bigg(\int_{(0,1)^N+y} |u_k|^q\Bigg)^{1-p/q}. \tag{1.7}$$

Let now $y_k \in \mathbb{Z}^N$ be such that

$$\sup_{y\in\mathbb{Z}^N} \int_{(0,1)^N+y} |u_k|^q \le 2 \int_{(0,1)^N+y_k} |u_k|^q = \int_{(0,1)^N} |u_k(\cdot - y_k)|^q.$$

Note that the right-hand side here converges to zero since $u_k(\cdot - y_k) \rightharpoonup 0$ and the embedding $H^{m,p}(\mathbb{R}^N) \hookrightarrow L^q((0,1)^N)$ is compact. Thus, by (1.7), $u_k \to 0$ in $L^q(\mathbb{R}^N)$. □

Example 1.1.10. Let $N > 2$, and let

$$\mathcal{G}^r = \{u \mapsto 2^{rj} u(2^j(\cdot - y))\}_{y\in\mathbb{R}^N, j\in\mathbb{Z}}, \quad r = \frac{N-2}{2}. \tag{1.8}$$

Elements of $\mathcal{G}^r$ are isometries on $\dot{H}^{1,2}(\mathbb{R}^N)$ and on $L^{\frac{2N}{N-2}}(\mathbb{R}^N)$. With this choice of $\mathcal{G}$ the limiting Sobolev embedding $\dot{H}^{1,2}(\mathbb{R}^N) \hookrightarrow L^{\frac{2N}{N-2}}(\mathbb{R}^N)$ is $\mathcal{G}^r$-cocompact. This is a particular case of Theorem 3.2.1 presented later in the book.

1.2 Profile decomposition

We give here a definition of profile decomposition in Hilbert space. The Banach space version, presented in Chapter 4, requires to define additional notions, which are the subject of Chapter 2.

Definition 1.2.1 (Concentration family). Let H be a Hilbert space and let $\mathcal{G}$ be a group of linear isometries of H. One says that a countable set of pairs

$$\{w^{(n)}, (g_k^{(n)})_{k\in\mathbb{N}}\}_{n\in\mathbb{N}} \subset H \times \mathcal{G}^{\mathbb{N}}$$

is a *concentration family* for a bounded sequence $(u_n)_{n\in\mathbb{N}}$ in H, if $g_k^{(1)} = \mathrm{id}$,

$$g_k^{(n)^{-1}} u_k \rightharpoonup w^{(n)}, \tag{1.9}$$

and

$$g_k^{(n)^{-1}} g_k^{(m)} \rightharpoonup 0 \quad \text{whenever } m \ne n. \tag{1.10}$$

The functions $w^{(n)}$ are called *concentration profiles* of $(u_k)_{k\in\mathbb{N}}$, associated with *scaling sequences* $(g_k^{(n)})_{k\in\mathbb{N}}$, and sequences $(g_k^{(n)} w^{(n)})_{k\in\mathbb{N}} \subset H$ are called *elementary concentrations* (or *blowups*, or *cores*) for the sequence $(u_k)_{k\in\mathbb{N}}$. Property (1.10) is called (asymptotic) *decoupling*.

Remark 1.2.2. Note that, since $\mathcal{G}$ consists of isometries of H, $g^{-1} = g^*$ for each $g \in \mathcal{G}$ and, therefore, $g_k^{(n)^{-1}} g_k^{(m)} \rightharpoonup 0$ if and only if $(g_k^{(m)} v, g_k^{(n)} w) \to 0$ for any $v, w \in H$. Thus, in the context of Hilbert space, decoupling property may be also called *asymptotic orthogonality*.

Example 1.2.3. Let $H = L^2(\mathbb{R}^N)$. If $\mathcal{G}$ is a group of shifts

$$\mathcal{G}_{\mathbb{R}^N} \stackrel{\text{def}}{=} \{g_y : u \mapsto u(\cdot - y)\}_{y\in\mathbb{R}^N} \tag{1.11}$$

relation (1.10) is equivalent to $|y_k^{(n)} - y_k^{(m)}| \to \infty$, since $g_{y_k} \rightharpoonup 0$ if and only if $|y_k| \to \infty$.

If $\mathcal{G}$ is a group of rescalings $\{g_{s,y}u \mapsto 2^{rs}u(2^s(\cdot - y))\}_{y\in\mathbb{R}^N, s\in\mathbb{R}}$ on $L^2(\mathbb{R}^N)$, with $r = N/2$ so that it preserves the L^2-norm, asymptotic orthogonality (1.10) is expressed by

$$|s_k^{(n)} - s_k^{(m)}| + (2^{s_k^{(n)}} + 2^{s_k^{(m)}})|y_k^{(n)} - y_k^{(m)}| \to \infty, \quad m \neq n, \tag{1.12}$$

since $g_{s_k,y_k} \rightharpoonup 0$ in $L^2(\mathbb{R}^N)$ if and only if $|s_k| \to \infty$ or $|y_k| \to \infty$.

Definition 1.2.4 (Profile decomposition). Let $(u_n)_{n\in\mathbb{N}}$ be a bounded sequence in a Hilbert space H. One says that (u_n) admits a *profile decomposition* if it has a concentration family $\{w^{(n)}, (g_k^{(n)})_{k\in\mathbb{N}}\}_{n\in\mathbb{N}} \subset H \times \mathcal{G}^{\mathbb{N}}$ such that the series

$$S_k \stackrel{\text{def}}{=} \sum_n g_k^{(n)} w^{(n)} \tag{1.13}$$

called *defect of compactness* converges in H unconditionally (in n) and uniformly with respect to k, and

$$u_k - S_k \stackrel{\mathcal{G}}{\rightharpoonup} 0. \tag{1.14}$$

Such concentration family is called *complete*.

Remark 1.2.5. If H is $\mathcal{G}$-cocompactly embedded into a Banach space F, then from (1.14) it follows that $u_k - S_k \to 0$ in the norm of F.

The following statement is an analog of Parseval identity in presence of asymptotic orthogonality.

Proposition 1.2.6. *Let $\{w^n, (g_k^{(n)})_k \in \mathbb{N}\}_{n\in\mathbb{N}} \subset H \times \mathcal{G}^{\mathbb{N}}$ be a complete concentration family for a bounded sequence $(u_k)_{k\in\mathbb{N}} \subset H$. Then*

$$\|u_k\|^2 = \sum_n \|w^{(n)}\|^2 + \|u_k - S_k\|^2 + o(1), \tag{1.15}$$

Proof. By convergence properties of the series (1.13), we may without loss of generality assume that the concentration family for (u_k) has finitely many, say M, nonzero concentration profiles $w^{(n)}$. Then

$$\begin{aligned}\|u_k\|^2 &= \left\| u_k - S_k + \sum_{n=1}^M g_k^{(n)} w^{(n)} \right\|^2 \\ &= \|u_k - S_k\|^2 + 2\left(u_k - S_k, \sum_{n=1}^M g_k^{(n)} w^{(n)} \right) + \left\| \sum_{n=1}^M g_k^{(n)} w^{(n)} \right\|^2\end{aligned}$$

$$
\begin{aligned}
&= \|u_k - S_k\|^2 + 2\sum_{n=1}^{M}([g_k^{(n)}]^{-1}(u_k - S_k), w^{(n)}) \\
&+ \sum_{n=1}^{M}\|g_k^{(n)}w^{(n)}\|^2 + \sum_{m\neq n}(g_k^{(n)}w^{(n)}, g_k^{(n)}w^{(n)}) \\
&= \|u_k - S_k\|^2 + o(1) + \sum_{n=1}^{M}\|w^{(n)}\|^2 + o(1).
\end{aligned}
$$

In the transition to the last line we used the following properties: $[g_k^{(n)}]^{-1}u_k \rightharpoonup w^{(n)}$ and $[g_k^{(n)}]^{-1}S_k \rightharpoonup w^{(n)}$ (with (1.10) involved) in the second term, isometry of $g_k^{(n)}$ in the third term, and (1.10) in the last term. □

Example 1.2.7. Any bounded sequence in $\dot{H}^{1,2}(\mathbb{R}^N)$, $N \geq 3$, has a renamed subsequence that admits a profile decomposition relative to the group (1.8), that takes the form

$$
u_k - \sum_{n=1}^{\infty} 2^{\frac{N-2}{2}s_k^{(n)}} w^{(n)}(2^{s_k^{(n)}}(\cdot - y_k^{(n)})) \to 0 \quad \text{in } L^{\frac{2N}{N-2}}(\mathbb{R}^N), \tag{1.16}
$$

with

$$
2^{-\frac{N-2}{2}s_k^{(n)}} u_k(2^{-s_k^{(n)}} \cdot + y_k^{(n)}) \rightharpoonup w^{(n)} \quad \text{as } k \to \infty, n \in \mathbb{N},
$$

with the asymptotic orthogonality expressed by (1.12), and with (1.15) satisfied. Indeed, (1.16) follows from (1.13), (1.14), and cocompactness of the Sobolev embedding from Example 1.1.10. This example is a particular case of the profile decomposition of Solimini [112].

1.3 Brezis–Lieb lemma

We now address effects of asymptotic orthogonality on values of functionals in the Hilbert space, including the norm. We start with the asymptotic orthogonality produced by weak convergence in a Hilbert space: if $u_k \rightharpoonup u$, then $(u, u_k - u) \rightharpoonup 0$, and thus (u_k) has an asymptotically orthogonal decomposition into u and $u_k - u$. This leads to an "asymptotic Pythagoras theorem":

$$
\|u_k\|^2 = \|u\|^2 + \|u_k - u\|^2 + o(1), \tag{1.17}
$$

which follows from the asymptotic orthogonality in the obvious identity

$$
\|u_k\|^2 = \|u\|^2 + \|u_k - u\|^2 + 2(u_k - u, u).
$$

The Brezis–Lieb lemma gives a similar property for the quantity $\|u\|_p^p$, defined for a measure space, under assumption of convergence almost everywhere, which is a stronger assumption than weak convergence (cf. Lemma 1.3.1 below). Further in this section, we illustrate how the Brezis–Lieb lemma and relation (1.17), together with cocompactness, yield existence of minimizers in the Sobolev inequality.

Lemma 1.3.1. *Assume that $(u_n)_{n\in\mathbb{N}}$ is a bounded sequence in $L^p(\Omega,\mu)$, $p\in(1,\infty)$, where (Ω,μ) is a measure space. If $u_n\to u$ almost everywhere, then $|u_n-u|\rightharpoonup 0$, and, consequently, $u_n\rightharpoonup u$.*

Proof. Without loss of generality, we may assume that $u=0$. Let $v\in L^{p'}(\Omega,\mu)$, $p'=\frac{p}{p-1}$, and let

$$A_n=\{x\in\Omega;\ |u_n(x)|\le|v(x)|^{p'-1}\},\quad B_n=\Omega\setminus A_n.$$

By the Hölder inequality, we have

$$\begin{aligned}\int_\Omega |u_n||v|d\mu &\le \int_{A_n}|u_n||v|d\mu+\|u_n\|_p\Big(\int_{B_n}|v|^{p'}d\mu\Big)^{1/p'}\\ &\le \int_\Omega \mathbb{1}_{A_n}|u_n||v|d\mu + C\Big(\int_\Omega \mathbb{1}_{B_n}|v|^{p'}d\mu\Big)^{1/p'}.\end{aligned}$$

Both integrands are bounded by an integrable function $|v|^{p'}$ and converge to zero a. e., so the right-hand side vanishes by the Lebesgue dominated convergence theorem. □

Theorem 1.3.2 (Brezis–Lieb lemma – general nonlinearity). *Let $1\le q<\infty$ and let (Ω,μ) be a measure space. Let $F:\Omega\to\mathbb{R}$ be a continuous function satisfying*

$$|F(a+b)-F(a)|\le\varepsilon|a|^q+C_\varepsilon|b|^q,\quad a,b\in\mathbb{R},\varepsilon>0. \tag{1.18}$$

If $(u_n)_{n\in\mathbb{N}}$ is a sequence bounded in $L^q(\Omega,\mu)$ and convergent almost everywhere in Ω to a function u, then

$$\int_\Omega F(u_n)dx=\int_\Omega F(u)dx+\int_\Omega F(u_n-u)dx+o(1). \tag{1.19}$$

Proof. Note that (1.18) with $a=0$ gives that $|F(s)|\le\inf_\varepsilon C_\varepsilon|s|^q$. Let $\varepsilon>0$ and

$$v_k^\varepsilon \overset{\text{def}}{=} (|F(u_n)-F(u_n-u)-F(u)|-\varepsilon|u_k-u|^q)_+, \tag{1.20}$$

so that $v_k^\varepsilon\to 0$ almost everywhere in Ω. Combining (1.18) for $a=u_k-u$, $b=u_k$ with the estimate on $F(u)$ by $|u|^q$, we have

$$v_k^\varepsilon=(|F(u_n)-F(u_n-u)-F(u)|-\varepsilon|u_k-u|^q)_+\le C'_\varepsilon|u|^q. \tag{1.21}$$

Then by the Lebesgue dominated convergence theorem, $\int_\Omega v_k^\varepsilon d\mu \to 0$. This implies

$$\limsup \int_\Omega |F(u_k) - F(u_k - u) - F(u)| d\mu \le \varepsilon \limsup \int_\Omega |u_k - u|^q d\mu.$$

Since ε in the right-hand side above can be arbitrarily small, (1.19) follows. □

Most often the name Brezis–Lieb lemma is applied to the following particular case of Theorem 1.3.2.

Corollary 1.3.3 (Brezis–Lieb lemma). *Let $q \in [1, \infty)$ and let (Ω, μ) be a measure space. Assume that $(u_k)_{k\in\mathbb{N}}$ is a bounded sequence $L^q(\Omega, \mu)$, convergent to w almost everywhere. Then*

$$\int_\Omega |u_k|^q d\mu - \int_\Omega |u_k - w|^q d\mu - \int_\Omega |w|^q d\mu \to 0. \tag{1.22}$$

Remark 1.3.4. Since $\lim_{q\to\infty}(a^q + b^q)^{\frac{1}{q}} = \max\{a, b\}$ whenever $a, b > 0$, it is natural to ask if the following analog of the Brezis–Lieb lemma for $q = \infty$ is true:

$$u_k \rightharpoonup w \text{ a. e. } \Longrightarrow \|u_k\|_\infty = \max\{\|w\|_\infty, \|u_k - w\|_\infty\} + o(1).$$

The answer is, without additional conditions, negative. Consider $w(x) = \sin\frac{1}{x}$, $x \in (0, \frac{1}{\pi})$, and let $u_k(x) = w(x) + \varphi(k^2(x - \frac{1}{2k\pi+\pi/2}))$, where φ is a nonnegative smooth function, supported in $(-\frac{1}{4\pi}, \frac{1}{4\pi})$ with $\varphi(x) \le \varphi(0) = 1$. Then $u_k \to w$ pointwise in $(0, \frac{1}{\pi})$, $\|u_k\|_\infty = 2$ while $\|u_k - w\|_\infty = \|w\|_\infty = 1 \neq 2$.

If we heuristically understand the Brezis–Lieb lemma as a consequence of asymptotic separation of supports of w and $(u_k - w)$, then the counterexample above is the consequence of w and $(u_k - w)$ having their peak values at the same point. The following statement imposes a condition of local uniform convergence that separates the maximal values.

Lemma 1.3.5 (Brezis–Lieb lemma for $q = \infty$). *Let $(u_k)_{k\in\mathbb{N}}$ be a bounded sequence in $L^\infty(\Omega, \mu)$, where (Ω, μ) is a measure space. Assume that for every $\varepsilon > 0$,*

$$u_k \text{ converges to } w \text{ uniformly on the set } \Omega_\varepsilon = \{x \in \Omega : |w(x)| \ge \varepsilon\}. \tag{1.23}$$

Then

$$\|u_k\|_\infty = \max\{\|w\|_{\infty,\Omega}, \|u_k - w\|_\infty\} + o(1). \tag{1.24}$$

Proof. In two cases, when $w = 0$ a. e. or when $u_k \to w$ uniformly in Ω, the assertion is trivial. It suffices then to prove the lemma when $w \neq 0$ on a set of positive measure and that $\delta = \lim_{k\to\infty} \|u_k - w\|_{\infty,\Omega} > 0$. By (1.23), for every $\varepsilon > 0$, $\|u_k - w\|_{\infty,\Omega\setminus\Omega_\varepsilon} \to \delta$.

Fix $\varepsilon > 0$, $\varepsilon < \min\{\|w\|_{\infty,\Omega}, \frac{1}{2}\delta\}$. Then we will have $\|u_k - w\|_{\infty,\Omega_\varepsilon} < \varepsilon$ and $|\|u_k - w\|_{\infty,\Omega} - \|u_k\|_{\infty,\Omega\setminus\Omega_\varepsilon}| \le \varepsilon$ for all k sufficiently large, and thus

$$\begin{aligned}\|u_k\|_{\infty,\Omega} &= \max\{\|u_k\|_{\infty,\Omega_\varepsilon}, \|u_k\|_{\infty,\Omega\setminus\Omega_\varepsilon}\}\\ &\le \max\{\|w\|_{\infty,\Omega}, \|u_k - w\|_{\infty,\Omega}\} + 2\varepsilon.\end{aligned}$$

At the same time,

$$\begin{aligned}\|u_k\|_{\infty,\Omega} &= \max\{\|u_k\|_{\infty,\Omega_\varepsilon}, \|u_k\|_{\infty,\Omega\setminus\Omega_\varepsilon}\}\\ &\ge \max\{\|w\|_{\infty,\Omega_\varepsilon}, \|u_k\|_{\infty,\Omega\setminus\Omega_\varepsilon}\} - \varepsilon \ge \max\{\|w\|_{\infty,\Omega}, \|u_k - w\|_{\infty,\Omega}\} - 2\varepsilon.\end{aligned}$$

Since ε is arbitrary, (1.24) follows. □

Remark 1.3.6. The Brezis–Lieb lemma, as well as its generalizations, Theorem 1.3.2 and later Theorem 4.7.1, remain valid also in the case when u_k, $k \in \mathbb{N}$, are functions with values in $\mathbb{R}^m$, $m \in \mathbb{N}$. The proof remains verbally the same provided that one reads the notation $|\cdot|$ as a norm in $\mathbb{R}^m$.

Example 1.3.7. Let $N \ge 3$ and let $2^* = \frac{2N}{N-2}$. The minimum in

$$S = \inf_{u\in\dot{H}^{1,2}(\mathbb{R}^N),\ \|u\|_{2^*}=1} \int_{\mathbb{R}^N} |\nabla u|^2 dx \tag{1.25}$$

is attained. Moreover, for every minimizing sequence $(u_n)_{n\in\mathbb{N}}$ in $\dot{H}^{1,2}(\mathbb{R}^N)$ (i. e., such that $\|u_n\|_{2^*} = 1$ and $\|\nabla u_n\|_2^2 \to S$) there exists a renamed subsequence and sequences $(y_n)_{n\in\mathbb{N}}$ in $\mathbb{R}^N$ and $(s_n)_{n\in\mathbb{N}}$ in $\mathbb{R}$, such that sequence

$$v_n \overset{\text{def}}{=} 2^{\frac{N-2}{2}s_n} u_n(2^{s_n}(\cdot - y_n)), \quad n \in \mathbb{N}, \tag{1.26}$$

converges to a minimizer in the norm of $\dot{H}^{1,2}(\mathbb{R}^N)$. Indeed, if $(u_n)_{n\in\mathbb{N}}$ is a minimizing sequence, then for any choice of sequences (y_n) in $\mathbb{R}^N$ and (s_n) in $\mathbb{R}$, the corresponding rescaled sequence (1.26) will be also a minimizing sequence. It may not occur, however, that the corresponding rescaled sequence (v_n) will weakly converge to zero for all choices of (y_n) and (s_n), since by cocompactness of the embedding $\dot{H}^{1,2}(\mathbb{R}^N) \hookrightarrow L^{2^*}(\mathbb{R}^N)$ (Example 1.1.10), one would have $u_k \to 0$ in $L^{2^*}(\mathbb{R}^N)$, which is a contradiction. Let us therefore fix a renamed subsequence of (y_n), (s_n) and (u_n) such that corresponding rescaled sequence (v_n) converges weakly to some $v \ne 0$. It is easy to show that v is a minimizer. Indeed, by (1.22) and (1.17) we have

$$1 = \|v_n\|_{2^*}^{2^*} = \|v\|_{2^*}^{2^*} + \|v_n - v\|_{2^*}^{2^*} + o(1), \tag{1.27}$$

$$S = \|\nabla v_n\|_2^2 = \|\nabla v\|_2^2 + \|\nabla v_n - v\|_2^2 + o(1). \tag{1.28}$$

Let $t = \|v\|_{2^*}^{2^*}$. Then by (1.27) we have $\|v_n - v\|_{2^*}^{2^*} \to 1 - t$, and by (1.28) we have $S \ge St^{\frac{N-2}{N}} + S(1-t)^{\frac{N-2}{N}}$, which can hold, given that $t \ne 0$, only if $t = 1$. Consequently, $v_n \to v$

in $L^{2^*}(\mathbb{R}^N)$ and thus $\|v\|_{2^*} = 1$. From the weak semicontinuity of the $\dot{H}^{1,2}(\mathbb{R}^N)$-norm, it follows that v is a minimizer, and, since $v_n \rightharpoonup v$ and $\|\nabla v_n\|_2 \to \|\nabla v\|_2$, we have $v_n \to v$ in $\dot{H}^{1,2}(\mathbb{R}^N)$. By the Polia–Szegö inequality (or the symmetry argument in [60]), the minimizer is necessarily decreasing radial with respect to some point, and thus it satisfies an ordinary differential equation of second order. Condition $v \in \dot{H}^{1,2}(\mathbb{R}^N)$ together with normalization selects a unique, up to a scaling (1.1), nonsingular radial solution

$$v = \frac{[N(N-2)]^{\frac{N-2}{4}}}{(1+r^2)^{\frac{N-2}{2}}}. \tag{1.29}$$

1.4 Lions' lemma for the Moser–Trudinger functional

Let $\Omega \subset \mathbb{R}^N$, $N \geq 2$, be a bounded domain and let $\alpha_N = N\omega_N^{1/(N-1)}$ where $\omega_N = \frac{2\pi^{N/2}}{\Gamma(\frac{N}{2})}$ is the area of the unit $N-1$-dimensional sphere. In particular, $\alpha_2 = 4\pi$. The following inequality is known as the Moser–Trudinger inequality:

$$\sup_{u \in H_0^{1,N}(\Omega),\ \|\nabla u\|_N \leq 1} \int_\Omega e^{\alpha |u|^{\frac{N}{N-1}}} dx < \infty, \quad \alpha \leq \alpha_N. \tag{1.30}$$

It is known that, despite that the functional may lack a uniform bound on a bounded set,

$$\int_\Omega e^{\lambda |u|^{\frac{N}{N-1}}} dx < \infty \quad \text{for any } u \in H_0^{1,N}(\Omega) \text{ and } \lambda > 0. \tag{1.31}$$

Indeed, for any $\varepsilon > 0$ there exists $M_\varepsilon > 0$ such that $\|\nabla(|u| - M_\varepsilon)_+\|_N < \varepsilon$, so that, using the inequality $|a+b|^q \leq 2^{q-1}(a^q + b^q)$ for $a, b \geq 0$, $q = \frac{N}{N-1}$, we have

$$\begin{aligned}\int_\Omega e^{\lambda |u|^{\frac{N}{N-1}}} dx &\leq \int_\Omega e^{\lambda |M_\varepsilon + (|u|-M_\varepsilon)_+|^{\frac{N}{N-1}}} dx \\ &\leq e^{\lambda 2^{\frac{1}{N-1}} M_\varepsilon^{\frac{N}{N-1}}} \int_\Omega e^{\lambda 2^{\frac{1}{N-1}} (|u|-M_\varepsilon)_+^{\frac{N}{N-1}}} dx.\end{aligned}$$

We fix now $\varepsilon > 0$ small enough so that the integral in the right-hand side is bounded by (1.30).

Lemma 1.4.1 (Lions, [84]). *Let $\Omega \subset \mathbb{R}^N$, $N \geq 2$, be a bounded domain and let sequence $(u_n)_{n\in\mathbb{N}}$ in $H_0^{1,N}(\Omega)$, $\|\nabla u_n\|_N \leq 1$, converge weakly to a function u. Then for any*

$$\alpha < \frac{\alpha_N}{\limsup \|\nabla(u_n - u)\|_N^{\frac{N}{N-1}}},$$

one has

$$\limsup \int_{\Omega} e^{\alpha |u_n|^{\frac{N}{N-1}}} dx < \infty. \tag{1.32}$$

Proof. We give the proof for $N = 2$. For the general case see [84]. Let us use the following obvious inequality: for every $\varepsilon > 0$ and $t \in \mathbb{R}$,

$$(1+t)^2 - (1+\varepsilon)t^2 \le 1 + 1/\varepsilon. \tag{1.33}$$

Set $v_n \stackrel{\text{def}}{=} u_n - u$. Note that $\limsup \|\nabla v_n\|_2^2 \le 1 - \|\nabla u\|_2^2$ by (1.17). Then, with any $\varepsilon > 0$ such that $\alpha(1-\varepsilon)(1 - \|\nabla u\|_2^2)^2 < \alpha_2$ we have by (1.33),

$$|u_n|^2 = |u + v_n|^2 \le (1+\varepsilon)|v_n|^2 + (1 + 1/\varepsilon)|u|^2.$$

Applying Hölder inequality to $e^{\alpha |u_n|^2}$ and taking into account the estimate above, we get, with any $r \in (1, \infty)$,

$$\int_{\Omega} e^{\alpha |u_n|^2} dx \le \left(\int_{\Omega} e^{\alpha(1+\varepsilon) r |v_n|^2} dx \right)^{1/r} \left(\int_{\Omega} e^{\alpha(1+1/\varepsilon) r' |u|^2} dx \right)^{1/r'}.$$

Set r close enough to 1 and ε small enough so that $\alpha(1+\varepsilon) r \limsup \|\nabla v_n\|_2^2 < \alpha_2$. Then the first integral in the right-hand side will be bounded by the Moser–Trudinger inequality. The second integral will be bounded by inequality (1.31). This proves the lemma. □

Corollary 1.4.2. *The functional* $\int_{\Omega} e^{\alpha |u|^{\frac{N}{N-1}}} dx$, $\alpha \le \alpha_N$, *is weakly continuous at any point of* $\{u \in H_0^{1,N}(\Omega),\ \|\nabla u\|_N \le 1\}$, *unless* $\alpha = \alpha_N$ *and* $u = 0$.

Proof. Like in Lemma 1.4.1 we give the proof for the case $N = 2$, with $\alpha_2 = 4\pi$. Let $u_n \rightharpoonup u$, $\|\nabla u_n\|_2 \le 1$. Let $p \in (1, \frac{4\pi/\alpha}{1 - \|\nabla u\|_2^2})$, noting that the interval is nonempty whenever $\alpha < 4\pi$ or $u \ne 0$. By (1.17) $\limsup \|\nabla (u_n - u)\|_2^2 \le 1 - \|\nabla u\|_2^2$, so we have

$$p < \frac{4\pi/\alpha}{\limsup \|\nabla (u_n - u)\|_2^2}. \tag{1.34}$$

We have, using the derivative of $e^{\alpha(tu_n + (1-t)u)^2}$ with respect to $t \in (0,1)$, and applying Hölder inequality with exponent p,

$$\left| \int_{\Omega} e^{\alpha u_n^2} dx - \int_{\Omega} e^{\alpha u^2} dx \right|$$
$$= \left| 2\alpha \int_0^1 \int_{\Omega} e^{\alpha(tu_n + (1-t)u)^2} (tu_n + (1-t)u)(u_n - u)\, dx dt \right|$$

$$\le 8\pi \sup_{t\in[0,1]} \left(\int_\Omega e^{\alpha p(tu_n+(1-t)u)^2}\,dx \right)^{\frac{1}{p}} \left(\int_\Omega |tu_n + (1-t)u|^{p'} |u_n - u|^{p'}\,dx \right)^{\frac{1}{p'}} .$$

Let us apply Lemma 1.4.1, taking into account (1.34). It is easy to see, following the proof of Lemma 1.4.1, that the bound in (1.32) for the family of sequences $tu_n + (1-t)u \rightharpoonup u$, $t \in [0,1]$, is uniform in t, so the first multiple in the right- hand side is bounded, while by compactness of Sobolev embedding for 2-dimensional bounded domains the second multiple converges to zero. □

In Section 3.11, we address further weak continuity properties of the Moser–Trudinger functional.

1.5 Bibliographic notes

Example 1.1.6 is based on Proposition 1 of [68]. The term *profile decomposition* is due to Gallagher [53], and the term *cocompact embedding* was introduced by the author in [125]. The notions themselves have been in use well before the adopted terminology. An early proof of cocompactness for subcritical Sobolev embeddings, Theorem 1.1.9, can traced to a lemma by Lieb in [76]. The earliest proof of cocompactness of Sobolev embeddings on $\mathbb{R}^N$ relative to the group rescalings $\mathcal{G}^r$ (Example 1.1.10), known to the author, is due to Solimini [112].

Lemma 1.3.1 is quoted from [134, Proposition 5.4.7]. The Brezis–Lieb lemma is a simplified version of [25, Theorem 2], and Corollary (1.3.3) is [25, Theorem 1]. Existence of minimizer in the limiting Sobolev embedding was proved by Talenti [121], while the proof given in Example 1.3.7 follows [18] (a textbook version is [120, Theorem 4.2]). The same paper gives a profile decomposition (restricted to critical sequences of the semilinear elliptic functional) for the embedding $\dot{H}^{1,2}(\mathbb{R}^N) \hookrightarrow L^{2^*}(\mathbb{R}^N)$, which was extended to general sequences by [112].

The Moser–Trudinger inequality has been first proved, without the optimal constant α_N, by Yudovich [135], and independently reproduced by Pohozhaev, Peetre, and, finally, Trudinger [130]. The version with the optimal constant is due to Moser [94], who also introduced Moser functions as test functions for the optimality of the constant. Similar borderline embeddings for $H_0^{s,p}(\Omega)$ with $sp = N$ are known as well; see [1, 88, 97]. Lions lemma was proved in [84].

We make a brief mention of another weak convergence method, which like concentration compactness, is also applied to finding solutions to PDE. The method originates in works of F. Murat and L. Tartar who gave it a similarly sounding name *compensated compactness*. We will exemplify it by a modified version of Murat's lemma [124, p. 278]. Let $\Omega \subset \mathbb{R}^N$, $N > 2$, be a bounded domain. If $u_k \rightharpoonup u$ in $H^{2,1}(\Omega)$, then $\nabla u_k \to \nabla u$ in $L^1(\Omega)$ is generally false. A correct counterpart of this statement, based on the modified Calderón–Zygmund theorem, says that the assertion becomes true if one assumes in addition that $\nabla^2 u_k$ is bounded in the Hardy space $\mathcal{H}^1(\Omega) \subset L^1(\Omega)$ instead of being bounded only in $L^1(\Omega)$.

In other words, a missing compactness property of a Sobolev space is recovered in a suitably chosen large subspace.

The space $\mathcal{H}^1(\mathbb{R}^N)$ is characterized by equivalent norms

$$\max_{i=1,\dots N} \|\partial_i(-\Delta)^{-\frac{1}{2}} u\|_1 \quad \text{and} \quad \left\|\sup_{t>0} |h_t * u|\right\|_1,$$

where $h_t = t^{-N} h(t^{-1}\cdot)$ and $h \in C_0^\infty(\mathbb{R}^N)$ is a nonnegative function with $\int_{\mathbb{R}^N} h(x)dx = 1$. In a heuristic sense, functions in $\mathcal{H}^1$ have a less oscillatory character than functions in $L^1 \setminus \mathcal{H}^1$. Furthermore, nonnegative functions in $\mathcal{H}^1_{\mathrm{loc}}$ are characterized by improved integrability $u \log u \in L^1_{\mathrm{loc}}$ (a result by Elias Stein). The dual of $\mathcal{H}^1$ is the space BMO (see the Appendix, Section 10.2) and $\mathcal{H}^1 = \mathrm{VMO}^*$, where $\mathrm{VMO}(\mathbb{R}^N)$ is the closure of the Schwarz class of rapidly vanishing functions in the BMO-norm. An important property of $\mathcal{H}^1$, extending Lemma 1.3.1 where $p \in (1,\infty)$, is that a bounded sequence in $\mathcal{H}^1$ convergent almost everywhere is weakly*-convergent to the same limit; see [70].

An important paper [32] of Coifman, Lions, Meyer, and Semmes presents a range of cases, where functions in $\mathcal{H}^1$ emerge naturally (a local version of this statements is also true):

1. If $u_i \in H^{1,p_i}(\mathbb{R}^N, \mathbb{R}^N)$, $p_i \in (1,\infty)$, $\sum_{i=1}^N \frac{1}{p_i} = 1$, $i = 1, \dots N$, $N \geq 2$, and $u = (u_1, \dots, u_N)$, then Jacobian $\det \nabla u$ is in $\mathcal{H}^1(R^N)$ and not just in $L^1(\mathbb{R}^N)$.
2. If $u \in H^{1,2}(\mathbb{R}^N, \mathbb{R}^N)$, $N \geq 2$, and $\operatorname{div} u = 0$, then

$$\sum_{i,j=1}^N \frac{\partial u_i}{\partial x_j}\frac{\partial u_j}{\partial x_i} \in \mathcal{H}^1(\mathbb{R}^N).$$

3. If $E \in L^p(\mathbb{R}^N, \mathbb{R}^N)$, $B \in L^{p'}(\mathbb{R}^N, \mathbb{R}^N)$, $p \in (1,\infty)$, $N \geq 2$, $\operatorname{div} E = 0$, and $\operatorname{curl} B = 0$, then $E \cdot B \in \mathcal{H}^1(\mathbb{R}^N)$.

2 Delta-convergence and weak convergence

Delta-convergence was originally studied in the context of the fixed-point theory, but it has recently emerged as a technical tool for dealing with profile decompositions in Banach spaces.

Delta-convergence is a mode of convergence in metric spaces similar to weak convergence, and in Hilbert spaces it coincides with weak convergence. As it follows from Theorem 2.1.3 below, in L^p-spaces, $1 < p < \infty$, Delta-convergence of u_n to u is equivalent to $|u_n - u|^{(p-2)}(u_n - u) \rightharpoonup 0$ in $L^{p'}$, which is generally different from weak convergence unless $p = 2$. Note also that from Lemma 1.3.1 it follows that if $u_n \to u$ a. e. and is bounded in L^p; then u_n is both weakly and Delta-convergent to u. Similarly to the Banach–Alaoglu theorem (Theorem 10.1.1), every bounded sequence in a metric space (satisfying certain convexity conditions) has a Delta-convergent subsequence. Unlike weak convergence, which is a topological property, Delta-convergence depends on the norm, but weak and Delta-convergence may coincide under a suitable choice of an equivalent norm.

2.1 Definition of Delta-convergence

Definition 2.1.1. Let (E, d) be a metric space. One says that a sequence $(x_n)_{n\in\mathbb{N}}$ in E is Delta-convergent to a point x (to be written $x_n \rightharpoondown x$), if for any $y \in E$,

$$d(x_n, x) \le d(x_n, y) + o_{n\to\infty}(1). \tag{2.1}$$

(The remainder in (2.1) is not supposed to be uniform with respect to y.) Heuristically, a Delta-limit of a sequence can be understood as a point closest to the tail of the sequence, in the asymptotic sense. Delta-limit is not necessarily unique.

In Hilbert spaces, Delta-convergence and weak convergence coincide.

Theorem 2.1.2. *Let $(x_n)_{n\in\mathbb{N}}$ be a sequence in a Hilbert space H and let $x \in H$. Then $x_n \rightharpoondown x$ if and only if $x_n \rightharpoonup x$.*

Proof. Consider the following identity, which is immediate by expansion of the scalar product:

$$\|x_n - x\|^2 = \|x_n - y\|^2 - \|x - y\|^2 - 2(x_n - x, x - y), \quad y \in H. \tag{2.2}$$

Assume first that $x_n \rightharpoonup x$. Then from (2.2), it follows that $\|x_n - x\|^2 = \|x_n - y\|^2 - \|x - y\|^2 + o(1)$, which immediately implies $x_n \rightharpoondown x$.

Assume now that $x_n \rightharpoondown x$. Let $z \in H$ be a unit vector and set $y = x - tz$, $t > 0$. Applying the definition of Delta-convergence to (2.2), we have

$$-\|x - y\|^2 - 2(x_n - x, x - y) = \|x_n - x\|^2 - \|x_n - y\|^2 \le o(1),$$

https://doi.org/10.1515/9783110532432-002

which gives $2t(x_n - x, z) \le o(1) + t^2$ and, therefore, $\limsup(x_n - x, z) \le t/2$. Since t is arbitrary, $\limsup(x_n - x, z) \le 0$, and since z is an arbitrary unit vector, replacing z with $-z$ gives $\liminf(x_n - x, z) \ge 0$. Thus $(x_n - x, z) \to 0$ whenever $\|z\| = 1$ and, therefore, $x_n \rightharpoonup x$. □

The proof above can be generalized in a way that yields a characterization of Delta-convergence in a uniformly smooth Banach space. Recall that if the space E is uniformly smooth, then the function $N(x) \stackrel{\text{def}}{=} \frac{1}{2}\|x\|^2$ is Frechet-differentiable at any $x \neq 0$ with the derivative uniformly continuous on bounded sets bounded away from zero. Moreover, $x^* \stackrel{\text{def}}{=} N'(x)$ is the unique conjugate element of x, that is, $\|x^*\|_{E^*} = \|x\|$ and $\langle x^*, x\rangle = \|x\|^2$. See [80, Section 1e] for details.

Theorem 2.1.3. *Let $(x_n)_{n\in\mathbb{N}}$ be a bounded sequence in a uniformly smooth Banach space E, and let $x \in E$. Then $x_n \rightharpoondown x$ if and only if $(x_n - x)^* \rightharpoonup 0$ in E^*.*

Proof. Note that $N(x)$ is a convex function. Indeed, using convexity of the norm and a trivial inequality $(a^2 + b^2) \le 2a^2 + 2b^2$, we have

$$N\left(\frac{x+y}{2}\right) \le \frac{1}{2}\left(\frac{1}{2}\|x\| + \frac{1}{2}\|y\|\right)^2 \le \frac{1}{2}N(x) + \frac{1}{2}N(y).$$

Let $z_n = x_n - x$. We may assume that $\|z_n\|$ is bounded away from zero, since when $x_n \to x$ in norm the assertion of the theorem is trivial.

Assume first that $x_n \rightharpoondown x$. Then $z_n \rightharpoondown 0$, and for any unit vector w and $t > 0$, $N(z_n) \le N(z_n + tw) + o(1)$. This implies, by convexity of N, that $\langle N'(z_n + tw), tw\rangle \ge o(1)$. Then $\liminf\langle N'(z_n + tw), w\rangle \ge 0$, and since N' is uniformly continuous on the sequence (z_n), by taking $t \to 0$, we get $\liminf\langle N'(z_n), w\rangle \ge 0$. Replacing w with $-w$ we arrive at $\langle N'(z_n), w\rangle \to 0$, that is, $(x_n - x)^* \rightharpoonup 0$ in E^*.

Assume now the converse, that $(z_n)^* \rightharpoonup 0$ in E^*. By convexity of the function N, $N(z_n + v) \ge N(z_n) + \langle N'(z_n), v\rangle$ for any $v \in E$. Since the last term converges to zero, we have $N(x_n - x) \le N(x_n - x + v) + o(1)$, and thus $x_n \rightharpoondown x$. □

2.2 Chebyshev and asymptotic centers. Delta-completeness and Delta-compactness

Let (E, d) be a metric space.

Definition 2.2.1 (Chebyshev center and Chebyshev radius). Let $A \subset E$ be a non-empty set and let

$$I_A(y) = \sup_{x\in A} d(x, y), \quad y \in E, \tag{2.3}$$

and

$$\mathbf{rad}\,(A) \stackrel{\text{def}}{=} \inf_{y\in E} I_A(y). \tag{2.4}$$

Quantity (2.4) is called the Chebyshev radius of A, and a minimum point in (2.4), if it exists, is called Chebyshev center of A, **cen** (A).

In general, Chebyshev center is not unique.

Definition 2.2.2 (Asymptotic radius and asymptotic center). Let $(x_n)_{n\in\mathbb{N}}$ be a sequence in a metric space (E,d), and let

$$I_{as}(y) = \limsup_n d(x_n, y), \quad y \in E. \tag{2.5}$$

The asymptotic radius $\mathbf{rad}\, x_n$ is the infimum value of the functional (2.5) over $y \in E$, and asymptotic center of the sequence $\mathbf{cen}\, x_n$, if it exists, is a point of minimum of the functional (2.5).

Definition 2.2.3 (Asymptotic completeness). One calls a metric space (E,d) asymptotically complete if every bounded sequence in E has an asymptotic center. If, in addition, the asymptotic center of every bounded sequence is unique, (E,d) called strictly asymptotically complete.

We will show soon that strict asymptotic completeness of a complete metric space can be assured by the uniform rotundity condition, which in restriction to Banach spaces coincides with uniform convexity. To illustrate how the notions above are used in the fixed-point theory, we give below a version of the Browder fixed-point theorem for metric spaces.

Theorem 2.2.4. *Let (E,d) be a strictly asymptotically complete metric space and let $T: E \to E$ be a nonexpansive map, that is, $d(Tx,Ty) \le d(x,y)$ for all $x,y \in E$. Let $w \in E$, be such that the sequence $T^n w$ is bounded and let $c = \mathbf{cen}\, T^n w$. Then $Tc = c$.*

Proof. By definition of the asymptotic radius, $\mathbf{rad}\, T^n w = \limsup d(T^n w, c)$. Note, however, that since T is nonexpansive, $\limsup d(T^n w, Tc) \le \limsup d(T^{n-1} w, c) = \mathbf{rad}\, T^n w$, which means that Tc is also an asymptotic center of $(T^n w)$, but by the assumption of strict asymptotic completeness the asymptotic center is unique, and thus $Tc = c$. □

Note that this fixed-point theorem appears elementary only because the condition of strict asymptotic completeness may be hard to satisfy.

Proposition 2.2.5. *A sequence $(x_n)_{n\in\mathbb{N}}$ in the metric space (E,d) is Delta-convergent to x if and only if every subsequence of (x_n) has x as its asymptotic center.*

Proof. Let $x_n \rightharpoondown x$ and let $(v_n)_{n\in\mathbb{N}}$ be a subsequence of $(x_n)_{n\in\mathbb{N}}$. Then (2.1) holds for $(v_n)_{n\in\mathbb{N}}$, and by taking the upper limits in the both sides of (2.1) one has $I_{as}(x) \le I_{as}(y)$ for any $y \in E$, which implies that x is an asymptotic center of $(v_n)_{n\in\mathbb{N}}$.

Assume now the converse, namely that x is an asymptotic center of every subsequence of $(x_n)_{n\in\mathbb{N}}$, but $(x_n)_{n\in\mathbb{N}}$ is not Delta-convergent to x. Then there would exist a subsequence $(v_n)_{n\in\mathbb{N}}$ of $(x_n)_{n\in\mathbb{N}}$, a point $y \in E$ and an $\varepsilon > 0$ such that $d(v_n, x) \geq d(v_n, y) + \varepsilon$. Taking the upper limit in both sides, we have $\limsup d(v_n, y) \leq \limsup d(v_n, x) - \varepsilon$, which implies that x is not an asymptotic center of $(v_n)_{n\in\mathbb{N}}$, a contradiction. □

Remark 2.2.6. By Proposition 2.2.5 a Delta-limit of a sequence in a metric space is its asymptotic center, but the converse is not true. For example, the asymptotic center of the sequence $((-1)^n)_{n\in\mathbb{N}}$ in $\mathbb{R}$ is 0, but the sequence is not Delta-convergent. All its Delta-convergent subsequences are constant sequences $(1)_{n\in\mathbb{N}}$ or $(-1)_{n\in\mathbb{N}}$.

Theorem 2.2.7 (Delta-compactness theorem, T.-C. Lim [78]). *Let (E, d) be an asymptotically complete metric space. If $(x_n)_{n\in\mathbb{N}}$ is a bounded sequence in E, then it has a Delta-convergent subsequence.*

Proof. In this proof, we will write $(x'_k)_{k\in\mathbb{N}} \prec (x_k)_{k\in\mathbb{N}}$ if (x'_k) is a subsequence of (x_k).

Let $r_0 = \inf\{\mathbf{rad}\, v_n : (v_n) \prec (x_n)\}$. Choose a subsequence $(v_n^{(1)}) \prec (x_n)$ so that $\mathbf{rad}\, v_n^{(1)} < r_0 + \frac{1}{2}$. Set inductively, assuming that subsequences $(v_n^{(m)}) \prec \cdots \prec (v_n^{(1)}) \prec (x_n)$ are defined, $r_m = \inf\{\mathbf{rad}\, v_n : (v_n) \prec (v_n^{(m)})\}$ and choose $(v_n^{(m+1)}) \prec (v_n^{(m)})$ such that $\mathbf{rad}\, v_n^{(m+1)} < r_m + \frac{1}{2^m}$. Note that $(r_m)_{m\in\mathbb{N}}$ is a nondecreasing bounded sequence, and set $r = \lim r_m$. Let now $w_n = v_n^{(n)}$, $n \in \mathbb{N}$. Since for every $m \in \mathbb{N}$, $(w_n)_{n\geq m+1} \prec (v_n^{(m+1)})$, $\mathbf{rad}\, w_n \leq r_m + \frac{1}{2^m}$. This implies that $\mathbf{rad}\, w_n = r$, and the same conclusion applies to any subsequence of (w_n).

By asymptotic completeness sequence (w_n) has an asymptotic center, which we denote as x. Let $(v_n) \prec (w_n)$ and assume that x is not an asymptotic center of (v_n). By asymptotic completeness, (v_n) has then an asymptotic center different from x, which we denote by y. Since x is not an asymptotic center, $\limsup d(v_n, y) < \limsup d(v_n, x)$, but this implies $\mathbf{rad}\, v_n < r$, a contradiction. Thus $x = \mathbf{cen}\ v_n$. Since (v_n) was an arbitrary subsequence of (w_n), by Proposition 2.2.5 we have $w_n \rightharpoonup x$. □

Remark 2.2.8. Since Theorem 2.1.2 identifies Delta-convergence in Hilbert spaces with weak convergence, once we know that Hilbert spaces are asymptotically complete (see the next section), Theorem 2.2.7 proves the classical Banach–Alaoglu theorem (Theorem 10.1.1) in the case of Hilbert space.

2.3 Rotund metric spaces

We now consider a class of strictly asymptotically complete metric spaces, which includes all uniformly convex Banach spaces and exhibits an analogous weak (i. e., Delta-) compactness property.

Definition 2.3.1 (Uniformly rotund space, John Staples [114]). A metric space (E, d) is *uniformly rotund* if there exists a function $\eta : [0, \infty)^2 \to (0, \infty)$ such that for any $\delta > 0$

and for any $x, y \in E$ with $d(x,y) \geq \delta$ and for some $\eta = \eta(r,\delta)$

$$\mathbf{rad}(B_{r+\eta}(x) \cap B_{r+\eta}(y)) \leq r - \eta \quad \text{for any } r > 0. \tag{2.6}$$

As we show below, this property is a natural generalization of the property of uniform convexity in normed vector spaces, which is defined as follows (see [80, Definition 1.e.1])

$$\forall \delta > 0 \; \exists \eta > 0 : \; x, y \in E, \|x\| \leq 1, \|y\| \leq 1, \|x - y\| \geq \delta \Longrightarrow \left\| \frac{x+y}{2} \right\| \leq 1 - \eta. \tag{2.7}$$

Proposition 2.3.2. *A normed vector space E is uniformly convex if and only if it is uniformly rotund.*

Proof. Uniform rotundity of a uniformly convex normed vector space follows immediately from (2.7). Let us show the converse, namely that if the normed vector space is uniformly rotund, then it is uniformly convex. Let $\|x\| \leq 1$, $\|y\| \leq 1$, such that $\|x-y\| \geq \delta$. It follows easily that both 0 and $x + y$ belong to $\bar{B}_1(x) \cap \bar{B}_1(y)$. By uniform rotundity, setting $\eta = \eta(1,\varepsilon)$, we have $\mathbf{rad}(\bar{B}_1(x) \cap \bar{B}_1(y)) \leq 1 - \eta$. Therefore, $\|x + y - 0\| \leq 2\,\mathbf{rad}(\bar{B}_1(x) \cap \bar{B}_1(y)) \leq 2 - 2\eta$ and so $\|\frac{x+y}{2}\| \leq 1 - \eta$ follows, thus proving uniform convexity. □

Uniform rotundity assures uniqueness of asymptotic centers, and for complete metric spaces, existence of asymptotic centers as well.

Proposition 2.3.3. *Let (E, d) be a uniformly rotund metric space. Then every bounded sequence in E has at most one asymptotic center.*

Proof. Let $(x_n)_{n\in\mathbb{N}}$ be a bounded sequence in E. Assume that it has two asymptotic centers $x \neq y$ and an asymptotic radius r. Then $\limsup d(x_n, x) = \limsup d(x_n, y) = r$. From uniform rotundity, it follows immediately that there is $\eta > 0$ and a point $z \in B_{r+\eta}(x) \cap B_{r+\eta}(y)$ such that $\limsup d(x_n, z) < r$, which is in contradiction to r being the asymptotic radius of (x_n). □

Corollary 2.3.4. *If $(x_n)_{n\in\mathbb{N}}$ is bounded sequence in a uniformly rotund metric space, then it has at most one Delta-limit.*

Proof. Since Delta-limit of a sequence is its asymptotic center by Proposition 2.2.5, the assertion follows from uniqueness of the asymptotic center (Proposition 2.3.3). □

Theorem 2.3.5. *Let $(x_n)_{n\in\mathbb{N}}$ be a bounded sequence in a uniformly rotund metric space (E, d). Then every sequence $(y_k)_{k\in\mathbb{N}}$ that minimizes the functional* (2.5) *for (x_n) is a Cauchy sequence.*

Proof. It suffices to consider the case $r \overset{\text{def}}{=} \mathbf{rad}\, x_n > 0$, since if $r = 0$ the zero infimum value of the functional (2.5) can be attained only on a Cauchy sequence.

Assume that there is a minimizing sequence $(y_n)_{n\in\mathbb{N}}$ for (2.5), which is not Cauchy. Then there exists $\varepsilon > 0$ such that for any $N \in \mathbb{N}$ there exist integers $m, n \geq N$, such that $d(y_m, y_n) \geq \varepsilon$. Then, by uniform rotundity, for N large enough there exists $\eta > 0$ such that $\mathbf{rad}\, x_n \leq \mathbf{rad}\,(B_{r+\eta}(y_m) \cap B_{r+\eta}(y_n)) < r - \eta$, which is a contradiction. □

Combining this theorem with the definition of complete metric space and Proposition 2.3.3, we have the following statement.

Corollary 2.3.6. *Every complete uniformly rotund metric space is strictly asymptotically complete.*

Delta-convergent sequences in uniformly rotund metric spaces always satisfy a stronger relation than (2.1.1).

Proposition 2.3.7. *Let (E, d) be a uniformly rotund metric space. Let $(x_n)_{n\in\mathbb{N}}$ be a bounded sequence in E, Delta-convergent to some $x \in E$. Then, for each element $z \in E$, $z \neq x$, there exist positive constants n_0 and c depending on z such that*

$$d(x_n, x) \leq d(x_n, z) - c \quad \textit{for all } n \geq n_0, \tag{2.8}$$

Proof. If the assertion is false, we can find $z \neq x$ and a subsequence $(x_{k_n})_{n\in\mathbb{N}}$ such that $d(x_{k_n}, x) - d(x_{k_n}, z) \to 0$. Passing again to a subsequence, we can also assume that $d(x_{k_n}, x) \to r > 0$. Set $\eta = \eta(r, d(x, z))$. Since, for large n, $x_{k_n} \in B_{r+\eta}(x) \cap B_{r+\eta}(z)$, we can deduce from (2.6) existence of $y \in E$ such that $d(x_{k_n}, y) < r - \eta$, which contradicts the Delta-convergence of (x_n) to x. □

Boundedness of Delta-convergent sequences stated below, similar to that of weakly convergent sequences, is a consequence of the uniform boundedness principle, although this is not as immediate as in the case of weak convergence.

Theorem 2.3.8. *Every Delta-convergent sequence in a uniformly convex and uniformly smooth Banach space is bounded.*

Proof. Let $(x_k)_{k\in\mathbb{N}}$ be a Delta-convergent sequence in a uniformly convex and uniformly smooth Banach space E. Since $x_k \rightharpoonup x$ is equivalent to $x_k - x \rightharpoonup 0$, without loss of generality we may prove the theorem for the case $x_k \rightharpoonup 0$.

Since strongly convergent sequences are bounded, we may restrict the argument to the case $\inf \|x_k\| > 0$. Since E is uniformly smooth, there exists a continuous function η on $[0, 1]$ with nonnegative values, such that (see [80, p. 61]) $\lim_{t\to 0} \eta(t)/t = 0$, and

$$\big|\|u + v\| - \|u\| - \langle u^*, v\rangle\big| \leq \eta(\|v\|), \quad \text{whenever } \|u\| = 1 \text{ and } \|v\| \leq 1.$$

Then, using the notation $y(u, v) = \|u + v\| - \|u\| - \langle u^*, v\rangle$, $u, v \in E$, we have

$$\begin{aligned}\|u + v\|^2 - \|u\|^2 &= (\|u + v\| - \|u\|)(\|u + v\| - \|u\| + 2\|u\|)\\ &= (y(u, v) + \langle u^*, v\rangle)^2 + 2\|u\|(y(u, v) + \langle u^*, v\rangle).\end{aligned}$$

Substitute now $u = \frac{x_k}{\|x_k\|}$ and $v = \frac{z}{\|x_k\|}$ with an arbitrary vector z. Then, by Proposition 2.3.7 we have for all k sufficiently large (but not uniformly with respect to z)

$$0 \le \|x_k + z\|^2 - \|x_k\|^2 = \alpha_k^2 + 2\|x_k\|\alpha_k,$$

where

$$\alpha_k = \|x_k\|^2 \gamma\left(\frac{x_k}{\|x_k\|}, \frac{z}{\|x_k\|}\right) + \|x_k\| \langle x_k^* / \|x_k\|, z \rangle.$$

Consequently, either $\alpha_k \ge 0$ or $\alpha_k \le -2\|x_k\| \to -\infty$. The latter case can be easily ruled out, since $\|x_k^*\|/\|x_k\| = 1$, $\langle x_k^*/\|x_k\|, z\rangle$ is bounded, $\|x_k\||\omega(\frac{x_k}{\|x_k\|}, \frac{z}{\|x_k\|})| \to 0$ as $\|x_k\| \to \infty$, and so α_k is bounded. Therefore, for k sufficiently large, one has

$$\|x_k\|\gamma\left(\frac{x_k}{\|x_k\|}, \frac{z}{\|x_k\|}\right) + \langle x_k^*/\|x_k\|, z\rangle \ge 0,$$

and thus,

$$\langle x_k^*/\|x_k\|, z\rangle \ge -\eta(t_k)/t_k,$$

where $t_k = 1/\|x_k\| \to 0$. In other words, we have $|\langle \psi(\|x_k\|) x_k^*/\|x_k\|, z\rangle| \le 1$, for k sufficiently large (without, as we keep noting, uniformity with respect to z), where $\psi(s) = \frac{s^{-1}}{\eta(s^{-1})}$ satisfies $\psi(s) \to \infty$ when $s \to \infty$. By the uniform boundedness principle, sequence $\psi(\|x_k\|)$ is bounded, which implies that $\|x_k\|$ is bounded. □

2.4 Opial condition and Van Dulst norm

In this section, we discuss connections between Delta-convergence and weak convergence.

Definition 2.4.1 (Opial condition – [96, Condition (2)]). Let E be a normed vector space. One says that a sequence $(x_n)_{n\in\mathbb{N}}$ in E, which is weakly convergent to a point $x_0 \in E$, is an *Opial sequence* if

$$\liminf \|x_n - x_0\| \le \liminf \|x_n - x\| \quad \text{for every } x \in E. \tag{2.9}$$

One says that the space E satisfies the Opial condition if (2.9) holds for every weakly convergent sequence.

Remark 2.4.2. Every Opial sequence has a Delta-convergent subsequence, whose Delta-limit equals its weak limit: consider a subsequence that realizes the lower limit in the left hand side of (2.9).

The following statement applies, mainly, to uniformly convex Banach spaces.

Proposition 2.4.3. *Let E be a strictly asymptotically complete reflexive Banach space. Then E satisfies the Opial condition if and only if for any bounded sequence $(x_n)_{n\in\mathbb{N}}$ in E,*

$$x_n \rightharpoonup x \Longleftrightarrow x_n \rightharpoondown x. \tag{2.10}$$

Note that the condition that the sequence is bounded can be omitted by Theorem 2.3.8, provided that the space is uniformly smooth and uniformly convex.

Proof. 1. Necessity. Assume that the Opial condition is satisfied. Assume that $x_n \rightharpoonup x$, but x_n is not Delta-convergent to x. Then, by Delta-compactness, a further renamed extraction is Delta-convergent to some $y \neq x$. However, by Opial condition, there is a yet further extraction that is Delta-convergent to x, which is a contradiction, implying that $x_n \rightharpoondown x$.

Assume conversely that $x_n \rightharpoondown x$, but on a renamed subsequence x_n is not weakly convergent to x. Then, on a renamed further extraction, $x_n \rightharpoonup y \neq x$. By the Opial condition, on a further extraction, $x_n \rightharpoondown y$, which contradicts to uniqueness of Delta-limit. Consequently, $x_n \rightharpoonup x$.

2. Sufficiency. Assume (2.10) for every bounded sequence. If $x_n \rightharpoonup x_0$, but (2.9) does not hold, then (x_n) has a bounded subsequence, Delta-convergent to some point $y_0 \neq x_0$. Then by (2.10), this subsequence would weakly converge to y_0, a contradiction. □

Proposition 2.4.4. *Let E be a uniformly convex Banach space. All closed convex subsets of E are closed with respect to Delta-convergence if and only if every bounded sequence in E satisfies* (2.10).

Proof. 1. Assume first that every closed convex set in E is closed with respect to Delta-convergence. If $x_n \rightharpoondown x$ is a bounded sequence, and $x_n \rightharpoonup y \neq x$ is its extraction, let $f \in E^*$ be such that $\langle f, x-y\rangle > 0$. Since $\langle f, x_n - y\rangle \to 0$, from Delta-closedness of convex closed sets it follows that $\langle f, x-y\rangle = 0$, a contradiction. Thus $x_n \rightharpoonup x$.

If $x_n \rightharpoonup x$, but on a renamed subsequence one has $x_n \rightharpoondown y \neq x$, repeating the previous argument we arrive at a further subsequence that weakly converges to y, which is a contradiction.

2. Assume the converse, that every bounded sequence in E satisfies (2.10). Since every closed convex set in E is weakly closed, it will be thus closed with respect to Delta-convergence. □

Proposition 2.4.5. *Let (E_0, E_1) be compatible strictly convex Banach spaces with a common dense set E, satisfying the Opial condition. Then for any $\theta \in (0,1)$ and $p \in [1,\infty]$ spaces $(E_0, E_1)_{\theta,p}$, interpolated by the real method, with the norms* (10.5), *satisfy the Opial condition.*

Proof. Without loss of generality, it suffices to show that if $u_k \rightharpoonup 0$ in $(E_0, E_1)_{\theta,p}$, then for any $v \in E$, $\|u_k\|_{\theta,p} \le \|u_k + v\|_{\theta,p} + o(1)$. Note that it suffices in definition (10.4) of the K-functional to consider $u_k^0 \in E_0$ and $u_k^1 \in E_1$ such that $u_k^0 + u_k^1 = u_k$, $u_k^0 \rightharpoonup 0$ and $u_k^1 \rightharpoonup 0$. Indeed, if $u_k^0 \rightharpoonup w \neq 0$, then $u_k^1 \rightharpoonup -w$ and, using Delta-convergence, we have for k

sufficiently large $\|u_k^0\|_{E_0}+t\|u_k^1\|_{E_1} > \|u_k^0-w\|_{E_0}+t\|u_k^1+w\|_{E_1}$, so that for k large enough the pair u_k^0 and u_k^1 does not contribute to minimization. Therefore, given $v \in E$, $v \neq 0$, by the Opial condition for E_0 and E_1 and Proposition 2.3.7, $K(u_k, t) < K(u_k+v, t)-c$ for some $c = c(v)$ and k sufficiently large. This implies $\liminf \|u_k\|_{\theta,p} \leq \liminf \|u_k + v\|_{\theta,p} + o(1)$, that is, the Opial condition holds. □

Proposition 2.4.6. *Let (Ω, μ) be a measure space. If $(u_k)_{k\in\mathbb{N}}$ is a bounded sequence in $L^p(\Omega, \mu)$, $p \in (1, \infty)$, convergent to a function u almost everywhere, then $u_k \rightharpoonup u$ and $u_k \rightharpoondown u$ in $L^p(\Omega, \mu)$.*

Proof. By Lemma 1.3.1, $u_k \rightharpoonup u$. Since $(u_k - u)_{k\in\mathbb{N}}$ is bounded in $L^p(\Omega, \mu)$ and converges to zero a. e., the sequence $|u_k - u|^{p-2}(u_k - u)$ is bounded in $L^{p'}(\Omega, \mu)$. Then, by Lemma 1.3.1, $|u_k-u|^{p-2}(u_k-u) \rightharpoonup 0$ in $L^{p'}(\Omega, \mu)$. Assume without loss of generality that $\|u_k - u\|_p \geq \delta > 0$ for all $k \in \mathbb{N}$. Then by Theorem 2.1.3, $u_k \rightharpoondown u$. □

Corollary 2.4.7. *If $(u_k)_{k\in\mathbb{N}}$ is a bounded sequence in $H^{1,p}(\mathbb{R}^N)$, $1 < p < N$, and $q \in [p, p^*)$, then $u_k \rightharpoonup u$ in $L^q(\mathbb{R}^N)$ if and only if $u_k \rightharpoondown u$ in $L^q(\mathbb{R}^N)$.*

Proof. Assume that $u_k \rightharpoonup u$ in $L^q(\mathbb{R}^N)$. Then, since (u_k) is bounded in $H^{1,p}(\mathbb{R}^N)$, it necessarily converges to u almost everywhere. Indeed, if it were false, then by local compactness of Sobolev embeddings, there would exist a subsequence of (u_k) that converges almost everywhere, to a different limit than u, which contradicts Lemma 1.3.1. Then by Proposition 2.4.6, $u_k \rightharpoondown u$ in $L^q(\mathbb{R}^N)$. The same argument applies if we assume that $u_k \rightharpoondown u$ in $L^q(\mathbb{R}^N)$. □

Corollary 2.4.8. *Space ℓ^p, $p \in (1, \infty)$, satisfies the Opial condition.*

Proof. It suffices to verify (2.10) when $\|u_k-u\|_p \geq \delta > 0$. The assertion then follows from the following chain of equivalent, up to extraction of subsequence, the statements:
- $u_k \rightharpoondown 0$;
- $(u_k - u)^* \rightharpoonup 0$ in $\ell^{p'}$ (by Theorem 2.1.3);
- $|u_k - u|^{p-2}(u_k - u)$ is bounded in $\ell^{p'}$ and converges to zero pointwise;
- $u_k - u$ bounded in ℓ^p and converges to zero pointwise;
- $u_k \rightharpoonup u$ in ℓ^p (by Lemma 1.3.1). □

Example 2.4.9.
1. Hilbert spaces, since they are uniformly convex and uniformly smooth, satisfy the Opial condition by Theorem 2.1.2 and Proposition 2.4.3.
2. The space $L^p([0, 1])$, $p \in (1, \infty)$, equipped with the standard norm, does not satisfy the Opial condition unless $p = 2$. Indeed, consider the following sequence of functions on $L^p([0, 1])$. Let $\psi_0(t) = -2$ for $t \in [0, 1/3]$ and $\psi_0(t) = 1$ for $t \in [1/3, 1]$. Let $\psi_n(t) = \psi_0(2^n t)$, $t \in [0, 1/2^n]$, extended periodically to $[0, 1]$. By density in L^p, $p < \infty$, of the set of functions constant on subsequent intervals of length $1/2^m$ for all $m \in \mathbb{N}$, one easily sees that ψ_n is Delta-convergent to a nonzero constant λ_p

that minimizes $\int_0^1 |\psi_0(t) - \lambda|^p dt = \frac{1}{3}|\lambda + 2|^p + \frac{2}{3}|\lambda - 1|^p$ over $\lambda \in \mathbb{R}$, and $\lambda_p = 0$ if and only if $p = 2$. At the same time $\psi_n \rightharpoonup 0$ for any $p \in (1, \infty)$.

If the Banach space is separable, it always has an equivalent norm such that Delta-convergence and weak convergence coincide:

Theorem 2.4.10 (van Dulst). *Any separable Banach space E admits an equivalent norm $\|\cdot\|_1$ such that $(E, \|\cdot\|_1)$ satisfies the Opial condition and, moreover, for any sequence $(x_n)_{n\in\mathbb{N}}$ in E,*

$$x_n \rightharpoonup x \Longrightarrow \|x_n - x\| \le \|x_n - y\| + o(1), \quad \textit{for any } y \in E, \textit{ that is, } x_n \xrightarrow{\Delta} x. \tag{2.11}$$

Proof. By the Banach–Mazur theorem every separable Banach space is isometrically isomorphic to a closed subspace of $C([0,1])$, which has a Schauder basis. Consider, without distinguishing in notation elements of E and their images in $C([0,1])$, a basis $\{y_j\}_{j\in\mathbb{N}}$ of $C([0,1])$ with $\|y_j\| = 1$, $j \in \mathbb{N}$. Then the associated coefficient functionals y_i^* will be bounded, and if we set $P_j y = \sum_{i=1}^j \langle y_i^*, y\rangle y_i$, $y \in E$, $j \in \mathbb{N}$, and $P_0 = 0$, then $\{\|P_j\|\}_{j\in\mathbb{N}}$ is bounded (see the Appendix, Section 10.1). Define for every $x \in E$,

$$\|x\|_1 \stackrel{\text{def}}{=} \sup_{j=0,1,\dots} \|(I - P_j)x\|. \tag{2.12}$$

Clearly, this is an equivalent norm on E, since $\|x\| = \|x - P_0 x\| \le \|x\|_1 \le \sup_{j=0,1,\dots} \|I - P_j\|\|x\|$. Without loss of generality, in order to prove (2.11) it suffices to show that if $x_n \rightharpoonup 0$, then for every $y \in E$, $\|x_n\|_1 \le \|x_n + y\|_1 + o_{n\to\infty}(1)$. Let us fix $y \in E$. Let $\varepsilon > 0$, and let $j_\varepsilon \in \mathbb{N}$ be such that $\|(I - P_{j_\varepsilon})y\|_1 \le \varepsilon$. Note that since $x_n \rightharpoonup 0$, we have $\|P_{j_\varepsilon} x_n\|_1 \to 0$. Then

$$\begin{aligned}
\|x_n + y\|_1 &\ge \left\|(I - P_{j_\varepsilon})x_n + P_{j_\varepsilon} y\right\|_1 - \varepsilon - o(1) \\
&= \sup_{j=0,1,\dots} \left\|(1 - P_j)(I - P_{j_\varepsilon})x_n + (1 - P_j)P_{j_\varepsilon} y\right\| - \varepsilon - o(1) \\
&\ge \sup_{j=j_\varepsilon, j_\varepsilon+1,\dots} \left\|(1 - P_j)(I - P_{j_\varepsilon})x_n\right\| - \varepsilon - o(1) \\
&= \left\|(I - P_{j_\varepsilon})x_n\right\|_1 - \varepsilon - o(1) \ge \|x_n\|_1 - \left\|(P_{j_\varepsilon})x_n\right\|_1 - \varepsilon - o(1) \\
&= \|x_n\|_1 - o(1) - \varepsilon.
\end{aligned}$$

In other words, $\liminf(\|x_n + y\|_1 - \|x_n\|_1) \ge -\varepsilon$, and since ε is arbitrary, we have (2.11). □

Remark 2.4.11. The mere fact that the Opial condition can be achieved by choosing a different equivalent norm, in order to make weak convergence and Delta-convergence coincide, may not be satisfactory in applications. In the context of the Browder fixed-point theorem, a map that is nonexpansive in the original norm may not be such in the new norm. In the context of profile decompositions, the new norm may no longer be invariant with respect to the same group as the original norm.

Sobolev spaces with the standard norm, in view of Example 2.4.9, do not satisfy the Opial condition, but the Opial condition, as we show below in Theorem 4.2.1, is satisfied by Besov and Triebel–Lizorkin spaces with scale-invariant norms (3.23) and (3.24), and the identification of $\dot{H}^{s,p}$ as $\dot{F}^{s,p,2}$, yields an equivalent scale-invariant norm for Sobolev spaces meeting the Opial condition, without invoking Theorem 2.4.10.

2.5 Defect of energy. Brezis–Lieb lemma with Delta-convergence

The Brezis–Lieb lemma (Corollary 1.3.3) can be understood as a quantitative estimate for the defect of convergence in L^p. In this section, we would like to address similar "energy estimates" for the defect of compactness in general Banach spaces, not only for weakly convergent sequences, but also for Delta-convergent ones.

Weak semicontinuity of the norm assures that whenever $x_k \rightharpoonup x$, one has $\|x_k\| \geq \|x\| + o(1)$, while in uniformly convex spaces the gap between $\|x_k\|$ and $\|x\|$ can be estimated in terms of the modulus of convexity. Assuming that $\|x_k\| \leq 1$, we have

$$x_k \rightharpoonup x \implies \|x_k\| \geq \|x\| + \delta(\|x_k - x\|) + o(1). \tag{2.13}$$

Indeed, by uniform convexity, (see (10.2) in the Appendix) one has, with $t_k = \max\{\|x\|, \|x_k\|\}$,

$$\left\|\frac{x_k + x}{2}\right\| \leq t_k - t_k\delta(\|x_k - x\|/t_k). \tag{2.14}$$

Note that since the norm is weakly lower semicontinuous, $t_k = \|x_k\| + o(1)$, and $\|\frac{x_k+x}{2}\| \geq \|x\| + o(1)$. Substituting these two relations into (2.14), we have (2.13).

For Hilbert spaces, one has a stronger counterpart of (2.13), namely (1.17). Since weak convergence in Hilbert spaces coincides with Delta-convergence, it is natural to ask whether Delta-convergent sequences in a general Banach space satisfy some analog of (2.13). From the definition of Delta-convergence, one has immediately that $\|x_k\| \geq \|x_k - x\| + o(1)$ when $x_k \rightharpoondown x$. We see from (1.17), that $\|x_k\|$ dominates both $\|x\|$ and $\|x - x_k\|$, while in general Banach space $\|x_k\|$ dominates $\|x\|$ (with a remainder when the space is uniformly convex) in the case of weak convergence, and $\|x_k\|$ dominates $\|x - x_k\|$ in the case of Delta-convergence. It is natural to ask then if, in case of uniform convex space, $\|x_k\|$ dominates $\|x - x_k\|$ with a nontrivial remainder dependent on $\delta(\|x\|)$.

Lemma 2.5.1. *Let E be a uniformly convex Banach and let δ be the modulus of convexity of E. If (u_k) is a sequence in E, $\|u_k\| \leq 1$, $k \in \mathbb{N}$, and $u_k \rightharpoondown u$, then $\|u\| < 2$ and*

$$\|u_k\| \geq \|u_k - u\| + \delta(\|u\|) \tag{2.15}$$

for all k sufficiently large.

Proof. If $u = 0$, the assertion is immediate, so assume that $u \neq 0$. Note that for k sufficiently large, $\|u_k - u\| < \|u_k\|$. This inequality implies that $\|u\| < 2\|u_k\| \leq 2$ and it also implies that $u_k \neq 0$ for these values of k. Thus we may apply to u_k and $u_k - u$ relation (10.3) from the Appendix with $C_1 = \|u_k\|$ and $C_2 = 1$, getting

$$\left\|u_k - \frac{1}{2}u\right\| = \left\|\frac{u_k + (u_k - u)}{2}\right\| \leq \|u_k\| - \delta(\|u\|)\,.$$

Finally, since $u_k \rightharpoondown u$, one also has $\|u_k - u\| \leq \|u_k - \frac{1}{2}u\|$ for sufficiently large k and (2.15) follows. □

In Lebesgue spaces, the Brezis–Lieb lemma (Corollary 1.3.3) gives a more refined evaluation of the defect of energy.

Remark 2.5.2. Delta-convergence is *necessary* for the assertion of the Brezis–Lieb lemma: if a vector-valued sequence $(u_k)_{n\in\mathbb{N}}$ in $L^p(\Omega,\mu;\mathbb{R}^m)$, $p \in [1,\infty)$, $m \in \mathbb{N}$, and a function $u \in L^p(\Omega,\mu;\mathbb{R}^m)$ are such that for any $v \in L^p(\Omega,\mu;\mathbb{R}^m)$,

$$\int_\Omega |u_k - v|^p d\mu \geq \int_\Omega |u - v|^p d\mu + \int_\Omega |u_k - u|^p d\mu + o(1), \tag{2.16}$$

then by the definition of Delta-limit $u_k \rightharpoondown u$.

Let us consider a sufficient condition for (2.16) that will be weaker than convergence almost everywhere. Note that by (1.17) pointwise convergence is not required in the case $p = 2$.

Lemma 2.5.3. *Let $p \geq 3$. Then the following inequality holds true:*

$$F(t,\theta) \overset{\text{def}}{=} |1 + t^2 + 2t\theta|^{p/2} - 1 - |t|^p - p|t|^{p-2}t\theta - pt\theta \geq 0, \quad |t| \leq 1,\ |\theta| \leq 1. \tag{2.17}$$

Proof. For each $t \in [-1,1]$, the function $\theta \mapsto F(t,\theta)$ is convex on $[-1,1]$. By an elementary computation, one easily gets that $\frac{\partial F(t,\theta)}{\partial\theta} \neq 0$ for all $t \in [-1,1]$, and thus $F(t,\theta) \geq \min\{F(t,-1), F(t,1)\}$. Since $F(t,-1) = F(-t,1)$, it suffices to show that $F(t,1) \geq 0$ for all $t \in [-1,1]$, that is,

$$|1 + t|^p \geq 1 + |t|^p + p|t|^{p-2}t + pt, \quad |t| \leq 1. \tag{2.18}$$

This holds if the two following inequalities hold:

$$f_+(t) = (1+t)^p - 1 - t^p - pt^{p-1} - pt \geq 0, \quad t \geq 0,$$
$$f_-(t) = (1-t)^p - 1 - t^p + pt^{p-1} + pt \geq 0, \quad 0 \leq t \leq 1.$$

Since both functions above vanish at zero, it suffices to show that $f_+' \geq 0$ and $f_-' \geq 0$. We have

$$\frac{1}{p}f_+'(t) = (1+t)^{p-1} - t^{p-1} - 1 - (p-1)t^{p-2},$$

and since this is also a function vanishing at zero, so it suffices to show that its derivative is nonnegative, that is,

$$\frac{1}{p(p-1)}f_+''(t) = (1+t)^{p-2} - t^{p-2} - (p-2)t^{p-3} \geq 0.$$

With $s = t^{-1}$,

$$\frac{1}{p(p-1)}f_+''(t) = \frac{(1+s)^{p-2} - 1 - (p-2)s}{s^{p-2}}.$$

By assumption $p \geq 3$, so the first term in the numerator above is convex and, therefore, $f_+''(t) \geq 0$ for all $t \in [0,1]$.

Consider now the derivative of f_-:

$$\frac{1}{p}f_-'(t) = -(1-t)^{p-1} - t^{p-1} + 1 + (p-1)t^{p-2},$$

which is nonnegative since $(1-t)^{p-1} + t^{p-1} \leq 1$. □

Theorem 2.5.4. *Let (Ω, μ) be a measure space and let $p \in [3, \infty)$ and $m \in \mathbb{N}$. Assume that $u_k \rightharpoonup u$ and $u_k \rightharpoondown u$ in $L^p(\Omega, \mu; \mathbb{R}^m)$. Then inequality with $v = 0$ (2.16) holds.*

Proof. From (2.17), it easily follows that

$$|u_k|^p \geq |u_k - u|^p + |u|^p + p|u|^{p-2}u \cdot (u_k - u) + p|u_k - u|^{p-2}(u_k - u) \cdot u.$$

Consider the integral of the inequality above over Ω. The integral of the second term in the right-hand side vanishes since $u_k \rightharpoonup u$, the integral of the third term vanishes, taking into account Theorem 2.1.3, since $u_k \rightharpoondown u$, and (2.16) follows. □

2.6 Bibliographic notes

For the notions of asymptotic radius and asymptotic center in the context of fixed-point theory, see Edelstein [40], and, for further details, the book of Goebel and Reich [63, pp. 18–22]. The proof of Theorem 2.2.4 is found, as Proof 2, in [63, p. 23]. Definition of asymptotic completeness and the Delta-compactness theorem are due to Teck-Cheong Lim [78]. Shortly after, an independent proof was provided by Tadeusz Kuczumow [74]. The proof of Delta-compactness in this book is a trivial adaptation to metric spaces of the proof of [62, Lemma 15.2] by Goebel and Kirk written for the case of Banach spaces. Notably, while the proof of Banach–Alaoglu theorem is dependent on the axiom of choice, the proof of the Delta-compactness theorem is not. We refer the reader to the survey [38] for a number of attempts to extend the notion of weak convergence to metric spaces, which indicates that, apart from definitions made for very specific situations, most definitions for a counterpart of weak convergence in metric space amount to Delta-convergence or its close modifications. Finding fixed points of nonexpansive maps as asymptotic centers of iterative sequences, which was a new and simpler proof of the Browder fixed-point theorem, is due to Michael Edelstein [40], with a generalization to metric spaces given by John Staples [114]. The notion of uniform rotundity from [114] generalizes uniform convexity of Banach spaces. The former implies asymptotic completeness of the space, which in turn implies its Delta-compactness, while the latter implies reflexivity of the space, which in turn implies its weak sequential compactness. The proof of asymptotic completeness (which implies Delta-compactness) of complete uniformly metric rotund spaces is given in [114]. Boundedness of Delta-convergent sequences (Theorem 2.3.8) is proved in [113], which also gives a characterization of Delta-convergence in uniformly smooth Banach spaces in terms of weak convergence in the conjugate space, although this is probably not the earliest reference.

The Opial property (sometimes called in literature nonstrict Opial property) was introduced in [96, Condition (2)] and Examples 2.4.9 are also taken from [96]. Theorem 2.4.10 is a weaker version of a theorem by D. van Dulst in [131]. The original theorem produces an equivalent norm satisfying a slightly modified version of (2.11) that uses the strict (for $y \neq x$) inequality, as it also appears in the definition of polar convergence in the early version of [113] and in [38]. Polar convergence, in fact, was an independently rediscovered, with a slight modification, Delta convergence, in particular, polar and Delta-convergence coincide in uniformly convex Banach spaces by Proposition 2.3.7, quoted here from [113]).

Inequality (2.15) is found in [113]. Theorem 2.5.4 was proved for the scalar-valued functions in [113] and for vector-valued functions in [8]. A counterexample showing that its result does not extend to $p < 3$, unless $p = 2$, is given in [8].

3 Cocompact embeddings with the rescaling group

This chapter presents several cocompact embeddings of functional spaces, relative to the rescaling group acting on $\mathbb{R}^N$,

$$\mathcal{G}^r = \{g_{j,y} : u \mapsto 2^{rj} u(2^j(\cdot - y)),\ j \in \mathbb{Z}, y \in \mathbb{R}^N\}, \quad r \in \mathbb{R}, \tag{3.1}$$

or to its subgroup, the group of integer shifts $\mathcal{G}_{\mathbb{Z}^N}$.

3.1 Definitions and elementary properties of cocompactness

Let us define an analog of $\mathcal{G}$-weak convergence (Definition 1.1.1) based, instead of weak convergence, on Delta-convergence.

Definition 3.1.1 ($\mathcal{G}$-Delta convergence). Let $\mathcal{G}$ be a set of homeomorphisms of a Banach space E. One says that a sequence $(u_n)_{n\in\mathbb{N}}$ in E is $\mathcal{G}$-Delta convergent to a point $u \in E$ relative to the set $\mathcal{G}$, if for any sequence $(g_n)_{n\in\mathbb{N}}$ in $\mathcal{G}$, $g_n(u_n - u)$ is Delta-convergent to zero in E. In this case, we use the notation $u_n \overset{\mathcal{G}}{\rightharpoonup} u$.

Remark 3.1.2. In face of the definition above, we can also define by analogy with Definition 1.1.4 a $\mathcal{G}$-Delta-cocompact embedding, but in this book we study only the cases where $\mathcal{G}$-Delta-cocompactness follows from $\mathcal{G}$-cocompactness.

Definition 3.1.3. Let $E \hookrightarrow F$ be two Banach spaces and let $\mathcal{G}$ be a bounded set of bounded linear operators on E. One says that the norm of F provides a *local metrization* of $\mathcal{G}$-weak convergence in E, if any sequence vanishing in F and bounded in E vanishes $\mathcal{G}$-weakly in E.

Example 3.1.4 (cf. Example 1.1.6). Metrization of $\mathcal{G}$-weak convergence is not unique. For example, given $p \in [1,\infty)$, all $\ell^q(\mathbb{Z})$-norms with $q > p$ are equivalent on a ball of $\ell^p(\mathbb{Z})$. If (u_k) is a bounded sequence in $\ell^p(\mathbb{Z})$, convergent to zero in $\ell^q(\mathbb{Z})$, $q > p$, then, for any sequence of integers (j_k), sequence $(u_k(\cdot - j_k))$ converges to zero by components. Since $(u_k(\cdot - j_k))$ is bounded in $\ell^p(\mathbb{Z})$, it is weakly convergent in $\ell^p(\mathbb{Z})$. Therefore, convergence of a sequence (u_k) in $\ell^q(\mathbb{Z})$ implies $\mathcal{G}_{\mathbb{Z}}$-weak convergence. In other words, each of the spaces $\ell^q(\mathbb{Z})$ with $q > p$ provides a local metrization of the $\mathcal{G}_{\mathbb{Z}}$-weak convergence in $\ell^p(\mathbb{Z})$.

Lemma 3.1.5. *Let $E \hookrightarrow F$ be two Banach spaces, and assume that F^* is dense in E^* and that the embedding $E \hookrightarrow F$ is cocompact relative to a bounded set $\mathcal{G}$ of bounded linear operators on E. Assume that operators in $\mathcal{G}$ extend continuously as operators on F and that the set of these extensions is bounded. Then the norm of F provides a local metrization of $\mathcal{G}$-convergence in E.*

Proof. It suffices to show that if a bounded sequence $(u_k)_{k\in\mathbb{N}}$ in E converges to zero in the norm of F, then it is $\mathcal{G}$-weakly convergent to zero in E. Let $u_k \to 0$ in F. By

https://doi.org/10.1515/9783110532432-003

assumptions on $\mathcal{G}$, there exists $C > 0$ such that for any sequence (g_k) in $\mathcal{G}$, $\|g_k u_k\|_F \le C\|u_k\|_F \to 0$ in F. Thus $g_k u_k \rightharpoonup 0$ in F, and since F^* is dense in E^*, we also have $g_k u_k \rightharpoonup 0$ in E. □

Lemma 3.1.6. *Let $V \hookrightarrow E \hookrightarrow F$ be three Banach spaces.*
(i) *If $\mathcal{G}$ is a set of homeomorphisms of V and the embedding $V \hookrightarrow E$ is $\mathcal{G}$-cocompact, then the embedding $V \hookrightarrow F$ is $\mathcal{G}$-cocompact.*
(ii) *If $\mathcal{G}$ is a set of homeomorphisms of E, whose restrictions to V are homeomorphisms of V, and the embedding $E \hookrightarrow F$ is $\mathcal{G}$-cocompact, then the embedding $V \hookrightarrow F$ is $\mathcal{G}$-cocompact.*

Proof. Assume that $u_n \overset{\mathcal{G}}{\rightharpoonup} 0$ in V.

Case (i). Since the embedding $V \hookrightarrow E$ is cocompact, then $u_n \to 0$ in E, and since E is continuously embedded into F, $u_n \to 0$ in F.

Case (ii). Since $E^* \hookrightarrow V^*$, $g_n u_n \rightharpoonup 0$ in V for every sequence $(g_n)_{n\in\mathbb{N}}$ in $\mathcal{G}$ (restricted to V) implies $g_n u_n \overset{\mathcal{G}}{\rightharpoonup} 0$ in E. Since the embedding $E \hookrightarrow F$ is $\mathcal{G}$-cocompact, we have $u_n \to 0$ in F. □

3.2 Cocompactness of the limiting Sobolev embedding

Consider the limiting Sobolev embedding $\dot{H}^{s,p}(\mathbb{R}^N) \hookrightarrow L^{p_s^*}(\mathbb{R}^N)$, $p_s^* = \frac{pN}{N-sp}$ with $0 < s < N/p$, $p \in (1,\infty)$. In the argument below we use the refined Sobolev inequality (10.25) from the Appendix.

Theorem 3.2.1. *Let $p \in (1,\infty)$, $s \in (0, N/p)$. Then the embedding $\dot{H}^{s,p}(\mathbb{R}^N) \hookrightarrow L^{p_s^*}(\mathbb{R}^N)$ is cocompact relative to the rescaling group* (3.1) *with $r = N/p - s$. Moreover, the $L^{p_s^*}(\mathbb{R}^N)$-norm provides a local metrization of $\mathcal{G}^r$-weak convergence in $\dot{H}^{s,p}(\mathbb{R}^N)$.*

Proof. Let $g_n u_n \rightharpoonup 0$ for any sequence $(g_n)_{N\in\mathbb{N}}$ in $\mathcal{G}^r$. Let us first show that

$$\|u_n\|_{\dot{B}^{s-N/p,\infty,\infty}} = \sup_{j\in\mathbb{Z}} \|2^{(s-N/p)j} P_j u_n\|_\infty \to 0 \tag{3.2}$$

(see the definition (3.23) of the Besov norm in the Appendix) or, equivalently, that for any sequence $(j_n)_{n\in\mathbb{N}}$ in $\mathbb{Z}$,

$$\|2^{-rj_n} P_{j_n} u_n\|_\infty \to 0, \tag{3.3}$$

which, if we set $v_n = g_{j_n,0} u_n$ and note that $v_n(\cdot - y_n) \rightharpoonup 0$ for any sequence $(y_n)_{n\in\mathbb{N}} \subset \mathbb{R}^N$, would follow from

$$\|P_0 v_n\|_\infty \to 0. \tag{3.4}$$

Indeed, choose any $s_0 > N/p$, so that $\dot{H}^{s_0,p}(\mathbb{R}^N)$ is compactly embedded into $C(\mathbb{R}^N)$. Since $P_0 v_n$ is bounded in $\dot{H}^{s_0,p}(\mathbb{R}^N)$, we have $v_n(\cdot - y_n) \rightharpoonup 0$ in $\dot{H}^{s_0,p}(\mathbb{R}^N)$ for any sequence $(y_n)_{n\in\mathbb{N}}$ in $\mathbb{R}^N$, and (3.4) follows by compactness of the embedding $\dot{H}^{s_0,p}(\mathbb{R}^N) \hookrightarrow C(\mathbb{R}^N)$. This yields (3.2).

Then substituting (3.2) into the refined Sobolev inequality (10.25) we have

$$\|u_n\|_{p_s^*} \leq C\|(-\Delta)^{\frac{s}{2}} u_n\|_p^{p/p_s^*} \|u_n\|_{\dot{B}^{s-N/p,\infty,\infty}}^{1-p/p_s^*} \to 0.$$

The local metrization property follows from Lemma 3.1.5. □

For the case $s = 1$, we give another proof of Theorem 3.2.1 that does not involve tools of harmonic analysis.

Proof. We may assume without loss of generality that $u_k \in C_0^\infty(\mathbb{R}^N)$. Let $(u_k)_{k\in\mathbb{N}}$ be a sequence in $\dot{H}^{1,p}(\mathbb{R}^N)$ and assume that for any $(j_k)_{k\in\mathbb{N}}$ in $\mathbb{Z}$ and any $(y_k)_{k\in\mathbb{N}}$ in $\mathbb{R}^N$, $g_{j_k,y_k}u_k \rightharpoonup 0$. Let $\chi \in C_0^\infty((\frac{1}{2},4),[0,3])$, such that $|\chi'| \leq 2$ for all t and $\chi(t) = t$ for $t \in [1,2]$. By continuity of the embedding $\dot{H}^{1,p}(\mathbb{R}^N) \hookrightarrow L^{p^*}(\mathbb{R}^N)$, we have for every $y \in \mathbb{Z}^N$,

$$\left(\int_{(0,1)^N+y} \chi(|u_k|)^{p^*} \mathrm{d}x\right)^{p/p^*} \leq C \int_{(0,1)^N+y} (|\nabla u_k|^p + \chi(|u_k|)^p)\mathrm{d}x,$$

from which follows, if we take into account that $\chi(t)^{p^*} \leq Ct^p$ for $t \geq 0$,

$$\begin{aligned}\int_{(0,1)^N+y} &\chi(|u_k|)^{p^*} \mathrm{d}x \\ &\leq C \int_{(0,1)^N+y} (|\nabla u_k|^p + \chi(u_k)^p)\mathrm{d}x \left(\int_{(0,1)^N+y} \chi(|u_k|)^{p^*} \mathrm{d}x\right)^{1-p/p^*} \\ &\leq C \int_{(0,1)^N+y} (|\nabla u_k|^p + \chi(|u_k|)^p)\mathrm{d}x \left(\int_{(0,1)^N+y} |u_k|^p \mathrm{d}x\right)^{1-p/p^*}.\end{aligned}$$

Adding the above inequalities over $y \in \mathbb{Z}^N$ and taking into account that $\chi(t)^p \leq Ct^{p^*}$ for $t \geq 0$, so that

$$\int_{\mathbb{R}^N} \chi(|u_k|)^p \mathrm{d}x \leq C\left(\int_{\mathbb{R}^N} |\nabla u_k|^p \mathrm{d}x\right)^{p^*/p} \leq C,$$

we get

$$\int_{\mathbb{R}^N} \chi(|u_k|)^{p^*} \mathrm{d}x \leq C \sup_{y\in\mathbb{Z}^N}\left(\int_{(0,1)^N+y} |u_k|^p \mathrm{d}x\right)^{1-p/p^*}. \tag{3.5}$$

Let $y_k \in \mathbb{Z}^N$ be such that

$$\sup_{y\in\mathbb{Z}^N}\left(\int_{(0,1)^N+y} |u_k|^p \mathrm{d}x\right)^{1-p/p^*} \leq 2\left(\int_{(0,1)^N+y_k} |u_k|^p \mathrm{d}x\right)^{1-p/p^*}.$$

Since $u_k(\cdot - y_k) \rightharpoonup 0$ in $\dot{H}^{1,p}(\mathbb{R}^N)$ and by the local compactness of subcritical Sobolev embeddings,

$$\int_{(0,1)^N + y_k} |u_k|^p \mathrm{d}x = \int_{(0,1)^N} |u_k(\cdot - y_k)|^p \mathrm{d}x \to 0.$$

Substituting this into (3.5), we get

$$\int_{\mathbb{R}^N} \chi(|u_k|)^{p^*} \mathrm{d}x \to 0.$$

Let

$$\chi_j(t) = 2^{rj}\chi(2^{-rj}t)), \quad j \in \mathbb{Z}.$$

Note that we may substitute for the original sequence u_k a sequence $g_{j_k,0}u_k$, with arbitrary $j_k \in \mathbb{Z}$, and so we have

$$\int_{\mathbb{R}^N} \chi_{j_k}(|u_k|)^{p^*} \mathrm{d}x \to 0. \tag{3.6}$$

Note now that, with $j \in \mathbb{Z}$,

$$\left(\int_{\mathbb{R}^N} \chi_j(|u_k|)^{p^*} \mathrm{d}x \right)^{p/p^*} \leq C \int_{2^{r(j-1)} \leq |u_k| \leq 2^{r(j+2)}} |\nabla u_k|^p \mathrm{d}x,$$

which can be rewritten as

$$\int_{\mathbb{R}^N} \chi_j(|u_k|)^{p^*} \mathrm{d}x \leq C \int_{2^{r(j-1)} \leq |u_k| \leq 2^{r(j+2)}} |\nabla u_k|^p \mathrm{d}x \left(\int_{\mathbb{R}^N} \chi_j(|u_k|)^{p^*} \mathrm{d}x \right)^{1-\frac{p}{p^*}}. \tag{3.7}$$

Adding the inequalities (3.7) over $j \in \mathbb{Z}$ and taking into account that the sets $2^{r(j-1)} \leq |u_k| \leq 2^{r(j+2)}$ cover $\mathbb{R}^N$ with a uniformly finite multiplicity, we obtain

$$\int_{\mathbb{R}^N} |u_k|^{p^*} \mathrm{d}x \leq C \int_{\mathbb{R}^N} |\nabla u_k|^p \mathrm{d}x \sup_{j \in \mathbb{Z}} \left(\int_{\mathbb{R}^N} \chi_j(|u_k|)^{p^*} \mathrm{d}x \right)^{1-p/p^*}. \tag{3.8}$$

Let j_k be such that

$$\sup_{j \in \mathbb{Z}} \left(\int_{\mathbb{R}^N} \chi_j(|u_k|)^{p^*} \mathrm{d}x \right)^{1-p/p^*} \leq 2 \left(\int_{\mathbb{R}^N} \chi_{j_k}(|u_k|)^{p^*} \mathrm{d}x \right)^{1-p/p^*},$$

and note that the right-hand side converges to zero due to (3.6). Then from (3.8), it follows that $u_k \to 0$ in $L^{p^*}(\mathbb{R}^N)$, which yields the cocompactness. □

Corollary 3.2.2. *Let $1 < p < \infty$, $s_0 \in \mathbb{R}$, $s \in (0, N/p)$, and let the number p_s^* be defined by $1/p_s^* = 1/p - s/N$. Then the embedding $\dot{H}^{s_0+s,p}(\mathbb{R}^N) \hookrightarrow \dot{H}^{s_0,p_s^*}(\mathbb{R}^N)$ is cocompact relative to the group* (3.1) *with $r = N/p - s$.*

Proof. By definition of potential Sobolev spaces, operator $(-\Delta)^{-s_0/2}$ acts isometrically from $\dot{H}^{s_0+s,q}(\mathbb{R}^N)$ to $\dot{H}^{s,q}(\mathbb{R}^N)$; apply Lemma 3.1.6. □

We now consider the space $\dot{H}^{1,p}(\mathbb{R}^N)$ for $N < p$, defined, as in the case $N > p$, as the completion of $C_0^\infty(\mathbb{R}^N)$ with respect to the gradient norm, which is known to have no continuous embedding into $L^1_{\mathrm{loc}}(\mathbb{R}^N)$, or indeed, into the space of distributions. Let us consider first the space $\dot{C}^{0,\lambda}(\mathbb{R}^N)$, with $\lambda > 0$, whose elements are equivalence classes of continuous functions, taken up to an additive constant, whose norm, given by

$$\sup_{x,y\in\mathbb{R}^N, x\neq y} \frac{|u(x) - u(y)|}{|x-y|^\lambda}, \tag{3.9}$$

is finite. This space is complete by the Arzela–Ascoli theorem. A well-known inequality (see [2], p. 100),

$$\sup_{x,y\in\mathbb{R}^N, x\neq y} \frac{|u(x) - u(y)|}{|x-y|^{\frac{p-N}{p}}} \le C\|\nabla u\|_p, \quad u \in C_0^\infty(\mathbb{R}^N), \tag{3.10}$$

means that $\dot{H}^{1,p}(\mathbb{R}^N)$ is continuously embedded into $\dot{C}^{0,\frac{p-N}{p}}(\mathbb{R}^N)$ and its elements, identified as functions up to an additive constant, have well-defined weak derivatives that belong to $L^p(\mathbb{R}^N)$.

Definition 3.2.3 (rescaling group $\mathcal{G}^r$ with $r < 0$). Definition of operators (3.1) for $r < 0$ extends to equivalence classes of functions on $\mathbb{R}^N$ in the sense that for each $j \in \mathbb{R}$, $y \in \mathbb{R}^N$,

$$g_{j,y}\tilde{u} = \left\{2^{rj}u(2^j \cdot -y) : \quad u \text{ is a representative of } \tilde{u}\right\}. \tag{3.11}$$

Note that group $\mathcal{G}^{\frac{N-p}{p}}$ acts isometrically on $\dot{H}^{1,p}(\mathbb{R}^N)$ and on $\dot{C}^{0,\frac{p-N}{p}}(\mathbb{R}^N)$.

Theorem 3.2.4. *Let $N < p < \infty$. The embedding $\dot{H}^{1,p}(\mathbb{R}^N) \hookrightarrow \dot{C}^{0,\frac{p-N}{p}}(\mathbb{R}^N)$ is cocompact relative to the group $\mathcal{G}^r$ with $r = \frac{N-p}{p} < 0$, and the norm* (3.9) *provides metrization of $\mathcal{G}^r$-weak convergence on $\dot{H}^{1,p}(\mathbb{R}^N)$.*

Proof. Let (u_k) be a sequence in $\dot{H}^{1,p}(\mathbb{R}^N)$ convergent $\mathcal{G}^r$-weakly to zero. In order to prove that it vanishes in the norm (3.9), it suffices to show that for any sequences (x_k) and (y_k) in $\mathbb{R}^N$, one has $|x_k - y_k|^{\frac{N-p}{p}}[u_k(x_k) - u_k(y_k)] \to 0$. Let $j_k \in \mathbb{Z}$ be such that $1 \le \frac{|x_k - y_k|}{2^{j_k}} \le 2$ and let $z_k = \frac{x_k - y_k}{2^{j_k}}$, $k \in \mathbb{N}$. Since the sequence (z_k) is bounded, we may assume without loss of generality that $z_k \to z \in \mathbb{R}^N$.

Let $(v_k)_{k\in\mathbb{N}}$ be a sequence in $\dot{H}^{1,p}(\mathbb{R}^N)$ represented by $2^{rj_k}u_k(y_k + 2^{j_k}\cdot)$. Then by definition of $\mathcal{G}^r$-weak convergence, we have $v_k \rightharpoonup 0$ in $\dot{H}^{1,p}(\mathbb{R}^N)$, and, in particular, $v_k(z) - v_k(0) \to 0$ in $\mathbb{R}$ for any sequence of representatives of v_k. Moreover, by (3.10)

$$|v_k(z_k) - v_k(z)| \le C\|\nabla v_k\|_p |z - z_k|^{\frac{p-N}{p}} \to 0,$$

and, therefore, $v_k(z_k) - v_k(0) \to 0$. Thus,

$$\begin{aligned}|x_k - y_k|^{\frac{N-p}{p}} &|u_k(x_k) - u_k(y_k)|\\ &\le 2\,2^{rj_k}|u_k(x_k) - u_k(y_k)|\\ &= 2\,2^{rj_k}[u_k(y_k + 2^{j_k}z_k) - u_k(y_k)]\\ &= 2[v_k(z_k) - v_k(0)] \to 0.\end{aligned}$$

Finally, since group $\mathcal{G}^r$ acts isometrically on $\dot{H}^{1,p}(\mathbb{R}^N)$ and on $\dot{C}^{0,-r}(\mathbb{R}^N)$ the last assertion of the theorem follows from Lemma 3.1.5. □

3.3 Embedding $\dot{H}^{1,p}(\mathbb{R}^N) \hookrightarrow L^{p^*,p}(\mathbb{R}^N)$ is not cocompact

Let $p \in (1, N)$. The Hardy inequality

$$\int_{\mathbb{R}^N} |\nabla u|^p \mathrm{d}x \ge \left(\frac{N-p}{p}\right)^p \int_{\mathbb{R}^N} \frac{|u|^p}{|x|^p}\mathrm{d}x \tag{3.12}$$

defines a continuous embedding $\dot{H}^{1,p}(\mathbb{R}^N) \hookrightarrow L^p(\mathbb{R}^N, \frac{\mathrm{d}x}{|x|^p})$ with norms invariant with respect to the dilation group

$$\mathcal{G} = \{u \mapsto 2^{s\frac{N-p}{p}} u(2^s\cdot)\}_{s\in\mathbb{R}}. \tag{3.13}$$

Proposition 3.3.1. *Embedding $\dot{H}^{1,p}_{\mathrm{rad}}(\mathbb{R}^N) \hookrightarrow L^p(\mathbb{R}^N, \frac{\mathrm{d}x}{|x|^p})$, $1 < p < N$, is not cocompact relative to the group* (3.13).

Proof. Let us define a sequence $(u_k)_{k\in\mathbb{N}}$ in $\dot{H}^{1,p}_{\mathrm{rad}}(\mathbb{R}^N)$ as follows. Let $\varphi \in C_0^\infty((1,2)) \setminus \{0\}$ and let

$$u_k(x) = \frac{1}{2k+1}\sum_{n=-k}^{n=k} 2^{n\frac{N-p}{p}} \varphi(2^n|x|). \tag{3.14}$$

Obviously, $u_k \in \dot{H}^{1,p}_{\mathrm{rad}}(\mathbb{R}^N)$, and the (positive) values of $\|\nabla u_k\|_p$ and of $\int_{\mathbb{R}^N} \frac{|u_k|^p}{|x|^p}\mathrm{d}x$ are independent of k. Once we show that $u_k \overset{\mathcal{G}}{\rightharpoonup} 0$, this will imply that the embedding is not cocompact. Indeed, for any $s_k \in \mathbb{R}$,

$$2^{s_k\frac{N-p}{p}} u_k(2^{s_k}x) = \frac{1}{2k+1}\sum_{n=-k}^{n=k} 2^{(n+s_k)\frac{N-p}{p}} \varphi(2^{(n+s_k)}|x|).$$

It suffices to consider two cases: $|s_k| \to \infty$ and (s_k) bounded. If $|s_k| \to \infty$, then for any compact set C there exists $k_0 > 0$ such that $2^{s_k \frac{N-p}{p}} u_k(2^{s_k} x) = 0$ for all $k \geq k_0$ and $x \in C$. If the sequence (s_k) is bounded, then $2^{s_k \frac{N-p}{p}} u_k(2^{s_k} \cdot)$ converges to zero uniformly on any compact set. Therefore, $2^{s_k \frac{N-p}{p}} u_k(2^{s_k} \cdot) \rightharpoonup 0$, and thus $u_k \overset{\mathcal{G}}{\rightharpoonup} 0$. □

Corollary 3.3.2. *Embedding $\dot{H}^{1,p}(\mathbb{R}^N) \hookrightarrow L^{p*,p}(\mathbb{R}^N)$, $1 < p < N$, is continuous but not cocompact relative to the rescaling group $\mathcal{G}^r$, (3.1), $r = \frac{N-p}{p}$.*

Proof. A quasinorm of a function u in the Lorentz space $L^{p*,p}(\mathbb{R}^N)$ can be given as the $L^p(\mathbb{R}^N, \frac{dx}{|x|^p})$-norm of the symmetric decreasing rearrangement $u^\star$ of u. By the Polya–Szegö inequality $\|\nabla u^\star\|_p \leq \|\nabla u\|_p$, (3.12) written for $u^\star$ yields a continuous embedding $\dot{H}^{1,p}(\mathbb{R}^N) \hookrightarrow L^{p*,p}(\mathbb{R}^N)$, $1 < p < N$.

Consider the sequence (3.14) and note that by the Hardy–Littlewood inequality the $L^p(\mathbb{R}^N, \frac{dx}{|x|^p})$-norm of $u_k^\star$ is bounded below by the $L^p(\mathbb{R}^N, \frac{dx}{|x|^p})$-norm of u_k, which is a positive number independent of k. On the other hand, it is easy to show that the sequence (3.14) (which we have seen to vanish $\mathcal{G}$-weakly relative to the group (3.13) so one has only regard the consequences of shifts) is $\mathcal{G}^r$-weakly convergent to zero. □

Remark 3.3.3. Embedding $\dot{H}^{1,p}(\mathbb{R}^N) \hookrightarrow L^{p*,q}(\mathbb{R}^N)$, $1 < p < N$, $q \in (p, p^*]$ is $\mathcal{G}^r$-cocompact, $r = \frac{N-p}{p}$. Indeed, if $u_k \overset{\mathcal{G}^r}{\rightharpoonup} 0$ in $\dot{H}^{1,p}(\mathbb{R}^N)$, then $u_k \to 0$ in $L^{p^*}(\mathbb{R}^N)$, and thus $u_k^\star \to 0$ in $L^{p^*}(\mathbb{R}^N)$. Since $(u_k^\star)$ is bounded in $L^p(\mathbb{R}^N, \frac{dx}{|x|^p})$, it will converge to zero by Hölder inequality in the norm of L^q, $q \in (p, p^*)$, with the corresponding weight, and thus $u_k \to 0$ in $L^{p^*,q}(\mathbb{R}^N)$.

3.4 Cocompactness and existence of minimizers

Cocompactness together with some convexity conditions allows to prove existence of extremal points in isoperimetric problems. We already gave two examples of use of cocompactness for finding extremal points, Example 1.1.8 and Example 1.3.7. The following existence result generalizes the latter. Note that the proof for general p cannot use (1.17), and resorts to a longer argument based on the Brezis–Lieb lemma for the gradient norm and a proof of a. e. convergence of the gradient.

Theorem 3.4.1. *Let $p \in (1, N)$, $p^* = \frac{pN}{N-p}$, let $f \in C^1_{\mathrm{loc}}(\mathbb{R})$ satisfy*

$$f(2^{\frac{N-p}{p} j} s) = 2^{(p^*-1)\frac{N-p}{p} j} f(s), \quad j \in \mathbb{Z}, s \in \mathbb{R}, \tag{3.15}$$

and set $F(s) = \int_0^s f(t) dt$. Assume that F satisfies (1.18) with $q = p^*$, and that, with some $\delta > 0$,

$$f(s)s \geq \delta F(s) > 0 \quad \textit{for all } s \in \mathbb{R} \setminus \{0\}. \tag{3.16}$$

Then the supremum in

$$c = \sup_{\|\nabla u\|_p \le 1} \int_{\mathbb{R}^N} F(u(x))\mathrm{d}x \tag{3.17}$$

is attained.

Note that (3.15) implies that $|f(s)| \le C|s|^{p^*-1}$, so $c < \infty$, and $|F(s)| \le C|s|^{p^*}$, and that $F(s) = |s|^{p^*}$ satisfies all the conditions of the theorem. Note also that it follows from (3.16) that $c > 0$.

Proof. Let $(u_n)_{n\in\mathbb{N}}$ in $\dot{H}^{1,p}(\mathbb{R}^N)$ be a maximizing sequence for (3.17), namely, $\|\nabla u_n\|_p \le 1$ and $\int_{\mathbb{R}^N} F(u_n(x))\mathrm{d}x \to c$. Note that the sequence $(g_n u_n)_{n\in\mathbb{N}}$ with any $g_n \in \mathcal{G}^{\frac{N-p}{p}}$ is also a maximizing sequence. Then there is a sequence $(g_n)_{n\in\mathbb{N}}$ in $\mathcal{G}^{\frac{N-p}{p}}$ such that $(g_n u_n)_{n\in\mathbb{N}}$ has a nonzero weak limit, since otherwise, by Theorem 3.2.1, $u_n \to 0$ in $L^{p^*}(\mathbb{R}^N)$, which implies

$$\int_{\mathbb{R}^N} F(u_n(x))\mathrm{d}x \le C \int_{\mathbb{R}^N} |u_n(x)|^{p^*} \mathrm{d}x \to 0,$$

a contradiction. We will now rename $(g_n u_n)$ with a nonzero weak limit u as (u_n).

It follows from the Ekeland's variational principle [42] (repeat the argument in [120, Corollary 5.3] replacing the whole Banach space with the differentiable manifold $\|\nabla u\|_p = 1$), that $(u_n)_{n\in\mathbb{N}}$ satisfies an asymptotic Lagrange multiplier relation: for some $\lambda_n \ge 0$, $n \in \mathbb{N}$,

$$\sup_{\|\nabla v\|_p = 1} \int_{\mathbb{R}^N} (|\nabla u_n(x)|^{p-2}\nabla u_n(x)\cdot\nabla v(x) - \lambda_n f(u_n(x))v(x))\mathrm{d}x \to 0. \tag{3.18}$$

Note that $\int_{\mathbb{R}^N} f(u_n)u_n \mathrm{d}x$ is bounded as $n \to \infty$, and that by (3.16) it is also bounded away from zero. Thus λ_n is also bounded and bounded away from zero so we may assume without loss of generality that $\lambda_n \to \lambda > 0$. Considering (3.18) for two different values of n, say $n = k$ and $n = j$, and taking $v = u_k - u_j$ (up to normalization), we get that, when $\min\{j, k\} \to \infty$,

$$\begin{aligned}(|\nabla u_k(x)|^{p-2}\nabla u_k - |\nabla u_j(x)|^{p-2}\nabla u_j(x))\cdot(\nabla u_k(x) - \nabla u_j(x))&\\ - \lambda(f(u_k(x)) - f(u_j(x)))(u_k(x) - u_j(x)) \to 0 \quad \text{in measure.}&\end{aligned} \tag{3.19}$$

Since (u_n) is a bounded sequence in $\dot{H}^{1,p}(\mathbb{R}^N)$, which is compactly embedded into $L^1(B)$ for any open ball $B \subset \mathbb{R}^N$, a renamed subsequence (u_n) converges almost everywhere, and thus from (3.19) we have, on a renamed subsequence, as $\min\{j, k\} \to \infty$,

$$(|\nabla u_k(x)|^{p-2}\nabla u_k - |\nabla u_j(x)|^{p-2}\nabla u_j(x))\cdot(\nabla u_k(x) - \nabla u_j(x)) \to 0 \quad \text{a.e. in } \mathbb{R}^N. \tag{3.20}$$

Then $(\nabla u_n(x))_{n\in\mathbb{N}}$ is a Cauchy sequence for a. e. $x \in \mathbb{R}^N$, and thus, (∇u_n) converges almost everywhere. Then, since $(u_n)_{n\in\mathbb{N}}$ has a weak limit $u \in \dot{H}^{1,p}(\mathbb{R}^N)$, we have necessarily that $\nabla u_n \to \nabla u$ a. e. in $\mathbb{R}^N$. By Remark 1.3.6 and Theorem 1.3.2, we have now

$$1 = \|\nabla u_n\|_p^p = \|\nabla u\|_p^2 + \|\nabla u_n - \nabla u\|_p^p + o(1), \tag{3.21}$$

$$c = \int_{\mathbb{R}^N} F(u_n)\mathrm{d}x + o(1) = \int_{\mathbb{R}^N} F(u)\mathrm{d}x + \int_{\mathbb{R}^N} F(u_n - u)\mathrm{d}x + o(1). \tag{3.22}$$

Let $t = \|\nabla u\|_p^p$. Then by (3.21), we have $\|\nabla u_n - \nabla u\|_p^p \to 1 - t$. Note that $\int_{\mathbb{R}^N} F(u(sx))\mathrm{d}x = s^{-N}\int_{\mathbb{R}^N} F(u(x))dx$ while $\|\nabla u(s\cdot)\|_p^p = s^{N-p}t$, so setting $s = t^{-\frac{1}{N-p}}$ we have $\int_{\mathbb{R}^N} F(u(sx))\mathrm{d}x \le ct^{\frac{N}{N-p}}$. Similarly, $\int_{\mathbb{R}^N} F(u_n(sx) - u(sx))\mathrm{d}x \le c(1-t)^{\frac{N}{N-p}} + o(1)$ By (3.22) this implies $1 \le t^{\frac{N}{N-p}} + (1-t)^{\frac{N}{N-p}}$ which can be true only if $t = 0$ or $t = 1$. Since $u \ne 0$, we have necessarily $t = 1$, which means that $u_n \to u$ in $\dot{H}^{1,p}(\mathbb{R}^N)$, and thus u is the maximizer. □

3.5 Cocompact embeddings of Besov and Triebel–Lizorkin spaces

Homogeneous Besov spaces $\dot{B}^{s,p,q}(\mathbb{R}^N)$ and homogeneous Triebel–Lizorkin spaces $\dot{F}^{s,p,q}\mathbb{R}^N$ are characterized by the respective equivalent norms,

$$\|u\|_{\dot{B}^{s,p,q}} = \big\|\big(\|2^{js}P_j u\|_{L^p}\big)_{j\in\mathbb{Z}}\big\|_{\ell^q}, \quad s \in \mathbb{R},\ 1 \le p \le \infty,\ 1 \le q \le \infty, \tag{3.23}$$

$$\|u\|_{\dot{F}^{s,p,q}} = \big\|\big\|(2^{js}P_j u)_{j\in\mathbb{Z}}\big\|_{\ell^q}\big\|_{L^p}, \quad s \in \mathbb{R},\ 1 \le p < \infty,\ 1 \le q \le \infty, \tag{3.24}$$

where operators P_j are defined in the Appendix, (10.22).

Like homogeneous Sobolev spaces these spaces have continuous embeddings into function spaces only when $r = \frac{N}{p} - s > 0$, while in general they are identified with classes of equivalence of functions modulo polynomials. An equivalent Besov norm (10.32) is defined in terms of wavelet coefficients.

Rescalings $\mathcal{G}^r$ (3.1) act on $\dot{B}^{s,p,q}(\mathbb{R}^N)$ and $\dot{F}^{s,p,q}$ isometrically with $r = \frac{N}{p} - s$. Indeed, norm (3.23) is invariant with respect to shifts, while actions of pure dilations $g_{i,0}$, $i \in \mathbb{Z}$, on $2^{sj}P_j u$ (with the appropriate value of r) yield shifts in $\ell^q(\mathbb{Z})$ that preserve the ℓ^q-norm. Similar reasoning applies to the norm (3.24). Note that elements of $\mathcal{G}^r$, when $r \le 0$, are well-defined on classes of equivalence modulo polynomials of a given degree, since translations and dilations on $\mathbb{R}^N$ map a polynomial into a polynomial of the same degree.

Theorem 3.5.1. *Let $s > t$ and assume that $\frac{N}{p} - s = \frac{N}{q} - t = r$. Then continuous embedding $\dot{B}^{s,p,a}(\mathbb{R}^N) \hookrightarrow \dot{B}^{t,q,b}(\mathbb{R}^N)$, $a, b, p, q \in [1, \infty]$, $b \ge a$, is cocompact relative to the group $\mathcal{G}^r$.*

Proof. Let $(u_n)_{n\in\mathbb{N}}$ be a sequence in $\dot{B}^{s,p,a}(\mathbb{R}^N)$, $\mathcal{G}^r$-weakly convergent to zero. Let $2^{rj}c_{j,k}(u_n)$ be the wavelet coefficients of u_n, given by (10.31). By Remark 10.2.2, given that $\frac{N}{p} - s = \frac{N}{q} - t = r$, the normalized basis and thus the wavelet expansion are the same in $\dot{B}^{s,p,a}$, $B^{t,q,a}$, and $B^{-r,\infty,\infty}$. Since the wavelet basis is a Schauder basis in $\dot{B}^{s,p,a}$ (see the Appendix, Section 10.2), the wavelet coefficients define continuous linear functionals $u \mapsto 2^{r_j} c_{j,k}(u)$ on $\dot{B}^{s,p,a}$, and thus $c_{0,0}(g_n u_n) \to 0$ for every sequence $(g_n)_{n\in\mathbb{N}}$ in $\mathcal{G}^r$. Therefore, $c_{j_n,k_n} \to 0$ for any sequence $(j_n, k_n) \subset \mathbb{Z} \times \mathbb{Z}^N$. Then from the wavelet representation of the Besov norm (10.32) we have

$$\|u_n\|_{B^{-r,\infty,\infty}} = \sup_{j\in\mathbb{Z},k\in\mathbb{Z}^N} |c_{j,k}(u_n)| \to 0. \tag{3.25}$$

Then, using again the equivalent Besov norm (10.32), we have, noting that $s > t$ implies $q > p$,

$$\begin{aligned}
\|u_n\|^a_{\dot{B}^{t,q,a}} &= \sum_{j\in\mathbb{Z}}\Big(\sum_{k\in\mathbb{Z}^N} |c_{j,k}(u_n)|^q\Big)^{\frac{a}{q}} \\
&\le \sum_{j\in\mathbb{Z}}\Big(\sum_{k\in\mathbb{Z}^N} |c_{j,k}(u_n)|^p\Big)^{\frac{a}{q}} \sup_{j\in\mathbb{Z},k\in\mathbb{Z}^N} |c_{j,k}(u_n)|^{a(1-p/q)} \\
&= s\|u_n\|^{ap/q}_{\dot{B}^{s,p,a}} \|u_n\|^{a(1-p/q)}_{\dot{B}^{-r,\infty,\infty}} \to 0.
\end{aligned}$$

Thus the embedding $\dot{B}^{s,p,a}(\mathbb{R}^N) \hookrightarrow \dot{B}^{t,q,a}(\mathbb{R}^N)$ is $\mathcal{G}^r$-cocompact. Then cocompactness of the embedding $\dot{B}^{s,p,a}(\mathbb{R}^N) \hookrightarrow \dot{B}^{t,q,b}(\mathbb{R}^N)$ follows from (10.27) and Lemma 3.1.6. □

$\mathcal{G}^r$-cocompactness of further embeddings involving homogeneous Besov and Triebel–Lizorkin spaces follows from the transitivity properties stated in Lemma 3.1.6 by combining Theorem 3.5.1 with known continuous embeddings (10.24)–(10.29).

Corollary 3.5.2. *Let $p, q \in [1,\infty)$, $s > t$, $r = N/p - s = N/q - t$, $1 \le a < q$, and $1 \le b \le \infty$. Then the embedding $\dot{B}^{s,p,a}(\mathbb{R}^N) \hookrightarrow \dot{F}^{t,q,b}(\mathbb{R}^N)$ is cocompact relative to the group $\mathcal{G}^r$.*

Proof. By using, in sequence, embeddings (10.28), (10.27), (10.26), and (10.29), we have the following chain of embeddings, fixing $q' \in (\max\{a,p\}, q)$ and setting t' satisfying $N/q' - t' = r$ so that $t' \in (t,s)$:

$$\begin{aligned}
\dot{B}^{s,p,a}(\mathbb{R}^N) &\hookrightarrow \dot{B}^{t',q',a}(\mathbb{R}^N) \hookrightarrow \dot{B}^{t',q',q'}(\mathbb{R}^N) \\
&= \dot{F}^{t',q',q'}(\mathbb{R}^N) \hookrightarrow \dot{F}^{t,q,b}(\mathbb{R}^N).
\end{aligned}$$

Note that $\mathcal{G}^r$ acts isometrically on each of these spaces. Since the first embedding in the chain is cocompact by Theorem 3.5.1, the resulting embedding is cocompact by Lemma 3.1.6. □

Corollary 3.5.3. *Let $p, q \in [1,\infty)$, $s > t$, $r = N/p - s = N/q - t$, $b \in (p,\infty]$ and $a \in [1,\infty]$. Then embedding $\dot{F}^{s,p,a}(\mathbb{R}^N) \hookrightarrow \dot{B}^{t,q,b}(\mathbb{R}^N)$ is cocompact relative to the group $\mathcal{G}^r$.*

Proof. Let $p < q' < \min\{q, b\}$ and let t' satisfy $N/q' - t' = r$, so that $t' \in (t,s)$. Invoking, in a sequence, embeddings (10.29), (10.26), (10.28), and (10.27), we have

$$\dot{F}^{s,p,a}(\mathbb{R}^N) \hookrightarrow \dot{F}^{t',q',q'}(\mathbb{R}^N) = \dot{B}^{t',q',q'}(\mathbb{R}^N) \hookrightarrow \dot{B}^{t,q,b}(\mathbb{R}^N).$$

Since the last embedding is cocompact by Theorem 3.5.1, the resulting embedding is cocompact by Lemma 3.1.6. □

Corollary 3.5.4. *Let $p, q \in [1, \infty)$, $s > t$, $r = N/p - s = N/q - t$ and $a, b \in [1, \infty]$. Then the embedding $\dot{F}^{s,p,a}(\mathbb{R}^N) \hookrightarrow \dot{F}^{t,q,b}(\mathbb{R}^N)$ is cocompact relative to the group $\mathcal{G}^r$.*

Proof. Let $q' \in (p, q)$ and let $N/q' - t' = r$ so that $t' \in (t, s)$. By Corollary 3.5.3 and then by Corollary 3.5.2, we have

$$\dot{F}^{s,p,a}(\mathbb{R}^N) \hookrightarrow \dot{B}^{t',q',q'}(\mathbb{R}^N) \hookrightarrow \dot{F}^{t,q,b}(\mathbb{R}^N). \tag{3.26}$$

Since both embeddings are cocompact, so is the resulting one by Lemma 3.1.6. □

Corollary 3.5.5. *Let $s > 0$, $p \in [1, N/s)$ and let $p_s^* = \frac{pN}{N-ps}$. Then embedding $\dot{F}^{s,p,a}(\mathbb{R}^N) \hookrightarrow L^{p_s^*}(\mathbb{R}^N)$, $a \in [1, \infty]$, is cocompact relative to the group $\mathcal{G}^r$, $r = N/p - s$.*

Proof. We use the identification (10.23) of Sobolev (and Lebesgue) spaces as Triebel–Lizorkin spaces and apply Corollary 3.5.4. □

Corollary 3.5.6. *Let $s > 0$, $p \in [1, N/s)$, $a, b \in [1, \infty]$, $a \leq b$, and let $p_s^* = \frac{pN}{N-ps}$. Then embedding $\dot{B}^{s,p,a}(\mathbb{R}^N) \hookrightarrow L^{p_s^*,b}(\mathbb{R}^N)$, is cocompact relative to the group $\mathcal{G}^r$, $r = N/p - s$.*

Proof. Let $t < s$ and let $r = N/p - s = N/q - t$. Combine the cocompact embedding from Theorem 3.5.1 with the known continuous embedding (10.30):

$$\dot{B}^{s,p,a}(\mathbb{R}^N) \hookrightarrow \dot{B}^{t,q,a}(\mathbb{R}^N) \hookrightarrow L^{q_t^*,a}(\mathbb{R}^N). \tag{3.27}$$

Note now that $q_t^* = p_s^*$ and that $L^{p_s^*,a}(\mathbb{R}^N) \hookrightarrow L^{p_s^*,b}(\mathbb{R}^N)$ for all $b \geq a$. The resulting embedding is $\mathcal{G}^r$-cocompact by Lemma 3.1.6. □

Corollary 3.5.7. *Let $s > 0$, $p \in [1, N/s)$ and let $p_s^* = \frac{pN}{N-ps}$. Then embedding $\dot{B}^{s,p,a}(\mathbb{R}^N) \hookrightarrow L^{p_s^*}(\mathbb{R}^N)$, $1 \leq a \leq p_s^*$ is cocompact relative to the group $\mathcal{G}^r$, $r = N/p - s$.*

Proof. The statement follows from Corollary 3.5.6 and the identification of $L^{p_s^*}$ as the Lorentz space $L^{p_s^*,p_s^*}$. □

Corollary 3.5.8. *For all $a, p \in [1, \infty)$, embeddings $\dot{B}^{N/p,p,a} \hookrightarrow \mathrm{BMO}$ and $\dot{F}^{N/p,p,a} \hookrightarrow \mathrm{BMO}$ are $\mathcal{G}^0$-cocompact. In particular, embedding $\dot{H}^{s,N/s}(\mathbb{R}^N) \hookrightarrow \mathrm{BMO}$, $0 < s < N$, is $\mathcal{G}^0$-cocompact.*

Proof. Apply, respectively, Corollary 3.5.2 and Corollary 3.5.4 with the target space $\dot{F}^{0,\infty,2} = BMO$. □

Remark 3.5.9. by Lemma 3.1.5, each of the norms of target spaces of embeddings in Theorem 3.5.1 and its corollaries provides a local metrization of the $\mathcal{G}^r$-weak convergence.

3.6 Cocompactness and interpolation

In this section, we show that cocompactness of embeddings under some general conditions is inherited under interpolation of spaces. We deal here only with the case when the scaling operators do not change from space to space. We refer the reader for the definitions of interpolation spaces to the Appendix, Section 10.1.

Definition 3.6.1. Let (E_0, E_1) be a Banach couple with E_1 continuously embedded in E_0 and let $\mathcal{G}$ be a set of linear operators $g : E_0 + E_1 \to E_0 + E_1$ which satisfies

$$g(A_j) \subset A_j \text{ and } g : A_j \to A_j \text{ is an isometry for } j = 0, 1. \tag{3.28}$$

Let E_1 be continuously embedded into some Banach space F_1. A family of bounded operators $\{M_t\}_{t\in(0,1)}$ from E_0 to E_1 is said to be a *family of $\mathcal{G}$-covariant mollifiers* (relative to a space F_1) if it satisfies the following conditions:

(i) For $j = 0, 1$, the norm of M_t as a continuous map from E_j into itself is bounded independently of $t \in (0,1)$, that is, $\sup_{t\in(0,1)} \|M_t\|_{E_j \to E_j} < \infty$.
(ii) The function $\sigma(t) \stackrel{\text{def}}{=} \|I - M_t\|_{E_1 \to F_1}$ satisfies $\lim_{t\to 0} \sigma(t) = 0$.
(iii) For each $g \in \mathcal{G}$, and $t \in (0,1)$, there exists an element $h_{g,t} \in \mathcal{G}$ such that $gM_t = M_t h_{g,t}$.

For the present, we show that Definition 3.6.1 is satisfied by the classical mollifiers.

Lemma 3.6.2. *Let* $(E_0, E_1) = (L^p(\mathbb{R}^N), H^{1,p}(\mathbb{R}^N))$, $1 \le p < N$, *and* $F_1 = L^r(\mathbb{R}^N)$ *with* $r \in (p, p^*)$. *Let $\mathcal{G}$ be the group of integer shifts*

$$\mathcal{G}_{\mathbb{Z}^N} = \{g_y : u \mapsto u(\cdot - y)\}_{y\in\mathbb{Z}^N}. \tag{3.29}$$

Let $\rho : \mathbb{R}^N \to [0, \infty)$ *be a smooth function supported in the open unit ball* $\{z \in \mathbb{R}^N : |z| < 1\}$ *which satisfies* $\int_{\mathbb{R}^N} \rho(x)dx = 1$.

Then, for each fixed $t \in (0,1)$ *the operator* M_t, *which is defined by*

$$(M_t u)(x) = \int\limits_{|z|<1} \rho(z)u(x + tz)dz\,, \tag{3.30}$$

is a bounded map of E_0 *into* E_1, *and the family* $\{M_t\}_{t\in(0,1)}$ *satisfies properties* (i), (ii), *and* (iii) *of Definition* 3.6.1.

Proof. The boundedness of M_t from E_0 into E_1 for each fixed t is simply the well- known mollification property. It is also obvious that $M_t : E_j \to E_j$ is bounded with $\|M_t\|_{E_j\to E_j} \le 1$ for $j = 0, 1$ and all $t \in (0,1)$, which gives property (i).

Property (iii) is an immediate consequence of the fact that $(M_t u)(\cdot - y) = M_t(u(\cdot - y))$ for each $y \in \mathbb{R}^N$. In fact, here we can take $h_{g,t} = g$ for each $g \in \mathcal{G}_{\mathbb{Z}^N}$ and each t.

It remains to prove property (ii). Consider the following identity:

$$u(x) - M_t u(x) = \int_{|z|<1} \rho(z)[u(x) - u(x+tz)]\mathrm{d}z = -\int_{|z|<1} \rho(z) \int_0^t z \cdot \nabla u(x+sz)\mathrm{d}s\mathrm{d}z.$$

Then

$$\left|u(x) - M_t u(x)\right|^p \leq \sup_{|y|<1} \rho(y)^p \left| \int_0^t \int_{|z|<1} \left|\nabla u(x+sz)\right| \mathrm{d}z\,\mathrm{d}s \right|^p.$$

By Hölder's inequality, we then have

$$\left|u(x) - M_t u(x)\right|^p \leq C t^{p/p'} \int_0^t \int_{|z|<1} \left|\nabla u(x+sz)\right|^p \mathrm{d}z\,\mathrm{d}s.$$

Integrating with respect to x, we obtain

$$\begin{aligned} \int_{\mathbb{R}^N} \left|u(x) - M_t u(x)\right|^p \mathrm{d}x &\leq C t^{p/p'} \int_0^t \int_{|z|<1} \int_{\mathbb{R}^N} \left|\nabla u(x+sz)\right|^p \mathrm{d}x\,\mathrm{d}z\,\mathrm{d}s \\ &= C t^{p/p'} \int_0^t \int_{|z|<1} \int_{\mathbb{R}^N} \left|\nabla u(x)\right|^p \mathrm{d}x\,\mathrm{d}z\,\mathrm{d}s \\ &= C t^{1+p/p'} \int_{\mathbb{R}^N} \left|\nabla u(x)\right|^p \mathrm{d}x. \end{aligned} \tag{3.31}$$

Here, and also later, we will use the following immediate consequence of Hölder's inequality: The inclusion $L^{p_0} \cap L^{p_1} \subset L^p$ holds whenever $1 \leq p_0 < p < p_1 \leq \infty$. Furthermore, the estimate

$$\|f\|_p \leq \|f\|_{p_0}^{1-\theta} \|f\|_{p_1}^{\theta} \tag{3.32}$$

holds for each $f \in L^{p_0} \cap L^{p_1}$, where $\theta = \frac{\frac{1}{p_0} - \frac{1}{p}}{\frac{1}{p_0} - \frac{1}{p_1}} \in (0,1)$.

Let s be some number satisfying $r < s < p^*$. Then $p < r < s$ and so (3.32) gives us that

$$\|u - M_t u\|_r \leq \|u - M_t u\|_p^{1-\theta} \|u - M_t u\|_s^{\theta}, \quad \text{where } \theta = \frac{\frac{1}{p} - \frac{1}{r}}{\frac{1}{p} - \frac{1}{s}} \in (0,1). \tag{3.33}$$

We estimate $\|u - M_t u\|_p$ and $\|u - M_t u\|_s$ using, respectively, (3.31) and the Sobolev embedding theorem. Substituting these estimates in (3.33), and noting that $1 + p/p' = p$, we obtain that

$$\|u - M_t u\|_r \le C\left(t^p \|u\|_{H^{1,p}}\right)^{1-\theta}\left(\|u - M_t u\|_{H^{1,p}}\right)^{\theta}$$
$$\le C t^{(1-\theta)p} \|u\|_{H^{1,p}} .$$

This establishes property (ii) and completes the proof of the lemma. □

An abstract statement concerning cocompactness of embeddings under interpolation of spaces is as follows.

Theorem 3.6.3. *Let (E_0, E_1) and (F_0, F_1) be two compatible couples of Banach spaces with E_j continuously embedded in F_j for $j = 0, 1$. Suppose, further, that E_1 is continuously embedded in E_0. Let $\mathcal{G}$ be a set of linear operators $g : F_0 + F_1 \to F_0 + F_1$ which satisfies (3.28) with respect to both of the couples (E_0, E_1) and (F_0, F_1). Assume that there exists a $\mathcal{G}$-covariant mollifier family $\{M_t : E_0 \to E_1\}_{t\in(0,1)}$ (see Definition 3.6.1). If, furthermore, E_1 is $\mathcal{G}$-cocompactly embedded into F_1, then, for every $\theta \in (0, 1)$ and $q \in [1, \infty]$, the space $(E_0, E_1)_{\theta,q}$ is $\mathcal{G}$-cocompactly embedded into $(F_0, F_1)_{\theta,q}$ and the space $[E_0, E_1]_\theta$ is $\mathcal{G}$-cocompactly embedded into $[F_0, F_1]_\theta$.*

Proof. We consider the case of real interpolation. The proof for the complex case is completely analogous.

In view of the continuous embedding $(E_0, E_1)_{\theta,q} \hookrightarrow E_0 + E_1 = E_0$, it follows that, for each fixed t, the operator M_t is bounded from $(E_0, E_1)_{\theta,q}$ into E_1. Suppose that $u_k \overset{\mathcal{G}}{\rightharpoonup} 0$ in $(E_0, E_1)_{\theta,q}$. Let $(g_k)_{k\in\mathbb{N}}$ be an arbitrary sequence in $\mathcal{G}$. Then

$$g_k M_t u_k = M_t h_{g_k,t} u_k \tag{3.34}$$

by property (iii). Since $h_{g_k,t} u_k \rightharpoonup 0$ in $(E_0, E_1)_{\theta,q}$, we deduce that $M_t h_{g_k,t} u_k \rightharpoonup 0$ in E_1 for each fixed $t \in (0, 1)$. The cocompactness of the embedding $E_1 \hookrightarrow F_1$ and (3.34) now imply that

$$\lim_{k\to\infty} \|M_t u_k\|_{F_1} = 0 . \tag{3.35}$$

In view of the continuous inclusions $E_j \hookrightarrow F_j$ and property (i), we have that $M_t : E_j \to F_j$ is bounded with

$$S_j \overset{\text{def}}{=} \sup_{t\in(0,1)} \|M_t\|_{E_j \to F_j} < \infty , \text{ for } j = 0, 1 . \tag{3.36}$$

Since $M_t u_k \in F_0 \cap F_1$, we can invoke (10.11) in the Appendix and then (3.36) to obtain that

$$\|M_t u_k\|_{(F_0,F_1)_{\theta,q}} \le c_{\theta,q} \|M_t u_k\|_{F_0}^{1-\theta} \|M_t u_k\|_{F_1}^{\theta}$$
$$\le c_{\theta,q} \left(S_0 \|u_k\|_{E_0}\right)^{1-\theta} \|M_t u_k\|_{F_1}^{\theta} .$$

Since $(u_k)_{k\in\mathbb{N}}$ is necessarily a bounded sequence in the space $(E_0, E_1)_{\theta,q}$ and is therefore also bounded in the space E_0, we can use (3.35) to obtain that

$$\lim_{k\to\infty} \|M_t u_k\|_{(F_0,F_1)_{\theta,q}} = 0 . \tag{3.37}$$

We now consider the operator $I - M_t$ in more detail. By (3.36), we of course have $I - M_t : E_0 \to F_0$ with $\|I - M_t\|_{E_0\to F_0} \le \|I\|_{E_0\to F_0} + S_0$. Using this estimate, property (ii) and Theorem 10.1.6 from the Appendix, we obtain that $I - M_t$ is a bounded operator from $(E_0, E_1)_{\theta,q}$ into $(F_0, F_1)_{\theta,q}$ and that

$$\begin{aligned}\|I - M_t\|_{(E_0,E_1)_{\theta,q}\to(F_0,F_1)_{\theta,q}} &\le \|I - M_t\|^{1-\theta}_{E_0\to F_0} \|I - M_t\|^{\theta}_{E_1\to F_1} \\ &\le \left(\|I\|_{E_0\to F_0} + S_0\right)^{1-\theta} \sigma(t)^{\theta} .\end{aligned}$$

Therefore, with the help of (3.37), we have

$$\begin{aligned}\limsup_{k\to\infty} \|u_k\|_{(F_0,F_1)_{\theta,q}} &\le \limsup_{k\to\infty} \|M_t u_k\|_{(F_0,F_1)_{\theta,q}} + \limsup_{k\to\infty} \left\|(I - M_t)u_k\right\|_{(F_0,F_1)_{\theta,q}} \\ &\le 0 + \limsup_{k\to\infty} \left\|(I - M_t)u_k\right\|_{(F_0,F_1)_{\theta,q}} \\ &\le \limsup_{k\to\infty} \left(\|I\|_{E_0\to F_0} + S_0\right)^{1-\theta} \sigma(t)^{\theta} \|u_k\|_{(E_0,E_1)_{\theta,q}} .\end{aligned}$$

We now use the boundedness of the sequence $(\|u_k\|_{(E_0,E_1)_{\theta,q}})_{k\in\mathbb{N}}$ once more, together with property (ii), to obtain that this last expression is bounded by a quantity which tends to 0 as t tends to 0. Since we can choose t as small as we please, this shows that $\lim_{k\to\infty} \|u_k\|_{(F_0,F_1)_{\theta,q}} = 0$ and completes the proof of the theorem. □

We will apply Theorem 3.6.3 to prove cocompactness of embeddings of inhomogeneous Sobolev spaces in the next section. We would like to remark, however, that if E_0 and E_1 are functional spaces with rescalings groups $\mathcal{G}^{r_0}$ and $\mathcal{G}^{r_1}$, respectively, with $r_0 \neq r_1$, their interpolations will also be scale-invariant.

Proposition 3.6.4. *Let X_0, X_1 be two compatible Banach spaces of functions from $\mathbb{R}^N$ to $\mathbb{R}^m$, $m \in \mathbb{N}$, equipped with respective norms $\|\cdot\|_0$, $\|\cdot\|_1$. Assume that there exist $r_0, r_1 \in \mathbb{R}$, such that the norm $\|\cdot\|_i$ is invariant with respect to the action of the respective group $\mathcal{G}^{r_i}$, $i = 0, 1$. Then for any $q \in (0, \infty]$, $\theta \in (0, 1)$, the norms of the interpolated, by the real method, space $X_{\theta,q}$, and of that by the complex method, X_θ, are invariant with respect to the group $\mathcal{G}^{r_\theta}$, $r_\theta =: (1 - \theta)r_0 + \theta r_1$.*

Proof. 1. Real method of interpolation. Let us calculate the value of the K-functional (10.4) under rescalings $g_{j,y} \in \mathcal{G}^0$, $j \in \mathbb{Z}$, $y \in \mathbb{R}^N$, noting first that $\|g_{j,y} x_i\|_{X_i} = 2^{-r_i j} \|x_i\|_{X_i}$, $i = 0, 1$, and thus,

$$\begin{aligned}&K(g_{j,y}x, t; X_0, X_1) \\ &\quad = \inf\{2^{-r_0 j} \|x_0\|_{X_0} + t2^{-r_1 j} \|x_1\|_{X_1} \ : \ x = x_0 + x_1,\ x_0 \in X_0,\ x_1 \in X_1\}\end{aligned}$$

$$= 2^{-r_0 j} K(x, 2^{(r_0-r_1)j} t; X_0, X_1).$$

Substituting this into the definition of the interpolated norm (10.5), we get for $0 < \theta < 1$, $1 \le q < \infty$, with $g_{j,y} \in \mathcal{G}^0$, $j \in \mathbb{Z}$, $y \in \mathbb{R}^N$,

$$\begin{aligned}
\|g_{j,y}x\|_{\theta,q} &= \left(\int_0^\infty (t^{-\theta} K(g_{j,y}x, t; X_0, X_1))^q \frac{dt}{t} \right)^{\frac{1}{q}} \\
&= 2^{-r_0 j} \left(\int_0^\infty (t^{-\theta} K(x, 2^{(r_0-r_1)j} t; X_0, X_1))^q \frac{dt}{t} \right)^{\frac{1}{q}} \\
&= 2^{-r_\theta j} \left(\int_0^\infty (s^{-\theta} K(x, s; X_0, X_1))^q \frac{ds}{s} \right)^{\frac{1}{q}} \\
&= 2^{-r_\theta j} \|x\|_{\theta,q},
\end{aligned}$$

which proves the real method case of the proposition for $q < \infty$. For $q = \infty$, we have similarly from (10.6)

$$\begin{aligned}
\|g_{j,y}x\|_{\theta,\infty} &= \sup_{t>0} t^{-\theta} K(g_{j,y}x, t; X_0, X_1) \\
&= 2^{-r_0 j} \sup_{t>0} t^{-\theta} K(x, 2^{(r_0-r_1)j} t; X_0, X_1) = 2^{-r_\theta j} \|x\|_{\theta,\infty}.
\end{aligned}$$

2. Complex method of interpolation. Let $G_{j,\eta} : \mathcal{F} \to \mathcal{F}$ (for the definition of the space $\mathcal{F}$ see the Appendix), $j \in \mathbb{Z}$, $\eta \in \mathbb{R}^N$, be given by $G_{j,\eta} f(z) = 2^{(1-z)r_0 j + z r_1 j} g_{j,\eta} f(z)$, $\operatorname{Re} z \in (0,1)$, where $g_{j,\eta} \in \mathcal{G}^0$. Let us show that operators $G_{j,\eta}$ are isometries on $\mathcal{F}$. Indeed, using scaling properties of the norms of X_0 and of X_1, we have

$$\begin{aligned}
\|G_{j,\eta} f\|_{\mathcal{F}} &= \max\left\{ \sup_{y\in\mathbb{R}} \|G_{j,\eta} f(iy)\|_{X_0}, \sup_{y\in\mathbb{R}} \|G_{j,\eta} f(1+iy)\|_{X_1} \right\} \\
&= \max\left\{ \sup_{y\in\mathbb{R}} \|2^{(1-iy)r_0 j} g_{j,\eta} f(iy)\|_{X_0}, \sup_{y\in\mathbb{R}} \|2^{(1+iyz)r_1 j} g_{j,\eta} f(1+iy)\|_{X_1} \right\} && (3.38) \\
&= \|f\|_{\mathcal{F}}. && (3.39)
\end{aligned}$$

Thus, with $\theta \in (0,1)$,

$$\begin{aligned}
\|2^{r_\theta j} g_{j,\eta} u\|_{X_\theta} &= \inf\{\|G_{j,\eta} f\|_{\mathcal{F}} : G_{j,\eta} f(\theta) = 2^{r_\theta j} g_{j,\eta} u\} \\
&= \inf\{\|f\|_{\mathcal{F}} : f(\theta) = u\} = \|u\|_{X_\theta}. && (3.40)
\end{aligned}$$

□

3.7 Cocompact embeddings of inhomogeneous Besov spaces

We first apply results of the previous section to prove cocompactness of Sobolev–Peetre embeddings of inhomogeneous Sobolev spaces $H^{s,p}(\mathbb{R}^N)$ relative to the group of integer shifts $\mathcal{G}_{\mathbb{Z}^N}$, followed by analogous result for inhomogeneous Besov spaces. Later, in Section 3.8, we give an alternative proof of cocompactness of general inhomogeneous spaces handled as intersections of corresponding homogeneous spaces with Lebesgue spaces.

Theorem 3.7.1. *Let $s \in (0,\infty)$, $p \in (1, N/s)$ and $p < q < p_s^* \stackrel{\text{def}}{=} \frac{pN}{N-sp}$. Then embedding $H^{s,p}(\mathbb{R}^N) \hookrightarrow L^q(\mathbb{R}^N)$ is cocompact relative to the group of integer shifts $\mathcal{G}_{\mathbb{Z}^N}$. Moreover, the embedding $H^{s+\gamma,p}(\mathbb{R}^N) \hookrightarrow H^{\gamma,q}(\mathbb{R}^N)$ is $\mathcal{G}_{\mathbb{Z}^N}$-cocompact for every $\gamma > 0$.*

Proof. Let $q \in (p, p_s^*)$ be fixed. Case 1: $s = 1$. The assertion of the theorem is proved for this case in Theorem 1.1.9.

Case 2: $s \in (0,1)$. For the number $p \in (1,\infty)$ appearing in the statement of Theorem 3.7.1, and for some number r in (p,p^*) we let (E_0, E_1) and (F_0, F_1) be the same couples $(L^p(\mathbb{R}^N), H^{1,p}(\mathbb{R}^N))$ and $(L^p(\mathbb{R}^N), L^r(\mathbb{R}^N))$ which appear in Lemma 3.6.2. Let us also choose the group $\mathcal{G}$ and the family of operators $\{M_t\}_{t\in(0,1)}$ to be as in Lemma 3.6.2.

We know, using Theorem 1.1.9, that A_1 is $\mathcal{G}_{\mathbb{Z}^N}$-cocompactly embedded in B_1. This, together with Lemma 3.6.2, provides us with all the conditions required for applying Theorem 3.6.3 in this context. More specifically, if we invoke the statement about complex interpolation spaces in Theorem 3.6.3, we obtain that $[L^p(\mathbb{R}^N), H^{1,p}(\mathbb{R}^N)]_\theta$ is $\mathcal{G}_{\mathbb{Z}^N}$-cocompactly embedded in $[L^p(\mathbb{R}^N), L^r(\mathbb{R}^N)]_\theta$ for each $\theta \in (0,1)$. By standard results ((10.13) and (10.12) in the Appendix), these two spaces are $H^{\theta,p}(\mathbb{R}^N)$ and $L^\gamma(\mathbb{R}^N)$ respectively, where γ is the number in the interval (p,r) given by

$$\frac{1}{\gamma} = \frac{1-\theta}{p} + \frac{\theta}{r}. \tag{3.41}$$

Setting $\theta = s$, we see that this establishes our result for $q = \gamma$. It will now be easy to extend the proof to all $q \in (p, p_s^*)$:

Let $(u_k)_{k\in\mathbb{N}}$ be an arbitrary sequence in $H^{s,p}$ which converges $\mathcal{G}_{\mathbb{Z}^N}$-weakly to 0.

Given an arbitrary q in (p, p_s^*), we choose $r \in (p, p^*)$ sufficiently close to p so that the number γ given by (3.41), with $\theta = s$, satisfies $p < \gamma < q$. By the previous step of our argument, we also have that $\lim_{k\to\infty} \|u_k\|_{L^\gamma(\mathbb{R}^N)} = 0$. Now let us choose some number $\delta \in (q, p_s^*)$. By the Sobolev embedding theorem, the sequence $(u_k)_{k\in\mathbb{N}}$, which is bounded in $H^{1,p}(\mathbb{R}^N)$, must also be bounded in $L^\delta(\mathbb{R}^N)$. Finally, we use the Hölder inequality to bound $\|u_k\|_q$ by $\|u_k\|_\gamma^{1-t}\|u_k\|_\delta^t$ for a suitable number $t \in (0,1)$. This suffices to complete the proof of Theorem 3.7.1 for the case $s \in (0,1)$.

Case 3: $s > 1$. Let p and q be as in the statement of the theorem. Noting that we always have $p < p^*$, let us choose numbers q_0 and q_1 which satisfy $p < q_0 < \min\{p^*, q\}$ and $q < q_1 < p_s^*$. Consider an arbitrary sequence $(u_k)_{k\in\mathbb{N}}$ in $H^{s,p}(\mathbb{R}^N)$ which is $\mathcal{G}_{\mathbb{Z}^N}$-weakly convergent to zero. Since in this case $H^{s,p}(\mathbb{R}^N)$ is continuously embedded into $H^{1,p}(\mathbb{R}^N)$, we have that $u_k(\cdot - y_k) \rightharpoonup 0$ in $H^{1,p}(\mathbb{R}^N)$ for any sequence $(y_k)_{k\in\mathbb{N}}$ of elements of $\mathbb{Z}^N$, that is, (u_k) is $\mathcal{G}_{\mathbb{Z}^N}$-weakly convergent in $H^{1,p}(\mathbb{R}^N)$. Then, by Example 1.1.9, $\lim_{k\to\infty} \|u_k\|_{q_0} = 0$.

Since $q_0 < q < q_1$, by Hölder inequality we have

$$\|u_k\|_q \leq \|u_k\|_{q_0}^{1-\theta} \|u_k\|_{q_1}^{\theta}, \quad \text{where } \theta = \frac{\frac{1}{q_0} - \frac{1}{q}}{\frac{1}{q_0} - \frac{1}{q_1}} \in (0,1). \tag{3.42}$$

Then, since $H^{s,p}(\mathbb{R}^N)$ is continuously embedded into $L^{q_1}(\mathbb{R}^N)$, we have $\|u_k\|_q \leq C\|u_k\|_{q_0}^{1-\theta}\|u_k\|_{H^{s,p}}^{\theta}$. Since weakly convergent sequences are bounded, we obtain that $\|u_k\|_q \leq C\|u_k\|_{q_0}^{\theta} \to 0$. □

Since Sobolev–Slobodecky spaces $W^{s,p}(\mathbb{R}^N)$ can be obtained by interpolation of Sobolev spaces (see (10.17) in the Appendix), we have the following.

Corollary 3.7.2. *Suppose that $s \in [1,\infty)$ and $p \in (1, \frac{N}{\lfloor s \rfloor+1})$. Embedding $W^{s,p}(\mathbb{R}^N) \hookrightarrow L^q(\mathbb{R}^N)$ is cocompact relative to the group of integer shifts $\mathcal{G}_{\mathbb{Z}^N}$ whenever $p < q < p_s^* \stackrel{\text{def}}{=} \frac{pN}{N-sp}$.*

Proof. The case $s \in \mathbb{N}$ is already proved in Theorem 3.7.1. Let us fix $k \in \mathbb{N}$, $s \in (k, k+1)$, and apply Theorem 3.6.3, with the real method of interpolation, to couples $(H^{k,p}(\mathbb{R}^N), H^{k+1,p}(\mathbb{R}^N))$ and $(L^\alpha(\mathbb{R}^N), L^\beta(\mathbb{R}^N))$, where $\alpha \in (p, p_k^*)$ and $\beta \in (p, p_{k+1}^*)$. Note that conditions of Theorem 3.6.3 are verified by Lemma 3.6.2, and we have $W^{s,p}(\mathbb{R}^N) = (H^{k,p}(\mathbb{R}^N), H^{k+1,p}(\mathbb{R}^N))_{s-k,p}$, and $L^{s,r}(\mathbb{R}^N) = (L^\alpha(\mathbb{R}^N), L^\beta(\mathbb{R}^N))_{s-k,p}$, with $\frac{1}{r} = \frac{k+1-s}{\alpha} + \frac{s-k}{\beta}$. Thus $W^{s,p}(\mathbb{R}^N)$ is cocompactly embedded into $L^r(\mathbb{R}^N)$ with any r greater between the values corresponding to $\alpha = \beta = p$, to $\alpha = p_k^*$, $\beta = p_{k+1}^*$, that is, $p < r < p_s^*$. □

We now apply Theorem 3.6.3 to couples of Sobolev spaces, for which the real interpolation method yields Besov spaces (see (10.15) in the Appendix) The continuity of the embeddings considered in this theorem is due to Jawerth [69].

Theorem 3.7.3. *Suppose that $0 < t < s < \infty$ and $1 < p_0 < p_1 < \infty$ and $q \in [1,\infty]$. If $\frac{N}{p_0} - \frac{N}{p_1} < s - t$, then the continuous embedding $B^{s,p_0,q}(\mathbb{R}^N) \hookrightarrow B^{s,p_1,q}(\mathbb{R}^N)$ is $\mathcal{G}_{\mathbb{Z}^N}$-cocompact.*

Corollary 3.7.4. *Let s, t, p_0, p_1, and N be as in Theorem 3.7.3. Then the embedding $B^{s,p_0,q_0}(\mathbb{R}^N) \hookrightarrow B^{t,p_1,q_1}(\mathbb{R}^N)$ is $\mathcal{G}_{\mathbb{Z}^N}$-cocompact whenever $1 \leq q_0 \leq q_1 \leq \infty$.*

This corollary follows immediately from Lemma 3.1.6. We take $V = B^{s,p_0,q_0}$, $E = B^{t,p_1,q_0}$ and $F = B^{t,p_1,q_1}$. By Theorem 3.7.3, V is $\mathcal{G}_{\mathbb{Z}^N}$-cocompactly embedded into E. The continuous embedding $E \hookrightarrow F$ follows from (10.15) and (10.7).

Theorem 3.7.5. *Let $s > 0$, $1 < p < \infty$, $p < q_0 \leq q < p_s^*$. Then the embedding $B^{s,p,q_0}(\mathbb{R}^N) \hookrightarrow L^q(\mathbb{R}^N)$ is $\mathcal{G}_{\mathbb{Z}^N}$-cocompact.*

The following lemma will be the main component of the proof of Theorem 3.7.3.

Lemma 3.7.6. *Suppose that $m_0, m_1 \in \mathbb{R}$, $0 \le m_1 < m_0$, $1 < p_0 < p_1 < \infty$, and assume further that*

$$\frac{1}{p_0} - \frac{1}{p_1} < \frac{m_0 - m_1}{N}. \tag{3.43}$$

For each $t \in (0,1)$, the operator M_t defined by (3.30) *is a bounded map from $H^{m_0,p_0}(\mathbb{R}^N)$ to $H^{m_1,p_1}(\mathbb{R}^N)$ and satisfies*

$$\lim_{t\to 0} \|I - M_t\|_{H^{m_0,p_0}(\mathbb{R}^N)\to H^{m_1,p_1}(\mathbb{R}^N)} = 0. \tag{3.44}$$

Proof. We begin by observing that the conditions on p_0 and p_1 in the statement of the lemma are equivalent (also if $(p_0)^*_{m_0}$ below is infinite) to

$$1 < p_0 < p_1 < (p_0)^*_{m_0-m_1} = \frac{p_0 N}{N - (m_0 - m_1)p}. \tag{3.45}$$

We shall make use once more of the operator $\Lambda = I - \Delta$, noting that Λ and each of its powers all commute with all of the operators M_t. Since $\Lambda^{m_0/2}$ defines an isometry between $H^{m_0,p_0}(\mathbb{R}^N)$ and $L^{p_0}(\mathbb{R}^N)$ as well as between $H^{m_1,p_1}(\mathbb{R}^N)$ and $H^{m_1-m_0,p_1}(\mathbb{R}^N)$, it suffices to prove the lemma in the case where the two parameters m_0 and m_1 are replaced by $m_0' = m_0 - m_1$ and $m_1' = m_1 - m_1 = 0$, that is, we can suppose that $m_1 = 0$. Note that this "shift" of the values of m_0 and m_1 does not change the stated conditions on p_0 and p_1.

Case 1: Assume first that $m_0 \ge 1$. By Lemma 3.6.2, we have

$$\lim_{t\to 0} \|I - M_t\|_{H^{1,p_0}\to L^r} = 0 \quad \text{for each } r \in (p_0, (p_0)^*). \tag{3.46}$$

This also implies that

$$\lim_{t\to 0} \|I - M_t\|_{H^{m_0,p_0}\to L^r} = 0. \tag{3.47}$$

Subcase 1.1: If $p_1 = r$, the lemma is proved.

Subcase 1.2: If $p_1 > r$, then we can obtain (3.44) by using (3.42), namely, there exists $\theta =\in (0,1)$, such that for each $f \in H^{m_0,p_0}$ we have

$$\begin{aligned} \|(I - M_t)f\|_{L^{p_1}} &\le \|(I - M_t)f\|_{L^r}^{1-\theta} \|(I - M_t)f\|_{L^s}^{\theta} \\ &\le \left(\|I - M_t\|_{H^{m_0,p_0}\to L^r} \|f\|_{H^{m_0,p_0}}\right)^{1-\theta} \left(2\|f\|_{L^s}\right)^{\theta}. \end{aligned} \tag{3.48}$$

Since $p_0 < s < (p_0)^*_{m_0}$, we have that $\|f\|_{L^s}$ is bounded by a constant multiple of $\|f\|_{H^{m_0,p_0}}$ which we can substitute in (3.49) and then use (3.47) to obtain the required property (3.44) in this case.

Subcase 1.3: If $p_1 < r$, we use an argument similar to the one for Subcase 1.2, also based on (3.42), namely, with some $\theta \in (0,1)$, for each $f \in H^{m_0,p_0}$ we have that

$$\begin{aligned}\|(I - M_t)f\|_{L^{p_1}} &\leq \|(I - M_t)f\|_{L^{p_0}}^{1-\theta}\|(I - M_t)f\|_{L^r}^{\theta}\\ &\leq \left(2\|f\|_{L^{p_0}}\right)^{1-\theta}\left(\|I - M_t\|_{H^{m_0,p_0}\to L^r}\|f\|_{H^{m_0,p_0}}\right)^{\theta}. \qquad (3.49)\end{aligned}$$

Obviously, $\|f\|_{L^{p_0}} \leq \|f\|_{H^{m_0,p_0}}$ and so the proof is also complete in this case.

Case 2: If $0 < m_0 < 1$, then we apply Theorem 10.1.6 to the operator $T = I - M_t$ and the couples $(A_0, A_1) = (L^{p_0}, H^{1,p_0})$ and $(B_0, B_1) = (L^{p_0}, L^r)$ where $r \in (p_0, (p_0)^*)$. We choose $\theta = m_0$ and use the facts (see (10.13) and (10.12) in Appendix) that $H^{m_0,p_0} = [L^{p_0}, H^{1,p_0}]_{m_0}$ and $[L^{p_0}, L^r]_{m_0} = L^{s_0}$, where

$$\frac{1}{s_0} = \frac{1-m_0}{p_0} + \frac{m_0}{r}. \qquad (3.50)$$

Thus we obtain that

$$\begin{aligned}\|I - M_t\|_{H^{m_0,p_0}\to L^{s_0}} &\leq \|I - M_t\|_{L^{p_0}\to L^{p_0}}^{1-m_0}\|I - M_t\|_{H^{1,p_0}\to L^r}^{m_0}\\ &\leq 2^{1-m_0}\|I - M_t\|_{H^{1,p_0}\to L^r}^{m_0}. \qquad (3.51)\end{aligned}$$

Since we are free to choose r arbitrarily close to p_0, we see from (3.50) that we can also have s_0 arbitrarily close to p_0. So, keeping (3.45) in mind, let us choose r so that $s_0 < p_1$ and let us choose a second number $s_1 \in (p_1, (p_0)^*_{m_0})$. Now we use (3.42) once more: for some $\theta \in (0,1)$, and for each $f \in H^{m_0,p_0}$, we have

$$\begin{aligned}\|(I - M_t)f\|_{L^{p_1}} &\leq \|(I - M_t)f\|_{L^{s_0}}^{1-\theta}\|(I - M_t)f\|_{L^{s_1}}^{\theta}\\ &\leq \left(\|I - M_t\|_{H^{m_0,p_0}\to L^{s_0}}\|f\|_{H^{m_0,p_0}}\right)^{1-\theta}\left(2\|f\|_{L^{s_1}}\right)^{\theta}. \qquad (3.52)\end{aligned}$$

The fact that $s_1 \in (p_0, (p_0)^*_{m_0})$ ensures that $\|f\|_{L^{s_1}}$ is bounded by a constant multiple of $\|f\|_{H^{m_0,p_0}}$. After we substitute this in (3.52) and apply (3.51) and then (3.46), we obtain (3.44) in this final case, and so complete the proof of the lemma. □

After these preparations, the proof of Theorem 3.7.3 is almost immediate. Let $\epsilon \in (0, t/2)$ and let $s_0 = s + \epsilon$, $s_1 = s - \epsilon$, $t_0 = t + \epsilon$ and $t_1 = t - \epsilon$. Consider the Banach couples

$$(A_0, A_1) = (H^{s_0,p_0}(\mathbb{R}^N), H^{s_1,p_0}(\mathbb{R}^N)) \quad \text{and} \quad (B_0, B_1) = (H^{t_0,p_1}(\mathbb{R}^N), H^{t_1,p_1}(\mathbb{R}^N)).$$

Let $\lambda = \frac{N}{p_0} - \frac{N}{p_1}$. For $j = 0, 1$, since $s_j - t_j = s - t > \lambda$, we obtain from Theorem 3.7.1, that A_j is $\mathcal{G}_{\mathbb{Z}^N}$-cocompactly embedded in B_j. This, together with Lemma 3.7.6, shows that the conditions for applying Theorem 3.6.3 are fulfilled. So we can deduce that $(A_0, A_1)_{\theta,q}$ is $\mathcal{G}_{\mathbb{Z}^N}$-cocompactly embedded into $(B_0, B_1)_{\theta,q}$ for each $\theta \in (0, 1)$ and $q \in [1, \infty]$. In particular, if we choose $\theta = 1/2$ we obtain the assertion of the theorem. □

We now turn to the proof of Theorem 3.7.5. Obviously, in view of (10.15), (10.7), and Lemma 3.1.6, it suffices to give the proof when $q_0 = q$. Fix some $\theta \in (0,1)$ and define s_0 and r so that they satisfy $s = \theta s_0$ and

$$\frac{1}{q} = \frac{1-\theta}{p} + \frac{\theta}{r}. \tag{3.53}$$

We next want to show that

$$q < r < p_{s_0}^*. \tag{3.54}$$

The first inequality of (3.54) follows from (3.53) and the fact that $p < q$. The second inequality of (3.54) is equivalent to

$$\frac{1}{r} > \frac{1}{p} - \frac{s_0}{N},$$

which readily follows from $1/q > 1/p - s/N = 1/p - \theta s_0/N$ and (3.53).

In view of (3.54) and Theorem 3.7.1, we have that $H^{s_0,p}(\mathbb{R}^N)$ is $\mathcal{G}_{\mathbb{Z}^N}$-cocompactly embedded into $L^r(\mathbb{R}^N)$. Then, by Theorem 3.6.3 it follows that the embedding

$$(L^p, H^{s_0,p})_{\theta,r} \hookrightarrow (L^p, L^{r_0})_{\theta,r}$$

is $\mathcal{G}_{\mathbb{Z}^N}$-cocompact. Using (10.15) and (10.12), we identify the above embedding as $B^{s,p,r} \hookrightarrow L^r$. □

3.8 Cocompact embeddings of intersections with $L^p(\mathbb{R}^N)$

$\mathcal{G}_{\mathbb{Z}^N}$-cocompactness of embeddings of general inhomogeneous spaces can be derived from $\mathcal{G}^r$-cocompactness of embeddings of their homogeneous counterparts. In this section we consider general Banach spaces of locally integrable functions in $\mathbb{R}^N$. A measurable function v will be identified as an element of a dual space to E by relation $\langle v, u\rangle = \int_{\mathbb{R}^N} u(\xi)v(\xi)d\xi$ for all $u \in E$.

Lemma 3.8.1. *Let E be a Banach space of functions on $\mathbb{R}^N$ cocompactly embedded to $L^q(\mathbb{R}^N)$ relatively to $\mathcal{G}_{\mathbb{R}^N}$. Assume that $C_0^\infty(\mathbb{R}^N)$ is dense in E^* and that*

$$g_{\xi_k}^*\varphi \to g_\xi^*\varphi \quad \text{in } E^* \tag{3.55}$$

whenever $\xi_k \to \xi$ in $\mathbb{R}^N$ and $\varphi \in C_0^\infty(\mathbb{R}^N)$. Then the embedding is also cocompact relative to $\mathcal{G}_{\mathbb{Z}^N}$.

Proof. Assume that $g_{z_k}u = u_k(\cdot - z_k) \rightharpoonup 0$ in E for any sequence (z_k) in $\mathbb{Z}^N$. Let (y_k) be a sequence in $\mathbb{R}^N$ and let $z_k \in \mathbb{Z}^N$ be such that $(y_k - z_k)$ is bounded. Consider an arbitrarily renamed subsequence of $(y_k - z_k)$ convergent to some $\xi \in \mathbb{R}^N$. Then, with $\varphi \in C_0^\infty(\mathbb{R}^N)$,

$$\begin{aligned}\langle \varphi, g_{y_k} u_k \rangle = \langle \varphi, g_{z_k} u_k \rangle \\ &+ \langle (g^*_{y_k - z_k}\varphi - g^*_\xi \varphi), g_{z_k} u_k \rangle \\ &+ \langle (g^*_\xi \varphi - \varphi), g_{z_k} u_k \rangle \\ &\to 0,\end{aligned}$$

which proves that $u_k \overset{\mathcal{G}_{\mathbb{R}^N}}{\rightharpoonup} 0$ in E. □

Theorem 3.8.2. *Let $\dot{E}$ be a Banach space of functions on $\mathbb{R}^N$, such that $C_0^\infty(\mathbb{R}^N)$ is dense in $\dot{E}^*$, and assume that the rescaling group $\mathcal{G}^r$, $r > 0$, acts on $\dot{E}$ isometrically. Let $E = \dot{E} \cap L^p(\mathbb{R}^N)$, $1 < p < N/r$, with the standard norm for intersection of spaces. If there is a $\mathcal{G}^r$-cocompact embedding $\dot{E} \hookrightarrow L^{N/r}(\mathbb{R}^N)$, then for every $q \in (p, N/r)$, there is $\mathcal{G}_{\mathbb{Z}^N}$-cocompact embedding $E \hookrightarrow L^q(\mathbb{R}^N)$.*

Proof. Let $h_j u \overset{\text{def}}{=} 2^{rj} u(2^j \cdot)$, $j \in \mathbb{Z}$. Assume that $u_k(\cdot - y_k) \rightharpoonup 0$ in E for any sequence (y_k) in $\mathbb{R}^N$.

Let (y_k) be an arbitrary sequence in $\mathbb{R}^N$, let (j_k) in $\mathbb{Z}$, and consider three cases.

Case 1: $j_k \to \infty$. Since (u_k) is bounded in $L^p(\mathbb{R}^N)$, by rescaling under the integration in the L^p-norm one has $h_{j_k} u_k(\cdot - y_k) \to 0$ in $L^p(\mathbb{R}^N)$, and thus $h_{j_k} u_k(\cdot - y_k) \rightharpoonup 0$ in $L^p(\mathbb{R}^N)$ as well as in $\dot{E}$.

Case 2: $j_k \to -\infty$. With an arbitrary function $\varphi \in C_0^\infty(\mathbb{R}^N)$, we have

$$|\langle \varphi, h_{j_k} u_k(\cdot - y_k) \rangle_{L^{N/r}}| = |\langle h^*_{j_k} \varphi, u_k(\cdot - y_k) \rangle_{L^{N/r}}| \le \|h^*_{j_k} \varphi\|_{L^{p'}} \|u_k\|_{L^p} \to 0,$$

which implies $h_{j_k} u_k(\cdot - y_k) \rightharpoonup 0$ in $L^{N/r}$, and thus $h_{j_k} u_k(\cdot - y_k) \rightharpoonup 0$ in $\dot{E}$.

Case 3: (j_k) is bounded. Since it suffices to consider a constant subsequence, relations $h_{j_k} u_k(\cdot - y_k) \rightharpoonup 0$ and $u_k(\cdot - y_k) \rightharpoonup 0$ are equivalent.

Thus we have shown that for any choice of $g_k \in \mathcal{G}^r$, $g_k u_k \rightharpoonup 0$ in $\dot{E}$, and by cocompactness of the embedding $h_{j_k} \dot{E} \hookrightarrow L^{N/r}(\mathbb{R}^N)$, $u_k \to 0$ in $L^{N/r}$. Since (u_k) is bounded in L^p, from the Hölder inequality we have $u_k \to 0$ in L^q, $p < q < N/r$.

We conclude that the embedding $E \hookrightarrow L^q(\mathbb{R}^N)$ is cocompact relative to $\mathcal{G}_{\mathbb{R}^N}$. Note that (3.55) is satisfied, so by Lemma 3.8.1 this embedding is cocompact relative to $\mathcal{G}_{\mathbb{Z}^N}$. □

Corollary 3.8.3. *Let $s > 0$, $p \in (1, N/s)$, $p^*_s = \frac{pN}{N-ps}$. For every $q \in (p, p^*_s)$, embeddings $B^{s,p,a}(\mathbb{R}^N) \hookrightarrow L^q(\mathbb{R}^N)$, $1 \le a \le p^*_s$ are cocompact relatively to $\mathcal{G}_{\mathbb{Z}^N}$.*

Proof. Given the identification of $B^{s,p,a}(\mathbb{R}^N)$ as intersection $\dot{B}^{s,p,a}(\mathbb{R}^N) \cap L^p(\mathbb{R}^N)$, $s > 0$, $p, a \in [1, \infty]$ (see Appendix, Section 10.2), apply Theorem 3.8.2 to Corollaries 3.5.5 and 3.5.7. □

In Section 6.2 we study, in a more general setting, the consequences of intersection of two spaces on profile decompositions.

3.9 Cocompactness of trace embeddings

Let $\mathbb{R}^N_+ = \mathbb{R}^{N-1} \times (0, \infty)$, with coordinates denoted as (x, z), $x \in \mathbb{R}^{N-1}$, $z > 0$. When it does not cause ambiguity, we will abbreviate $\mathbb{R}^{N-1} \times \{0\}$ in the notation as $\mathbb{R}^{N-1}$.

In this section, we consider the space $\dot{H}^{1,p}(\mathbb{R}^N_+)$, $p \in [1, N)$, defined as the closed subspace of all functions from $\dot{H}^{1,p}(\mathbb{R}^N)$ satisfying $u(x, z) = u(x, -z)$. We equip it with an equivalent norm

$$\|u\| = \left(\int_{\mathbb{R}^N_+} |\nabla u|^p \mathrm{d}x\mathrm{d}z \right)^{\frac{1}{p}}, \tag{3.56}$$

(which is a scalar multiple of the gradient norm) and with a group

$$\mathcal{G} = \{u \mapsto 2^{\frac{N-p}{p}j} u(2^j(x+y), 2^j z)\}_{y \in \mathbb{R}^{N-1}, j \in \mathbb{Z}}. \tag{3.57}$$

This group consists of isometries on $\dot{H}^{1,p}(\mathbb{R}^N_+)$, which also extend to isometries on $L^{\bar{p}}(\mathbb{R}^{N-1} \times \{0\})$, where $\bar{p} = \frac{p(N-1)}{N-p}$. We would like to address cocompactness of the trace embedding $\dot{H}^{1,p}(\mathbb{R}^N_+) \hookrightarrow L^{\bar{p}}(\mathbb{R}^{N-1} \times \{0\})$.

Theorem 3.9.1. *Embedding $\dot{H}^{1,p}(\mathbb{R}^N_+) \hookrightarrow L^{\bar{p}}(\mathbb{R}^{N-1} \times \{0\})$ is cocompact relative to the group (3.57), and the $L^{\bar{p}}(\mathbb{R}^{N-1} \times \{0\})$-norm gives a local metrization of the $\mathcal{G}$-weak convergence in $\dot{H}^{1,p}(\mathbb{R}^N_+)$.*

Proof. 1. Let (u_k) be a sequence in $\dot{H}^{1,p}(\mathbb{R}^N_+)$ and assume that $u_k \overset{\mathcal{G}}{\rightharpoonup} 0$. We will reduce the question whether $u_k \to 0$ in $L^{\bar{p}}(\mathbb{R}^{N-1} \times \{0\})$ to compactness of the local Sobolev embedding, namely, that if $v_k \rightharpoonup 0$ in $\dot{H}^{1,p}(\mathbb{R}^N_+)$, then

$$\int_{(0,1)^{N-1}} |v_k(x, 0)|^p \mathrm{d}x \to 0. \tag{3.58}$$

By density, we may assume without loss of generality that $u_k \in C_0^\infty(\mathbb{R}^N)$. Let $\chi \in C_0^\infty((\frac{1}{2}, 4))$, such that $\chi(t) \le t$, $\chi(t) = t$ whenever $t \in [1, 2]$ and $|\chi'| \le 2$, and define

$$\chi_j(s) = 2^{\frac{N-p}{p}j} \chi(2^{-\frac{N-p}{p}j} s), \quad j \in \mathbb{Z}, s > 0.$$

Let $\psi \in C_0^\infty((-2, 2)^{N-1})$ satisfy $\psi(x) = 1$ for $x \in (0, 1)^{N-1}$. By continuity of the embedding $\dot{H}^{1,p}(\mathbb{R}^N_+) \hookrightarrow L^{\bar{p}}(\mathbb{R}^{N-1} \times \{0\})$ written for a function $\psi(\cdot - y)\chi(u_k)$, we have, for every $y \in \mathbb{Z}^{N-1}$,

$$\left(\int_{(0,1)^{N-1}+y} \chi(u_k(x, 0))^{\bar{p}} \mathrm{d}x \right)^{p/\bar{p}} \le C \int_{((-2,2)^{N-1}+y) \times (0,\infty)} (|\nabla u_k|^p + \chi(u_k)^p) \mathrm{d}x\mathrm{d}z.$$

Taking into account that $\chi(s)^{\bar{p}} \le C|s|^p$, we then have

$$\begin{aligned}
&\int\limits_{(0,1)^{N-1}+y} \chi(u_k(x,0))^{\bar{p}}\mathrm{d}x \\
&\le C \int\limits_{((-2,2)^{N-1}+y)\times(0,\infty)} (|\nabla u_k|^p + \chi(u_k)^p)\mathrm{d}x\mathrm{d}z\Big(\int\limits_{(0,1)^{N-1}+y} \chi(u_k(x,0))^{\bar{p}}\mathrm{d}x\Big)^{1-p/\bar{p}} \\
&\le C \int\limits_{((-2,2)^{N-1}+y)\times(0,\infty)} (|\nabla u_k|^p + \chi(u_k)^p)\mathrm{d}x\mathrm{d}z\Big(\int\limits_{(0,1)^{N-1}+y} |(u_k(x,0)|^p\mathrm{d}x\Big)^{1-p/\bar{p}}.
\end{aligned}$$

Adding the above inequalities over $y \in \mathbb{Z}^{N-1}$, we obtain

$$\begin{aligned}
\int\limits_{(0,1)^{N-1}+y} \chi(u_k(x,0))^{\bar{p}}\mathrm{d}x \le C \int\limits_{\mathbb{R}^N_+} (|\nabla u_k|^p + \chi(u_k)^p)\mathrm{d}x\mathrm{d}z \\
\times \sup_{y\in\mathbb{Z}^{N-1}}\Big(\int\limits_{(0,1)^{N-1}+y} |u_k(x,0)|^p\mathrm{d}x\Big)^{1-p/\bar{p}}.
\end{aligned}$$

Note that, by the definition of χ and the limiting Sobolev embedding,

$$\begin{aligned}
\int\limits_{\mathbb{R}^N_+} \chi(u_k)^p\mathrm{d}x\mathrm{d}z &\le C \int\limits_{\mathbb{R}^N_+} \chi(u_k)^{p^*}\mathrm{d}x\mathrm{d}z \\
&\le C\Big(\int\limits_{\mathbb{R}^N_+} |\nabla u_k|^p\mathrm{d}x\mathrm{d}z\Big)^{p^*/p} \le C,
\end{aligned}$$

which implies

$$\int\limits_{(0,1)^{N-1}+y} \chi(u_k(x,0))^{\bar{p}}\mathrm{d}x \le C \sup_{y\in\mathbb{Z}^{N-1}}\Big(\int\limits_{(0,1)^{N-1}+y} |u_k(x,0)|^p\mathrm{d}x\Big)^{1-p/\bar{p}}. \tag{3.59}$$

Let $y_k \in \mathbb{Z}^{N-1}$ be such that

$$\sup_{y\in\mathbb{Z}^{N-1}}\Big(\int\limits_{(0,1)^{N-1}+y} |u_k(x,0)|^p\mathrm{d}x\Big)^{1-p/\bar{p}} \le 2\Big(\int\limits_{(0,1)^{N-1}+y_k} |u_k(x,0)|^p\mathrm{d}x\Big)^{1-p/\bar{p}}. \tag{3.60}$$

Since $u_k \overset{\mathcal{G}}{\rightharpoonup} 0$, $u_k(\cdot - (y_k, 0)) \rightharpoonup 0$ in $\dot{H}^{1,p}(\mathbb{R}^N_+)$ and, by (3.58),

$$\int\limits_{(0,1)^{N-1}+y_k} |u_k(\cdot,0)|^p\mathrm{d}x = \int\limits_{(0,1)^{N-1}} |u_k(\cdot - y_k, 0)|^p\mathrm{d}x \to 0.$$

Substituting this into (3.59), and repeating the argument above for $-u_k$ we obtain

$$\int_{\mathbb{R}^{N-1}} \chi(|u_k(x,0)|)^{\bar{p}} \mathrm{d}x \to 0.$$

Moreover, since for any sequence $j_k \in \mathbb{Z}$,

$$2^{\frac{N-p}{p} j_k} u_k(2^{j_k}\cdot, 0) \overset{\mathcal{G}}{\rightharpoonup} 0,$$

we also have, with arbitrary $j_k \in \mathbb{Z}$, $k \in \mathbb{N}$,

$$\int_{\mathbb{R}^{N-1}} \chi_{j_k}(|u_k(x,0)|)^{\bar{p}} \mathrm{d}x \to 0. \tag{3.61}$$

2. Note now that, with $j \in \mathbb{Z}$ and $r = \frac{N-p}{p}$,

$$\left(\int_{\mathbb{R}^{N-1}} \chi_j(|u_k(x,0)|)^{\bar{p}} \mathrm{d}x \right)^{p/\bar{p}} \le C \int_{2^{r(j-1)} \le |u_k| \le 2^{r(j+2)}} |\nabla u_k|^p \mathrm{d}x\mathrm{d}z,$$

which can be rewritten as

$$\int_{\mathbb{R}^{N-1}} \chi_j(|u_k(x,0)|)^{\bar{p}} \mathrm{d}x \le C \int_{2^{r(j-1)} \le |u_k| \le 2^{r(j+2)}} |\nabla u_k|^p \mathrm{d}x\mathrm{d}z \left(\int_{\mathbb{R}^{N-1}} \chi_j(|u_k(x,0)|)^{\bar{p}} \mathrm{d}x \right)^{1-p/\bar{p}}. \tag{3.62}$$

Adding the inequalities (3.62) over $j \in \mathbb{Z}$ and taking into account that the sets $2^{j-1} \le |u_k| \le 2^{j+2}$ cover $\mathbb{R}^N$ with a uniformly finite multiplicity, we obtain

$$\int_{\mathbb{R}^{N-1}} |u_k(x,0)|^{\bar{p}} \mathrm{d}x \le C \int_{\mathbb{R}^N} |\nabla u_k|^p \mathrm{d}x\mathrm{d}z \sup_{j \in \mathbb{Z}} \left(\int_{\mathbb{R}^{N-1}} \chi_j(|u_k(x,0)|)^{\bar{p}} \mathrm{d}x \right)^{1-p/\bar{p}}. \tag{3.63}$$

Let j_k be such that

$$\sup_{j \in \mathbb{Z}} \left(\int_{\mathbb{R}^{N-1}} \chi_j(|u_k(x,0)|)^{\bar{p}} \mathrm{d}x \right)^{1-p/\bar{p}} \le 2 \left(\int_{\mathbb{R}^{N-1}} \chi_{j_k}(|u_k(x,0)|)^{\bar{p}} \mathrm{d}x \right)^{1-p/\bar{p}},$$

and note that the right-hand side converges to zero due to (3.61). Then from (3.63), it follows that $u_k(\cdot,0) \to 0$ in $L^{\bar{p}}(\mathbb{R}^{N-1} \times \{0\})$.

3. Since $\mathcal{G}$ consists of isometries on $L^{\bar{p}}(\mathbb{R}^{N-1} \times \{0\})$, by Lemma 3.1.5, the norm of $L^{\bar{p}}(\mathbb{R}^{N-1} \times \{0\})$ provides local metrization of $\mathcal{G}$-weak convergence on $\dot{H}^{1,p}(\mathbb{R}^N_+)$. □

3.10 Spaces cocompactly embedded into themselves

Theorem 3.10.1. *The space $C^k(\mathbb{R}^N)$, $k \in \mathbb{N}_0 = \{0, 1, 2, \dots\}$, is cocompactly imbedded into itself relatively to the group of shifts $\mathcal{G}_{\mathbb{R}^N}$.*

Proof. Let (u_n) be a bounded sequence in $C^k(\mathbb{R}^N)$ such that for any sequence (y_n) in $\mathbb{R}^N$, $u_n(\cdot + y_n) \rightharpoonup 0$. Since $v \mapsto \partial^\beta v(0)$ is a continuous functional in $C^k(\mathbb{R}^N)$ for each $\beta \in \mathbb{N}_0^N$, $|\beta| \le k$, it follows from $u_n(\cdot + y_n) \rightharpoonup 0$ that $\partial^\beta u_n(y_n) \to 0$ in $\mathbb{R}^N$ whenever $|\beta| \le k$. Choosing $y_n \in \mathbb{R}^N$ such that

$$\sum_{|\beta|\le k} |\partial^\beta u_n(y_n)| \ge \frac{1}{2} \sup_{y\in\mathbb{R}^N} \sum_{|\beta|\le k} |\partial^\beta u_n(y)| = \frac{1}{2}\|u_n\|_{C^k(\mathbb{R}^N)},$$

we have $\|u_n\|_{C^k(\mathbb{R}^N)} \to 0$. □

Let $\dot{C}^{0,\alpha}(\mathbb{R}^N)$, $\alpha \in (0, 1]$, denote the factor space of functions modulo additive constants, with the norm $\|u\| = \sup_{x\ne y\in\mathbb{R}^N} \frac{|u(x)-u(y)|}{|x-y|^\alpha}$.

Theorem 3.10.2. *The space $\dot{C}^{0,\alpha}(\mathbb{R}^N)$, $\alpha \in (0, 1]$, is cocompactly embedded into itself relative to the rescaling group*

$$\mathcal{G}_{\mathbb{R}}^{-\alpha} \overset{\text{def}}{=} \{u \mapsto t^{-\alpha} u(t(\cdot) - y)\}_{t>0, y\in\mathbb{R}^N}, \tag{3.64}$$

acting on $\dot{C}^{0,\alpha}(\mathbb{R}^N)$ in the sense of Definition 3.2.3.

Proof. Note that the group (3.64) acts isometrically on $\dot{C}^{0,\alpha}(\mathbb{R}^N)$. Let (u_k) be a bounded sequence in $C^{0,\alpha}(\mathbb{R}^N)$ such that $t_k^{-\alpha} u_k(t_k(\cdot + z_k)) \rightharpoonup 0$ for any sequence (z_k) in $\mathbb{R}^N$ and any sequence (t_k) of positive numbers. Let $x_k, y_k \in \mathbb{R}^N$ be such that

$$\frac{|u_k(x_k) - u_k(y_k)|}{|x_k - y_k|^\alpha} \ge \frac{1}{2} \sup_{x\ne y\in\mathbb{R}^N} \frac{|u_k(x) - u_k(y)|}{|x - y|^\alpha} = \frac{1}{2}\|u_k\|_{\dot{C}^\alpha(\mathbb{R}^N)}. \tag{3.65}$$

Set $t_k = |x_k - y_k|$, $z_k = t_k^{-1} x_k$, $\omega_k = \frac{y_k - x_k}{t_k}$ and assume, without loss of generality, that $\omega_k \to \omega$. Set $v_k(x) = t_k^{-\alpha} u_k(t_k(x + z_k))$ and note, since $v \mapsto v(\xi) - v(\eta)$, $\xi, \eta \in \mathbb{R}^N$, is a continuous functional in $\dot{C}^\alpha(\mathbb{R}^N)$, that $v_k(\omega) - v_k(0) \to 0$. Then

$$\begin{aligned}\frac{|u_k(x_k) - u_k(y_k)|}{|x_k - y_k|^\alpha} &= |v_k(\omega_k) - v_k(0)| \le |v_k(\omega_k) - v_k(\omega)| + |v_k(\omega) - v_k(0)| \\ &\le \|v_k\||\omega_k - \omega|^\alpha + o(1) = \|u_k\||\omega_k - \omega|^\alpha + o(1) \to 0,\end{aligned}$$

and thus, by (3.65), $\|u_k\| \to 0$. □

Remark 3.10.3. An analogous proof extends the assertion of the theorem above to the spaces $\dot{C}^{k,\alpha}$, $k \in \mathbb{N}$, which are cocompactly embedded into themselves relatively to the group $\mathcal{G}_{\mathbb{R}}^{-k-\alpha}$.

Corollary 3.10.4. *Let E be a $\mathcal{G}_{\mathbb{R}^N}$-invariant Banach space continuously embedded into $C^k(\mathbb{R}^N)$ for some $k \in \mathbb{N}_0$. If $C_0^\infty(\mathbb{R}^N)$ is dense in E^* and (3.55) is satisfied, then this embedding is $\mathcal{G}_{\mathbb{Z}^N}$-cocompact.*

Proof. Apply Lemma 3.1.6 and Lemma 3.8.1. □

The following corollary generalizes Theorem 3.2.4.

Corollary 3.10.5. *Let E be a $\mathcal{G}_{\mathbb{R}}^{-k-\alpha}$-invariant Banach space continuously embedded into $\dot{C}^{k,\alpha}(\mathbb{R}^N)$, $\alpha \in (0,1]$, $k \in \mathbb{N}_0$. If $C_0^\infty(\mathbb{R}^N)$ is dense in E^*, $t_k^\alpha \varphi(t_k^{-1} \cdot + y_k) \to \varphi$ in E^* whenever $\varphi \in C_0^\infty(\mathbb{R}^N)$, $t_k \to 1$ and $y_k \to 0$, then this embedding is cocompact relative to $\mathcal{G}^{-k-\alpha}$.*

Proof. Apply Lemma 3.1.6 and an argument analogous to Lemma 3.8.1. □

3.11 Cocompactness of the radial Moser–Trudinger embedding

Let $B = B_1(0) \subset \mathbb{R}^N$, $N \geq 2$. In this section, we prove cocompactness of an embedding of the Sobolev space $H^{1,N}_{0,\mathrm{rad}}(B)$ of radial functions, relative to the multiplicative group of isometries

$$\mathcal{G} = \{u \mapsto g_s(u) = s^{(1-N)/N} u(|x|^s), s > 0\}. \tag{3.66}$$

Let

$$\eta_r \overset{\text{def}}{=} \log \frac{1}{r}, \quad 0 < r < 1,$$

and define

$$V_p(r) \overset{\text{def}}{=} \begin{cases} r^{-N} \eta(r)^{-N-(p-N)\frac{N-1}{N}}, & N \leq p < \infty, \\ \eta(r)^{\frac{1-N}{N}}, & p = \infty, \end{cases} \tag{3.67}$$

The following family of functions was used by Moser [94] in the proof that the constant α_N in (1.30) is optimal:

$$m_t(r) \overset{\text{def}}{=} (\omega_{N-1})^{-\frac{1}{N}} \eta_t^{\frac{N-1}{N}} \min\left\{\frac{\eta_r}{\eta_t}, 1\right\}, \quad r, t \in (0,1). \tag{3.68}$$

It is easy to calculate that functions $m_t(|x|)$ on B satisfy $\|\nabla m_t\|_N = 1$. Let us define a continuous linear functional on $H^{1,N}_{0,\mathrm{rad}}(B)$ associated with the function m_t, $t \in (0,1)$:

$$\langle m_t^*, u \rangle \overset{\text{def}}{=} \int_B |\nabla m_t(|x|)|^{N-2} \nabla m_t(|x|) \cdot \nabla u \, dx.$$

Lemma 3.11.1. *Let $u \in H^{1,N}_{0,\mathrm{rad}}(B)$. Then for every $t \in (0,1)$,*

$$\langle m_t^*, u \rangle = \omega_{N-1}^{1/N} \eta_t^{(1-N)/N} u(t). \tag{3.69}$$

Proof. We have

$$\langle m_t^*, u\rangle = \omega_{N-1} \int_1^t u'(r)|m_t'(r)|^{N-1} r^{N-1} \mathrm{d}r$$
$$= \omega_{N-1}\omega_{N-1}^{\frac{1-N}{N}} \eta_t^{(1-N)/N} \int_1^t u'(r)\mathrm{d}r = \omega_{N-1}^{1/N}\eta_t^{(1-N)/N} u(t).$$ □

Corollary 3.11.2. *Every function $u \in H^{1,N}_{0,\mathrm{rad}}(B)$ satisfies the inequality*

$$\|u\|_{\infty,V_\infty} = \sup_{r\in(0,1)} |u(r)|\eta_r^{(1-N)/N} \le \omega_{N-1}^{-1/N}\|\nabla u\|_N \tag{3.70}$$

and the constant $\omega_{N-1}^{-1/N}$ in the right-hand side is optimal.

Proof. Apply Hölder inequality to (3.69). The best constant is attained at $u = m_t$. □

Lemma 3.11.3. *The space $H^{1,N}_{0,\mathrm{rad}}(B)$ is continuously embedded into $L^p(B, V_p)$, $p \in [N, \infty]$.*

Proof. Case $p = \infty$ is proved in Corollary 3.11.2. The case $p = N$ is a well-known Hardy-type inequality (see, e. g., [6])

$$\int_B \frac{|u|^N}{|x|^N \eta_{|x|}^N} \mathrm{d}x \le \left(\frac{N-1}{N}\right)^N \int_B |\nabla u|^N \mathrm{d}x. \tag{3.71}$$

Case $p \in (N, \infty)$ follows from the endpoint cases by the Hölder inequality. □

Note that the $L^p(B, V_p)$-norms are invariant with respect to the group (3.66).

Theorem 3.11.4. *Embedding $H^{1,N}_{0,\mathrm{rad}}(B) \hookrightarrow L^\infty(B, V_\infty)$ is cocompact relative to the group (3.66).*

Proof. Observe first, by direct computation, that for every $s > 0$ and $t \in (0,1)$,

$$g_s m_t = m_{t^{1/s}} \quad \text{and} \quad g_s^* m_t^* = m_{t^{1/s}}^* \quad . \tag{3.72}$$

Let $t_k \in (0,1)$ be an arbitrary sequence and let $s_k = \log\frac{1}{t_k}$. If $u_k \stackrel{\mathcal{G}}{\rightharpoonup} 0$, then $\langle m_{1/e}^*, g_{s_k} u_k\rangle \to 0$. By (3.72), we have

$$\langle m_{1/e}^*, g_{s_k} u_k\rangle = \langle m_{t_k}^*, u_k\rangle.$$

Then, using (3.69) we have

$$\sup_{r\in(0,1)} |u_k(r)|\eta_r^{(1-N)/N} \to 0. \tag{3.73}$$
□

Corollary 3.11.5. *Let $p \in (N, \infty)$. Embedding $H^{1,N}_{0,\mathrm{rad}}(B) \hookrightarrow L^p(B, V_p)$ is cocompact relatively to the group* (3.66).

Proof. Let $u_k \overset{\mathcal{G}}{\rightharpoonup} 0$. By (3.71) and (3), we have

$$\|u_k\|_{p,V_p} \leq \int_B \frac{|u_k|^N}{|x|^N \eta^N_{|x|}} \mathrm{d}x \left(\sup_{r \in (0,1)} \frac{|u_k(r)|}{\eta_r^{\frac{N-1}{N}}} \right)^{p-N} \to 0.$$ □

Remark 3.11.6. By Lemma 3.1.5, the $L^p(B, V_p)$-norm, $p \in (N, \infty]$, provides a local metrization of the embedding $H^{1,N}_{0,\mathrm{rad}}(B) \hookrightarrow L^p(B, V_p)$.

Proposition 3.11.7. *If $u_k \overset{\mathcal{G}}{\rightharpoonup} 0$ in $H^{1,N}_{0,\mathrm{rad}}(B)$, then for any $\lambda > 0$,*

$$\int_B (e^{\lambda |u_k|^{\frac{N}{N-1}}} - 1)\mathrm{d}x \to 0.$$

(In other words, embedding $H^{1,N}_{0,\mathrm{rad}}(B) \hookrightarrow \exp L^{\frac{N}{N-1}}(B)$ is cocompact relative to the group (3.66).*)*

Proof. By Theorem 3.11.4, there is a sequence $\epsilon_k \to 0$ such that $|u_k|^{N/(N-1)}(r) \leq \epsilon_k \eta_r$. Then

$$0 \leq \int_B (e^{\lambda |u_k|^{\frac{N}{N-1}}} - 1)\mathrm{d}x \leq \int_B (r^{\lambda \epsilon_k} - 1)\mathrm{d}x \to 0. \tag{3.74}$$ □

3.12 Bibliographic notes

Theorem 3.2.1 for $s = 1$, $1 \leq p < N$, was proved by Solimini [112]. We give here a second proof, for general $s > 0$, following Jaffard [68], and also quote a third, elementary, proof for the case $s = 1$ from [126]. A fourth proof, following the wavelet decompositions approach of [13], is provided for the embeddings of Besov spaces, Theorem 3.5.1 (its Corollary 3.5.5 applies to Sobolev embeddings). The range of parameters in embeddings handled here is larger than in [13] because the argument in the latter is dependent on a technical property [13, Assumption 1.1] which is stronger than cocompactness.

An example of non-cocompact embedding into the Lorentz space, $\dot{H}^{1,p}(\mathbb{R}^N) \hookrightarrow L^{p^*,p}(\mathbb{R}^N)$ in Corollary 3.3.2, based on the noncompactness of the embedding defined by the Hardy inequality (Proposition 3.3.1), is due to Solimini [112].

Interpolation of cocompact embeddings with applications to inhomogeneous Sobolev and Besov spaces, studied in [35], is only a tentative incursion into the subject. Same applications as those following from Theorem 3.6.3 can be handled in the case $p < N/s$ by Theorem 3.8.2 by means of reduction of known $\mathcal{G}^r$-cocompactness for corresponding homogeneous spaces. Theorem 3.6.3 is restricted to spaces E_0 and E_1 equipped with the *same* scaling group. An illustrative Proposition 3.6.4 studies invariance of interpolated spaces when the endpoint spaces E_0, E_1 have respective $\mathcal{G}^{r_0}$ and $\mathcal{G}^{r_1}$-invariance, but presently we do not know of any result on interpolation of $\mathcal{G}^r$-cocompactness.

Cocompactness of trace embeddings in Section 3.9 is an elementary generalization of an analogous statement for $\dot{H}^{1,2}(\mathbb{R}^N_+)$ in [127]. A further generalization to traces on hyperplanes of lower dimension seems to be elementary but we have not seen it in literature. Moreover, given that sharper trace embeddings involve Besov spaces (see, e. g., [2]), there is a whole range of significant trace embeddings expected to be cocompact.

Cocompactness of Moser–Trudinger embeddings and corresponding profile decompositions were first proved for the radial case by [4], followed by the nonradial case for $N = 2$ in [7]. Embedding $H^{1,N}_{0,\mathrm{rad}}(B) \hookrightarrow L^N(B, V_N)$, associated with inequality (3.71), is not $\mathcal{G}$-cocompact ([4]). The argument is similar to Proposition 3.3.1. The problem in full generality remains open, and in most remaining cases no suitable scaling set $\mathcal{G}$ is known. Nonlinear dilations similar to the group (3.66) have been also studied for the radial subspace of $H^{1,2}$ on the hyperbolic space $\mathbb{H}^N$ in [100].

4 Profile decomposition in Banach spaces

In this chapter, we prove existence of profile decomposition for general Banach spaces and its counterparts for dual spaces and for spaces of vector-valued functions, provide its realization for Besov and Triebel–Lizorkin spaces (which, suitably normed, satisfy the Opial condition), and discuss consequences of asymptotic decoupling of bubbles in nonlinearities. We conclude the chapter with an example of a profile decomposition in the context of the Moser–Trudinger inequality.

4.1 Profile decomposition

Definition 1.2.4 of profile decomposition can be extended from Hilbert spaces to Banach spaces verbatim, but when concentration profiles are obtained by (1.9), convergence of the sum of elementary concentrations (4.3) generally cannot be assured by known means. Only by replacing weak convergence with Delta convergence (which makes no difference in Hilbert space) one has a Banach space counterpart of (1.15), (4.7). Thus we extend Definition 1.2.4 to Banach spaces by invoking Delta-convergence instead of weak convergence, which allows to use "energy" estimate (2.15), which leads to (4.7), and eventually to convergence of the sum (4.3) representing defect of compactness.

Definition 4.1.1 (Concentration family). Let E be a Banach space and let $\mathcal{G}$ be a set of linear bijective isometries of E. One says that a countable set of pairs $(w^{(n)}, (g_k^{(n)})_{k\in\mathbb{N}})_{n\in\mathbb{N}} \subset B \times \mathcal{G}^{\mathbb{N}}$ is a *concentration family* for a bounded sequence $(u_n)_{n\in\mathbb{N}}$ in E, if $g_k^{(1)} = \mathrm{id}$,

$$g_k^{(n)^{-1}} u_k \rightharpoonup w^{(n)}, \tag{4.1}$$

and

$$g_k^{(n)^{-1}} g_k^{(m)} \rightharpoonup 0 \quad \text{whenever } m \neq n. \tag{4.2}$$

The functions $w^{(n)}$ are called *concentration profiles* of $(u_n)_{n\in\mathbb{N}}$, associated with *scaling sequences* $(g_k^{(n)})_{k\in\mathbb{N}}$, sequences $(g_k^{(n)} w^{(n)})_{k\in\mathbb{N}}$ in E are called *elementary concentrations* (or *blowups*, or *cores*) for the sequence $(u_n)_{n\in\mathbb{N}}$, and property (4.2) is called (asymptotic) *decoupling*.

Definition 4.1.2 (Profile decomposition). Let $(u_n)_{n\in\mathbb{N}}$ be a bounded sequence in a Banach space E and let $\mathcal{G}$ be a set of linear bijective isometries of E. One says that (u_n) admits a *profile decomposition* if it has a concentration family $(w^{(n)}, (g_k^{(n)})_{k\in\mathbb{N}})_{n\in\mathbb{N}} \subset E \times \mathcal{G}^{\mathbb{N}}$ such that the series

$$S_k \stackrel{\text{def}}{=} \sum_n g_k^{(n)} w^{(n)}, \tag{4.3}$$

https://doi.org/10.1515/9783110532432-004

called *defect of compactness*, converges in E unconditionally (with respect to n) and uniformly with respect to k, and

$$u_k - S_k \overset{\mathcal{G}}{\rightharpoonup} 0. \tag{4.4}$$

Such concentration family is called *complete*.

Definition 4.1.3. A group $\mathcal{G}_0$ of isometries on a Banach space E is called a *scaling group* (or *dislocation group*) if the following relations are satisfied:

$$g_k \in \mathcal{G}_0,\ k \in \mathbb{N},\ g_k \not\rightharpoonup 0 \Longrightarrow \exists (k_j) \subset \mathbb{N}:\ (g_{k_j}^{-1}) \text{ and } (g_{k_j}) \text{ converge operator-strongly} \tag{4.5}$$

(i. e., pointwise), and

$$u_k \rightharpoonup 0,\ w \in X,\ g_k \in \mathcal{G}_0, k \in \mathbb{N},\ g_k \rightharpoonup 0 \implies u_k + g_k w \rightharpoonup 0. \tag{4.6}$$

Remark 4.1.4. When weak convergence and Delta-convergence coincide, condition (4.6) trivially holds true, so this definition generalizes the definition of the dislocation group from [127] to Banach spaces. Furthermore, in this case, concentration profiles (4.1) become weak limits as in (1.9).

Remark 4.1.5. When $\mathcal{G}_0$ is a scaling group, profiles $w^{(n)}$ in (4.3) are unique, up to extraction of a subsequence and up to multiplication by an operator $g \in \mathcal{G}_0$. The argument is repetitive of that in Proposition 3.4 in [127], which considers the case of Hilbert space.

Theorem 4.1.6. *Let E be a uniformly convex and uniformly smooth Banach space equipped with a scaling group $\mathcal{G}_0$ and let $\mathcal{G} \ni \mathrm{id}$ be a subset of $\mathcal{G}_0$.Then every bounded sequence $(x_k)_{k\in\mathbb{N}}$ in E has a subsequence that admits a profile decomposition relative to $\mathcal{G}$. Moreover, if $\|x_k\| \le 1$, then $\|w^{(n)}\| \le 2$ for all $n \in \mathbb{N}$ and*

$$\limsup \|x_k - S_k\| + \sum_n \delta(\|w^{(n)}\|) \le 1, \tag{4.7}$$

where S_k is the sum (4.3), $w^{(n)}$ are concentration profiles as in Definition 4.1.1, and δ is the modulus of convexity of E.

Remark 4.1.7. Condition $\|x_k\| \le 1$ can be removed by applying Theorem 4.1.6 to a subsequence of $x_k/\|x_k\|$ with $\|x_k\| \to \nu > 0$. Theorem 4.1.6 remains valid with δ in (4.7) replaced by $\nu\delta(\frac{\cdot}{\nu})$.

Remark 4.1.8. The assumption of uniform convexity in Theorem 4.1.6 cannot be removed. Indeed, let $E = L^\infty(\mathbb{R})$ equipped with the group of integer shifts $\mathcal{G}_\mathbb{Z}$ and Let x_k be a characteristic function $\mathbb{1}_{A_k}$ of the set

$$A_k = \bigcup_{j=1}^{2^k} \left[a_k^{(j)}, a_k^{(j)} + \frac{j}{2^k}\right]$$

where $a_k^{(j)} \in \mathbb{R}$ satisfy $|a_k^{(j)} - a_k^{(\ell)}| \geq 2k$ whenever $j \neq \ell$. Then x_k will have concentration profiles $\mathbb{1}_{[a,a+t]}$ for every $t \in (0,1]$, contrary to Theorem 4.1.6 where the set of all concentration profiles of a sequence is countable.

Corollary 4.1.9. *If space E, in addition to the assumptions of Theorem 4.1.6, is $\mathcal{G}$-Delta-cocompactly embedded into another Banach space F (if E satisfies the Opial condition, this coincides with $\mathcal{G}$-cocompactness), then the remainder $r_k \overset{\text{def}}{=} u_k - S_k$ converges to zero in the norm of F. If, furthermore, E is a Hilbert space, one also has* (1.15).

4.2 Opial condition in Besov and Triebel–Lizorkin spaces

Given that Delta-cocompactness of embeddings, distinct from cocompactness, remains generally unknown, application of Theorem 4.1.6 relies on coincidence between Delta-convergence and weak convergence together with cocompact embeddings that yield a remainder vanishing in the target space. In particular, the two convergence modes coincide in Besov and Triebel–Lizorkin spaces equipped with norms (3.23) and (3.24), respectively. In the case of Sobolev spaces, which are a subfamily of Triebel–Lizorkin spaces, the equivalent norm (3.24) is different from the standard Sobolev norm, but this does not present any difficulties: once the profile decomposition is stated in terms of weak convergence, we may revert to the standard Sobolev norm as well as discard the general energy inequality (4.7) in favor of the stronger energy inequality (4.25), specific for Sobolev spaces and the group $\mathcal{G}^r$. In this way, profile decomposition of Solimini (Theorem (4.6.4)) is a corollary of Theorem 4.1.6.

Theorem 4.2.1. *Let $s \in \mathbb{R}$ and $p, q \in (1, \infty)$. Besov spaces $\dot{B}^{s,p,q}(\mathbb{R}^N)$ with the norm* (3.23) *and Triebel–Lizorkin spaces $\dot{F}^{s,p,q}(\mathbb{R}^N)$ with the norm* (3.24)*, satisfy the Opial condition, and in each of these spaces Delta-convergence coincides with weak convergence.*

Proof. 1. Note that the spaces in question are uniformly convex and uniformly smooth, so once we show that Delta-convergence and weak convergence coincide, we will have the Opial condition satisfied by Proposition 2.4.3. Furthermore, it suffices to prove only that $u_n \rightharpoonup u$ always implies $u_n \rightharpoondown u$. Indeed, if $u_n \rightharpoondown u$, then by Theorem 2.3.8, (u_n) is a bounded sequence. Then it has a weakly convergent subsequence, whose limit is necessarily the Delta-limit of (u_n), namely u. This shows, however, that every weakly convergent subsequence of (u_n) has weak limit u, and thus $u_n \rightharpoonup 0$.

2. We may now apply characterization of Delta-convergence by Theorem 2.1.3, so it suffices to show that $u_n \rightharpoonup 0$ implies $u_n^* \rightharpoonup 0$. Finally, we may assume that the norm of u_n is bounded away from zero, since $u_n \to 0$ implies both weak and Delta-convergence.

Consider first Besov spaces. We have, for any $\varphi \in \mathcal{S}(\mathbb{R}^N)$,

$$\langle u_n^*, \varphi\rangle = \Big(\sum_{j\in\mathbb{Z}} \|2^{js}P_j u_n\|_{L^p}^q\Big)^{2/q-1} \\ \times \sum_{j\in\mathbb{Z}} \Big(\int_{\mathbb{R}^N} |2^{js}P_j u_n|^p \mathrm{d}x\Big)^{q/p-1} \int_{\mathbb{R}^N} |2^{js}P_j u_n|^{p-2} 2^{js}(P_j u_n) 2^{js} P_j \varphi \mathrm{d}x. \tag{4.8}$$

The first factor equals $\|u_n\|^{2-q}$, while by assumption $\|u_n\|$ is bounded and bounded away from zero, so it suffices to show that the remaining product vanishes. By density and linearity, it suffices to use as test functions φ such that $P_j\varphi = 0$ for all values of j except one. For any $j \in \mathbb{Z}$, $\int_{\mathbb{R}^N} |2^{js}P_j u_n|^p \mathrm{d}x \le \|u_n\|^p$, which is bounded as $n \to \infty$, and $u_n \rightharpoonup 0$ implies that $P_j u_n \to 0$ in L^∞, which means that, for the right-hand side of (4.8), reduced to a single term of the summation, we have

$$\Big(\int_{\mathbb{R}^N} |2^{js}P_j u_n|^p \mathrm{d}x\Big)^{q/p-1} \int_{\mathbb{R}^N} |2^{js}P_j u_n|^{p-2} 2^{js}(P_j u_n) 2^{js} P_j \varphi \mathrm{d}x \to 0.$$

Similar calculations for the Triebel–Lizorkin norm give

$$\langle u_n^*, \varphi\rangle \\ = \|u_n\|^{2-p} \int_{\mathbb{R}^N} \Big[\Big(\sum_{j\in\mathbb{Z}} |2^{js}P_j u_n|^q\Big)^{p/q-1} \sum_{j\in\mathbb{Z}} |2^{js}P_j u_n|^{q-2} 2^{js}(P_j u_n)\Big] 2^{js} P_j \varphi \mathrm{d}x. \tag{4.9}$$

With our particular choice of test functions φ, we have

$$|\langle u_n^*, \varphi\rangle| \le C(i) \int_{\mathbb{R}^N} \Big(\sum_{j\in\mathbb{Z}} |2^{js}P_j u_n|^q\Big)^{p/q-1} |P_i u_n|^{q-1} P_i \varphi \mathrm{d}x. \tag{4.10}$$

We now consider two cases. If $p \ge q$, then by Hölder inequality we have from (4.10)

$$|\langle u_n^*, \varphi\rangle| \le C(i)\|u_n\|^{1-q/p} \Big[\int_{\mathbb{R}^N} (|P_i u_n|^{q-1} P_i\varphi)^{p/q} \mathrm{d}x\Big]^{q/p} \to 0.$$

If, on the other hand, $p < q$, by substituting a trivial inequality $\sum_{j\in\mathbb{Z}} |2^{js}P_j u_n|^q \ge 2^{is}|P_i u_n|^q$, into (4.10) we get

$$|\langle u_n^*, \varphi\rangle| \le C(i) \int_{\mathbb{R}^N} |P_i u_n|^{p-1} P_i \varphi \mathrm{d}x \to 0.$$

We conclude that $u_n^* \rightharpoonup 0$, and thus $u_n \rightharpoonup 0$, which proves the theorem. □

4.3 Proof of Theorem 4.1.6

Throughout this section, it is assumed that E is a uniformly smooth and uniformly convex Banach space, equipped with a scaling group $\mathcal{G}_0$ and that $\mathcal{G}$ is a subset of $\mathcal{G}_0$ containing id. We will use the notation $(v_k) \prec (u_k)$ or $(u_k) \succ (v_k)$ whenever (v_k) is a subsequence of (u_k).

Our proof of Theorem 4.1.6 will start with several technical statements.

Lemma 4.3.1. *Let (g_k) be a sequence in $\mathcal{G}_0$. If $g_k \rightharpoonup 0$ then $g_k^{-1} \rightharpoonup 0$.*

Proof. If $g_k^{-1} \not\rightharpoonup 0$, then by (4.5) the sequence (g_k) has a strongly convergent subsequence, whose limit g is necessarily an isometry and, therefore, $g \neq 0$, a contradiction. □

Lemma 4.3.2. *Let (g_k) be a sequence in $\mathcal{G}$ such that g_k^{-1} is operator-strongly convergent. If $x_k \rightharpoonup 0$, then $g_k x_k \rightharpoonup 0$.*

Proof. By operator-strong convergence, there is a linear isometry h, such that $g_k^{-1} y \to hy$ for all $y \in E$,

$$\langle (g_k x_k)^*, y\rangle = \langle x_k^*, g_k^{-1} y\rangle = \langle x_k^*, hy\rangle + o(1) \to 0. \qquad \square$$

The next lemma describes how scaling sequences (g_k) become asymptotically decoupled.

Lemma 4.3.3. *Let (u_k) be a bounded sequence in E. If two sequences $(g_k^{(1)})_{k\in\mathbb{N}}$ and $(g_k^{(2)})_{k\in\mathbb{N}}$ in $\mathcal{G}$, satisfy $(g_k^{(1)})^{-1} u_k \rightharpoonup w^{(1)}$ and $(g_k^{(2)})^{-1}(u_k - g_k^{(1)} w^{(1)}) \rightharpoonup w^{(2)} \neq 0$, then $(g_k^{(1)})^{-1}(g_k^{(2)}) \rightharpoonup 0$.*

Proof. Assume that $(g_k^{(1)})^{-1}(g_k^{(2)})$ does not converge weakly to zero. Then by (4.5), on a renamed subsequence, $(g_k^{(1)})^{-1}(g_k^{(2)})$ converges operator-strongly to some isometry h. Then by Lemma 4.3.2,

$$(g_k^{(1)})^{-1}(g_k^{(2)})[(g_k^{(2)})^{-1}(u_k - g_k^{(1)} w^{(1)}) - w^{(2)}] \rightharpoonup 0,$$

which, taking into account (4.6), gives

$$(g_k^{(1)})^{-1} u_k - w^{(1)} - hw^{(2)} \rightharpoonup 0.$$

This, however, contradicts the definition of $w^{(1)}$ and the assumption that $w^{(2)} \neq 0$. □

In the next statement, we obtain decoupled scaling sequences by iteration.

Lemma 4.3.4. *Let u_k be a bounded sequence in E and let sequences $(g_k^{(n)})_k$ in $\mathcal{G}$, and $w^{(n)} \in E$, $n = 1, \dots, M$, satisfy $g_k^{(1)} = \mathrm{id}$, $(g_k^{(n)})^{-1} u_k \rightharpoonup w^{(n)}$, $n = 1, \dots M$, and $(g_k^{(n)})^{-1}(g_k^{(m)}) \rightharpoonup 0$ whenever $n < m$. Assume that there exists a sequence $(g_k^{(M+1)})$ in $\mathcal{G}$ such that, on a renumbered subsequence, $(g_k^{(M+1)})^{-1}(u_k - w^{(1)} - g_k^{(2)} w^{(2)} - \dots - g_k^{(M)} w^{(M)}) \rightharpoonup w^{(M+1)} \neq 0$. Then $(g_k^{(n)})^{-1}(g_k^{(M+1)}) \rightharpoonup 0$ for $n = 1, \dots M$.*

Proof. We apply Lemma 4.3.3, with 1 replaced by n, 2 by $M+1$, and u_k by $u_k - \sum_{m\neq n} g_k^{(m)} w^{(m)}$, taking into account (4.6). □

We now start the construction needed for the proof of Theorem 4.1.6.

The following partial strict order relation between sequences in E will be denoted as $>$. First, given two sequences (x_k) and (y_k) in E, we shall say that $(x_k) \succ (y_k)$ if there exists a sequence (g_k) in $\mathcal{G}$, an element $w \in E \setminus \{0\}$, and a renumeration (n_k) such that $g_{n_k}^{-1} x_{n_k} \rightharpoonup w$ and $y_k = x_{n_k} - g_{n_k} w$. By Lemma 2.5.1, if $(x_k) \succ (y_k)$ and $\|x_k\| \le 1$, then $\|y_k\| \le 1$ for all k sufficiently large. Then, by Delta-compactness of bounded sequences, every sequence (x_k) in E, $\|x_k\| \le 1$, which is not $\mathcal{G}$-Delta-convergent to 0, there is a sequence (y_k) in E, such that $\|y_k\| \le 1$ and $(x_k) \succ (y_k)$.

Then we shall say that $(x_k) > (y_k)$ in one step, if $(x_k) \succ (y_k)$ and in m steps, $m \ge 2$, if there exist m sequences $(x_k^1) \succ (x_k^2) \succ \cdots \succ (x_k^m)$, such that $(x_k^1) = (x_k)$ and $(x_k^m) = (y_k)$. Note that, for every sequence (x_k) in E, $\|x_k\| \le 1$, either there exists a finite number of steps $m_0 \in \mathbb{N}$ such that $(x_k) > (y_k)$ in m_0 steps for some (y_k) in E, $\|y_k\| \le 1$, and $p((y_k)) = 0$, or for every $m \in \mathbb{N}$ there exists a sequence (y_k) in E, $\|y_k\| \le 1$, such that $(x_k) > (y_k)$ in m steps. We will say that $(x_k) \ge (y_k)$ if either $(x_k) > (y_k)$ or $(x_k) = (y_k)$

Define now

$$\sigma((x_k)) = \inf_{(y_k)\ge(x_k)} \sup_{k\in\mathbb{N}} \|y_k\|$$

and observe that if $(x_k) \ge (z_k)$, then $\sigma((x_k)) \le \sigma((z_k))$, since the set of sequences (y_k) dominating (z_k) is a subset of sequences dominating (x_k).

Lemma 4.3.5. *Let $(x_k) > (y_k)$ in m steps, $\|x_k\| \le 1$ and $\eta > 0$. Then there exist $w^{(1)}, \dots, w^{(m)} \in E$, sequences $(g_k^{(1)}), \dots, (g_k^{(m)})$ in $\mathcal{G}$, and a renumeration (n_k) such that*

$$y_k = x_{n_k} - \sum_{n=1}^{m} g_{n_k}^{(n)} w^{(n)},$$

$(g_{n_k}^{(p)})^{-1} g_{n_k}^{(q)} \rightharpoonup 0$ for $p \neq q$, and for any set $J \subset J_m = (1, \dots, m)$,

$$\delta\Big(\sum_{n\in J} g_{n_k}^{(n)} w^{(n)}\Big) \le \sup \|x_{n_k}\| - \sigma((x_{n_k})) + \eta, \quad \textit{for all } k \textit{ sufficiently large.} \tag{4.11}$$

Proof. The first assertion follows from Lemma 4.3.4. Define

$$\alpha_k \stackrel{\text{def}}{=} x_{n_k} - \sum_{n\in J_m\setminus J} g_k^{(n)} w^{(n)},$$

$$\beta_k \stackrel{\text{def}}{=} x_{n_k} - \sum_{n\in J_m\setminus J} g_k^{(n)} w^{(n)} - \frac{1}{2}\sum_{n\in J} g_k^{(n)} w^{(n)} = \frac{1}{2}(\alpha_k + y_k).$$

By Lemma 2.5.1, $\|y_k\| \le \|\alpha_k\| \le \|x_k\| \le 1$ and $\beta_k \le 1$ for all k large. Note that, as in the construction above, we can take k large enough so that $\sup \|\beta_k\| \le \inf \|\beta_k\| + \eta$. By uniform convexity, we have for all k sufficiently large

$$\|\beta_k\| \le \|\alpha_k\| - \delta(\alpha_k - y_k).$$

Therefore,

$$\delta\Big(\Big\|\sum_{n\in J} g_k^{(n)} w^{(n)}\Big\|\Big) \le \|\alpha_k\| - \|\beta_k\| \le \sup \|x_k\| - \sigma((x_k)) + \eta. \qquad \square$$

Let us now define the following value associated with sequences in E:

$$p((u_k)_{k\in\mathbb{N}}) \overset{\text{def}}{=} \sup\{\|w\| : \exists (v_k) \prec (v_k) \text{ and } (g_k) \text{ in } \mathcal{G}, \text{ such that } g_k^{-1}(u_{n_k}) \rightharpoonup w\}.$$

Note that by the definition of $\mathcal{G}$-Delta convergence $u_k \overset{\mathcal{G}}{\rightharpoonup} u$ if and only if $p((u_k - u)_{k\in\mathbb{N}}) = 0$.

Proof of Theorem 4.1.6. Let $x_k \in E$, $\|x_k\| \le 1$, $k \in \mathbb{N}$. For every $j \in \mathbb{N}$, define $\epsilon_j = \delta(\frac{1}{2^j})$. Let $(x_k^{(1)}) \subset E$ be such that $(x_k) > (x_k^{(1)})$ and $\sup \|x_k^{(1)}\| < \sigma((x_k)) + \epsilon_1$. Consider the following iterations. Given $(x_k^{(j)})_k$, either $p((x_k^{(j)})_k) = 0$, in which case there is a profile decomposition with $r_k = x_k^{(j)}$, or there exists a sequence $(x_k^{(j+1)})_k < (x_k^{(j)})_k$, such that $\sup_k \|x_k^{(j+1)}\| < \sigma((x_k^{(j)})_k) + \frac{\epsilon_j}{2}$, $j \in \mathbb{N}$. Let us denote as n_k^j the cumulative enumeration of the original sequence that arises at the jth iterative step, and denote as m_{j+1} the number of elementary concentrations that are subtracted at the transition from $(x_k^{(j)})_k$ to $(x_k^{(j+1)})_k$ (using the convention $x_k^{(0)} \overset{\text{def}}{=} x_k$). Set $M_j = \sum_{i=1}^{j} m_i$, $M_0 = 0$. Then the sequence $(x_k^{(j)})_k$ admits the following representation:

$$x_k^{(j)} = x_{n_k^j} - \sum_{n=1}^{M_j} g_{n_k^j}^{(n)} w^{(n)}, \quad k \in \mathbb{N}.$$

By Lemma 4.3.5, under an appropriate renumeration such that (4.11) holds for all k,

$$\delta\Big(\Big\|\sum_{n=M_{j-1}+1}^{M_j} g_{n_k^j}^{(n)} w^{(n)}\Big\|\Big) \le \sup\|x_k^{(j+1)}\| - \sigma((x_k^{(j)})) + \frac{\epsilon_j}{2} < \epsilon_j, \quad k \in \mathbb{N},$$

and thus

$$\Big\|\sum_{n=M_{j-1}+1}^{M_j} g_{n_k^j}^{(n)} w^{(n)}\Big\| \le 2^{-j}, \quad j \in \mathbb{N}.$$

Let us now diagonalize the double sequence $x_k^{(j)}$ by considering

$$x_k^{(k)} = x_{n_k^k} - \sum_{n=1}^{M_k} g_{n_k^k}^{(n)} w^{(n)}.$$

Let us show that $x_k^{(k)} \xrightarrow{\mathcal{G}} 0$. Indeed, by definition of functional p and Lemma 4.3.5, $\delta(p(x_k)) \le \sup \|x\_k\| - \sigma(x_k)$ and, therefore, for any $j \in \mathbb{N}$ and all $k \ge j$,

$$p(x_k^{(k)}) \le p(x_k^{(j)}) \le \sup\|x_k^{(j)}\| - \sigma(x_k^{(j)}) \le \epsilon_j.$$

Since j is arbitrary, this implies $p(x_k^{(k)}) = 0$. Furthermore, denoting an arbitrary subset of $\{M_j + 1, \dots, M_{j+1}\}$, $j \in \mathbb{N}$, as J_j, we have

$$\left\| \sum_{n=M_k+1}^{\infty} g_{n_k^k}^{(n)} w^{(n)} \right\| \le \sum_{j=k}^{\infty} \left\| \sum_{n \in J_j} g_{n_k^k}^{(n)} w^{(n)} \right\| \le \frac{1}{2^{k-1}}.$$

We have therefore

$$x_{n_k^k} - \sum_{n=1}^{\infty} g_{n_k^k}^{(n)} w^{(n)} \xrightarrow{\mathcal{G}} 0,$$

understanding the series as a sum $S_k + S_k'$, of $S_k = \sum_{n=1}^{M_k} g_{n_k^k}^{(n)} w^{(n)}$ – a finite, not a priori bounded, sum – and of a series $S_k' = \sum_{n=M_k+1}^{\infty} g_{n_k^k}^{(n)} w^{(n)}$ that converges unconditionally and uniformly in k.

Note, however, that S_k is a sum of a bounded sequence $x_{n_k^k}$, a $\mathcal{G}$-Delta-vanishing (and thus bounded) sequence, and of the convergent series S_k' is bounded with respect to k. Therefore, the sum S_k' is bounded with respect to k and, consequently, the series $S_k + S_k'$ is convergent in norm, unconditionally and uniformly in k.

Finally, our construction can be carried out without further modifications if at the beginning, if $x_k \not\rightharpoonup 0$, one sets $g_k^{(1)} = \mathrm{id}$, or, if $x_k \rightharpoonup 0$ one starts the sum S_k with the zero term $g_k^{(1)} w^{(1)}$. □

4.4 Profile decomposition in the dual space

Given a continuous embedding $E \hookrightarrow F$, cocompact relative to a set $\mathcal{G}$ of scaling operators, this section deals with a dual embedding $F^* \hookrightarrow E^*$ and with existence of profile decomposition in F^* relative to a dual set of scalings

$$\mathcal{G}^{\#} \stackrel{\text{def}}{=} \{g^{*-1}, g \in \mathcal{G}\}.$$

Theorem 4.4.1. *Let F be a uniformly convex and uniformly smooth Banach space that satisfies the Opial condition. Let $\mathrm{id} \in \mathcal{G} \subset \mathcal{G}_0$ where $\mathcal{G}_0$ is a group of linear isometries on both E and F, which satisfies (4.5). If $E \overset{\mathcal{G}}{\hookrightarrow} F$ and E is dense in F, then any bounded sequence in F^* has a subsequence that admits a profile decomposition relative to $\mathcal{G}^{\#}$, and $F^* \overset{\mathcal{G}^{\#}}{\hookrightarrow} E^*$.*

Proof. 1. Note first that condition (4.5) holds for $\mathcal{G}_0^\#$ in F^*. Indeed, if $g_k^{*-1} \not\rightharpoonup 0$, then $\langle v, g_k^{-1}u\rangle \not\to 0$ for some $u \in F$, $v \in F^*$, and thus $g_k^{-1} \not\rightharpoonup 0$ in F, and, by density, $g_k^{-1} \not\rightharpoonup 0$ in E. Then, by (4.5), on a renamed subsequence, $g_k^{-1} \to g^{-1}$ in the strong operator sense in E. By continuity of the embedding, $g_k^{-1} \to g^{-1}$ in the strong operator sense in F. This implies that g^{-1} is a linear isometry on E and on F. Then, by a simple duality argument, g^{*-1} is a linear isometry on E^* and on F^*. Then, for any $v \in F^*$, $g_k^{*-1}v \rightharpoonup g^{*-1}v$, and $\|g_k^{*-1}v\|_{F^*} = \|g^{*-1}v\|_{F^*} = \|v\|_{F^*}$. Since by assumption F^* is uniformly convex, we have $g_k^{*-1} \to g^{*-1}$ in the strong sense.

2. Note also that F^* satisfies the Opial condition. Indeed, since F is uniformly convex and uniformly smooth, so is F^*. In each of F and F^*, by Proposition 2.4.3, the Opial condition is equivalent to the condition that weak convergence and Delta-convergence coincide, or by Theorem 2.1.3 that $u_k \rightharpoonup 0 \Leftrightarrow u_k^* \rightharpoonup 0$ for any sequence, where u_k^* is the dual conjugate of u_k (with $\langle u_k^*, u_k\rangle = \|u_k\|^2$). However, the latter holds in F if and only if it holds in F^*. Therefore, conditions of Theorem 4.1.6 are satisfied, which proves the first assertion of the theorem.

3. Consider now a sequence $(v_k)_{k\in\mathbb{N}}$, $v_k \overset{\mathcal{G}^\#}{\rightharpoonup} 0$ in F^*, as a sequence in E^*, and let v_k^* be a dual conjugate of v_k in E. Consider a profile decomposition for a renamed subsequence of (v_k^*) in E. Then, with $r_k \overset{\text{def}}{=} v_k^* - \sum_n g_k^{(n)} w^{(n)}$, we have

$$\begin{aligned}\|v_k\|_{E^*}^2 &= \sum_n \langle v_k, g_k^{(n)} w^{(n)}\rangle_E + \langle v_k, r_k\rangle_E \\ &\le \sum_n \langle g_k^{(n)*} v_k, w^{(n)}\rangle_E + \|v_k\|_{F^*}\|r_k\|_F.\end{aligned}$$

Note now that each term $\langle g_k^{(n)*} v_k, w^{(n)}\rangle_E$, $n \in \mathbb{N}$, vanishes by the assumption on (v_k), and that their sum is uniformly convergent relative to k and therefore vanishes as well. Sequence (v_k) is bounded in F^* (as a weakly convergent sequence), and $r_k \to 0$ in F since embedding $E \hookrightarrow F$ is $\mathcal{G}$-cocompact. This yields $v_k \to 0$ in E^*. □

The following two statements follow immediately from the definition of profile decomposition.

Proposition 4.4.2. *Let E be a Banach space where weak convergence and Delta-convergence coincide, equipped with a group of isometries $\mathcal{G}$, and let $\psi : E \to E$ be a continuous map satisfying $\psi(gu) = g\psi(u)$ for all $g \in \mathcal{G}$ and $u \in E$ and, moreover, let ψ be continuous as a map from E, equipped with the weak topology, to E equipped with the weak topology (i. e., "weak-to-weak" continuous). If $(u_k)_{k\in\mathbb{N}}$ has a profile decomposition on E relative to $\mathcal{G}$, with a concentration family $(w^{(n)}, (g_k^{(n)})_{k\in\mathbb{N}})$, $n \in \mathbb{N}$, then $\psi(u_k)$ has a profile decomposition on E whose concentration family is $(\psi(w^{(n)}), (g_k^{(n)})_{k\in\mathbb{N}})$, $n \in \mathbb{N}$.*

Proposition 4.4.3. *Let E be a Banach space where weak convergence and Delta-convergence coincide, equipped with a group of isometries $\mathcal{G}$, and let $\psi : E \to E^*$ be a continuous map satisfying*

$$\psi(gu) = g^{*-1}\psi(u) \quad \textit{for all } g \in \mathcal{G} \textit{ and } u \in E, \tag{4.12}$$

and, moreover, let ψ be continuous as a map from E, equipped with the weak topology, to E^* equipped with the weak topology ("weak-to-weak" continuous). If $(u_k)_{k\in\mathbb{N}}$ has a profile decomposition on E relative to $\mathcal{G}$ with a concentration family $(w^{(n)}, (g_k^{(n)})_{k\in\mathbb{N}})$, $n \in \mathbb{N}$, then $\psi(u_k)$ has a profile decomposition on E^* whose concentration family is $(\psi(w^{(n)}), (g_k^{*(n)^{-1}})_{k\in\mathbb{N}})$, $n \in \mathbb{N}$.

Example 4.4.4. Let $1 < p < N$, and consider $E = \dot{H}^{1,p}(\mathbb{R}^N)$ equipped with the group of rescalings $\mathcal{G}^{\frac{N-p}{p}}$. Let $f : \mathbb{R} \to \mathbb{R}$ be a continuous function satisfying

$$f(2^{\frac{N-p}{p}j}s) = 2^{(N-\frac{N-p}{p})j}f(s), \quad s \in \mathbb{R}, j \in \mathbb{Z}.$$

Then the map ψ defined by

$$\langle\psi(u), v\rangle = \int_{\mathbb{R}^N} f(u)v\,dx$$

satisfies (4.12).

Example 4.4.5. Let $1 < p < N$, and consider $E = H^{1,p}(\mathbb{R}^N)$ equipped with the group of shifts $\mathcal{G}_{\mathbb{Z}^N}$. Let $f : \mathbb{R} \times \mathbb{R}^N \to \mathbb{R}$ be a continuous function, satisfying

$$f(x + y, s) = f(x, s), \quad x \in \mathbb{R}^N, y \in \mathbb{Z}^N,$$

and

$$|f(x, s)| \le C(|s|^p + |s|^{p^*})$$

Then the map ψ defined by

$$\langle\psi(u), v\rangle = \int_{\mathbb{R}^N} f(x, u(x))v(x)\,dx$$

satisfies (4.12).

4.5 Profile decomposition for vector-valued functions

Let E be a Banach space equipped with a bounded set $\mathcal{G}$ of bounded linear operators on E. We consider now profile decompositions on a product space E^ν, $\nu = 2, 3, \dots$, relative to the diagonal action

$${}^\nu\mathcal{G} = \{(u_1, \dots, u_\nu) \mapsto (gu_1, \dots, gu_\nu)\}_{g\in\mathcal{G}} \tag{4.13}$$

as distinct from the product action $\{(u_1, \dots, u_\nu) \mapsto (g_1u_1, \dots, g_\nu u_\nu)\}_{g_1,\dots,g_\nu\in\mathcal{G}}$. Profile decompositions relative to the latter, larger, group follow immediately from profile decompositions in the scalar case, but elementary concentrations in the former case take a more specific form $(g_kw_1, \dots, g_kw_\nu)$ without weakening the remainder.

Proposition 4.5.1. *Let E be a Banach space equipped with a bounded set $\mathcal{G}$ of bounded linear operators on E. A sequence of vector-valued functions in E^ν, $\nu \in \mathbb{N}$, is ${}^\nu\mathcal{G}$-weakly vanishing if and only if its every component is $\mathcal{G}$-weakly vanishing.*

Proof. Sufficiency in the statement is immediate. Let us prove necessity. Let $(u_k^{(1)}, \dots, u_k^{(\nu)}) \overset{{}^\nu\mathcal{G}}{\rightharpoonup} 0$. Then, for any fixed $i = 1, \dots, \nu$, using test functions with only the ith component nonzero, we get $u_k^{(i)} \overset{\mathcal{G}}{\rightharpoonup} 0$. □

Corollary 4.5.2. *If a continuous embedding $E \hookrightarrow F$ is $\mathcal{G}$-cocompact, then $E^\nu \hookrightarrow F^\nu$ is ${}^\nu\mathcal{G}$-cocompact.*

Let us fix the norm on E^ν as $\sqrt{\|u_1\|^2 + \dots + \|u_\nu\|^2}$.

Proposition 4.5.3. *Let E be a Banach space equipped with a bounded set $\mathcal{G}$ of bounded linear operators on E. A sequence of vector-valued functions in E^ν, $\nu \in \mathbb{N}$, is ${}^\nu\mathcal{G}$-Delta vanishing if and only if its every component is $\mathcal{G}$-Delta vanishing.*

Proof. Sufficiency is immediate. Let us prove necessity. Let $(u_k^{(1)}, \dots, u_k^{(\nu)}) \overset{{}^\nu\mathcal{G}}{\rightharpoondown} 0$. Then, for any $g_k \in \mathcal{G}$ and any $v \in E$,

$$\|(g_k u_k^{(1)}, \dots, g_k u_k^{(\nu)})\|_{E^\nu}^2 \le \|(g_k u_k^{(1)}, \dots, g_k u_k^{(\nu)}) + (v, 0, \dots, 0)\|_{E^\nu}^2 + o(1).$$

This implies that $\|g_k u_k^{(1)}\| \le \|g_k u_k^{(1)} + v\| + o(1)$, that is, $g_k^{(1)} u_k \rightharpoondown 0$. Since the choice of the index 1 is arbitrary, we have $g_k u_k^{(i)} \rightharpoondown 0$ for any $i = 1, \dots, \nu$. □

Remark 4.5.4. We may now apply Theorem 4.1.6 to the uniformly convex and uniformly smooth Banach space E^ν equipped with the group ${}^\nu\mathcal{G}_0$ and its subset ${}^\nu\mathcal{G}$, where E, the group $\mathcal{G}_0$ and its subset $\mathcal{G}$ are as in Theorem 4.1.6. It is easy to see that ${}^\nu\mathcal{G}_0$ is a scaling group for E^ν. Then every sequence in E^ν has a profile decomposition relative to ${}^\nu\mathcal{G}$, and if E is $\mathcal{G}$- cocompactly embedded into F, then the remainder in (4.4) vanishes in F^ν. We may, however, prove a more general statement that does not explicitly require conditions of Theorem 4.1.6.

Theorem 4.5.5. *Let E be a Banach space where every bounded sequence has a subsequence that admits a profile decomposition relative to a group of bijective linear isometries $\mathcal{G}$. Then every bounded sequence in the product space E^ν, $\nu \in \mathbb{N}$, has a subsequence that admits a profile decomposition relative to the diagonal group ${}^\nu\mathcal{G}$.*

Proof. We will give here only a sketch of the proof, leaving to the reader to fill omitted details. Consider for simplicity the space $E \times E$ and consider without loss of generality two sequences (u_k) and $(\bar{u}_k)$ in E with respective complete concentration families $((g_k^{(n)})_{k\in\mathbb{N}}, w^{(n)})$ and $((\bar{g}_k^{(n)})_{k\in\mathbb{N}}, \bar{w}^{(n)})$.

Given $n \in \mathbb{N}$, if $g_k^{(n)^{-1}} \bar{g}_k^{(m)} \rightharpoonup 0$ for all m, we set $w^{(m_n)} = 0$. If, on the other hand, $g_k^{(n)^{-1}} \bar{g}_k^{(m)} \not\rightharpoonup 0$ for some m_n, then by (4.5), on a renamed subsequence, $g_k^{(n)^{-1}} \bar{g}_k^{(m_n)} \to$

$g_n \in \mathcal{G}$ (in the sense of strong operator convergence) and, furthermore, by Lemma 4.3.3, we have, on this subsequence, $g_k^{(n)^{-1}} \bar{g}_k^{(m')} \rightharpoonup 0$ for all $m' \neq m_n$. We then can replace the term $((\bar{g}_k^{(m_n)})_{k\in\mathbb{N}}, \bar{w}^{(m_n)})$ in the corresponding concentration family with $((g_k^{(n)})_{k\in\mathbb{N}}, g_n \bar{w}^{(m_n)})$ since

$$g_k^{(n)^{-1}} \bar{u}_k \rightharpoonup g_n \bar{w}^{(m_n)}.$$

Let us now apply this algorithm iteratively, extracting on the step $n+1$ a subsequence from the sequence obtained at the step n, and completing the extraction by the standard diagonalization argument. As a result, we get a renamed subsequence of $(u_k, \bar{u}_k)_{k\in\mathbb{N}}$ with (u_k) having the complete concentration family $((g_k^{(n)})_{k\in\mathbb{N}}, w^{(n)})_{n\in\mathbb{N}}$, and $(\bar{u}_k)$ having the complete concentration family, which we are free to order by the index n, $((g_k^{(n)})_{k\in\mathbb{N}}, g_n \bar{w}^{(n)})_{n\in\mathbb{N}}$. Thus we arrive at a complete concentration family for the sequence $(u_k, \bar{u}_k)$ in $E \times E$ that has the form

$$\{(g_k^{(n)}, g_k^{(n)})_{k\in\mathbb{N}}, (w^{(n)}, \tilde{w}^{(n)})\}_{n\in\mathbb{N}}, \tag{4.14}$$

where $\tilde{w}^{(n)} = g_n \bar{w}^{(m_n)}$. By Proposition 4.5.3, the remainder in (4.4) is ${}^V\mathcal{G}$-Delta vanishing. □

4.6 Profile decomposition in Besov, Triebel–Lizorkin, and Sobolev spaces

Remark 4.6.1. Profile decomposition in $\dot{B}^{s,p,q}(\mathbb{R}^N)$ and $\dot{F}^{s,p,q}(\mathbb{R}^N)$ relative to the group of rescalings is a particular case of the general profile decomposition theorem, Theorem 4.1.6, once we take into account that the Opial condition is verified by Theorem 4.2.1, and verify (4.5). (Note that this remark includes Sobolev spaces $\dot{H}^{s,p}(\mathbb{R}^N)$ as identified with $\dot{F}^{s,p,2}(\mathbb{R}^N)$ – see (10.23).) Indeed, (4.5) holds once we note that $g_{j_k,y_k} u = 2^{\frac{N-sp}{p} j_k}(2^{j_k}(\cdot - y_k)) \rightharpoonup 0$ if and only if $|j_k| + 2^{j_k}|y_k| \to \infty$. Therefore, if $g_{j_k,y_k} u \not\rightharpoonup 0$, then we have, on a renamed subsequence, $j_k = j \in \mathbb{Z}$ and $y_k \to y \in \mathbb{R}^N$. Then $g_{j_k,y_k} = g_{j,y_k} \to g_{j,y}$ in the strong operator topology, which is easy to verify directly from the definition of the respective norm. Furthermore, there is an explicit interpretation of the decoupling relation (4.2) in terms of the group (3.1):

Lemma 4.6.2. *Let E be $\dot{B}^{s,p,q}(\mathbb{R}^N)$ or $\dot{F}^{s,p,q}(\mathbb{R}^N)$ with $s > 0$ and $1 < p < N/s$. Let $r = \frac{N-sp}{p}$ and let $g_{j,y} u = 2^{rj} u(2^j(\cdot - y))$. Then $g_{j'_k,y'_k}^{-1} g_{j_k,y_k} \rightharpoonup 0$ if and only if*

$$|j_k - j'_k| + (2^{j_k} + 2^{j'_k})|y_k - y'_k| \to \infty. \tag{4.15}$$

Proof. Elementary calculations show that

$$g_{j',y'}^{-1} u = 2^{-rj'} u(2^{-j'} \cdot + y'),$$

and thus

$$g_{j',y'}^{-1} g_{j,y} u$$
$$= 2^{r(j-j')} u(2^{j-j'}(\cdot + 2^{j'}(y' - y))) = g_{j-j',2^{j'}(y-y')} u.$$

Note now that $g_{j_k,y_k} \rightharpoonup 0$ in E if and only if $|j_k| + 2^{j_k}|y_k| \to \infty$. Indeed, if $|j_k| + 2^{j_k}|y_k| \le C$, then on a renamed subsequence $j_k = j \in \mathbb{Z}$, $y_k \to y \in \mathbb{R}^N$, and $g_{j,y_k} \rightharpoonup g_{j,y}$. Conversely, if $|j_k| \to \infty$, we may without loss of generality test weak convergence on functions whose Fourier transform has compact support away from the origin, so that the support of $\mathcal{F}(u(\cdot - y_k))(2^{-j_k}\xi)\varphi(\xi)$ is disjoint from any annular domain for k sufficiently large. Finally, if j_k is bounded while $|y_k| \to \infty$, the support of $u(2^{j_k}(x - y_k))$ will become disjoint from any compact set for k sufficiently large. We conclude that $g_{j'_k,y'_k}^{-1} g_{j_k,y_k} \rightharpoonup 0$ if and only if $|j_k - j'_k| + 2^{j'_k}|y_k - y'_k| \to \infty$. It is easy to see that this is equivalent to (4.15). □

Remark 4.6.3. The same interpretation of decoupling applies to intersections of Besov and Triebel–Lizorkin spaces whose norms are $\mathcal{G}^r$ invariant with the same r. Furthermore, for the subgroup $\mathcal{G}_{\mathbb{R}^N}$ of $\mathcal{G}^r$, the same interpretation, that is,

$$g_{0,y'_k}^{-1} g_{0,y_k} \rightharpoonup 0 \iff |y_k - y'_k| \to \infty, \tag{4.16}$$

also holds for intersections of Besov and Triebel–Lizorkin spaces with different values of the parameter $r = \frac{N-sp}{p}$, in particular for the inhomogeneous spaces $B^{s,p,q}(\mathbb{R}^N)$ and $H^{s,p}(\mathbb{R}^N)$.

For subcritical embeddings of inhomogeneous Besov and Sobolev spaces, cocompact relative to the group of shifts $\mathcal{G}_{\mathbb{Z}^N}$ (as in Theorem 3.7.1, Theorem 3.7.3, Corollary 3.7.4, Corollary 3.7.5, or Corollary 3.8.3), profile decomposition follows from Theorem 4.1.6, with the decoupling relation defined by $|y_k^{(n)} - y_k^{(m)}| \to \infty$, $m \ne n$, and the remainder vanishing in the norm of the target space.

Profile decomposition in Sobolev spaces, $\dot{H}^{m,p}(\mathbb{R}^N)$, $m \in \mathbb{N}_0$, $p \in (1, N/m)$, in addition to the explicit decoupling relation (4.2) by Lemma 4.6.2, allows a sharper than (4.7) estimate for the norms of concentration profiles, set in terms of the gradient norm:

$$\sum_{n\in\mathbb{N}} \int_{\mathbb{R}^N} |\nabla^m w^{(n)}|^p \mathrm{d}x \le \int_{\mathbb{R}^N} |\nabla^m u_k|^p \mathrm{d}x + o(1). \tag{4.17}$$

We give an abbreviated proof here, referring to [112] for details. Without loss of generality, we may assume that the sum S_k is finite (the sum converges uniformly in k, so the tail amounts for a uniform arbitrarily small correction in (4.17)), and that concentration profiles $w^{(n)}$ are in $C_0^\infty(\mathbb{R}^N)$, by density of $C_0^\infty(\mathbb{R}^N)$ in $\dot{H}^{m,p}(\mathbb{R}^N)$. By convexity,

$$\int_{\mathbb{R}^N} |\nabla^m u_k|^p \mathrm{d}x \ge \int_{\mathbb{R}^N} |\nabla^m S_k|^p \mathrm{d}x + p\langle \mathcal{E}'(S_k), u_k - S_k\rangle, \tag{4.18}$$

where notation $\mathcal{E}'$ stands here for the Fréchet derivative of $\mathcal{E}(w) \stackrel{\text{def}}{=} \frac{1}{p}\int_{\mathbb{R}^N} |\nabla^m w|^p dx$. The second term in the right-hand side, since the individual blowup terms $g_k^{(n)} w^{(n)}$ in the sum S_k have asymptotically disjoint supports (details are left to the reader), is equal to a sum of terms of the form

$$p\langle [g_k^{(n)} w_k^{(n)}]^*, u_k - S_k\rangle + o(1) = p\langle [w_k^{(n)}]^*, g_k^{(n)}(u_k - S_k)\rangle + o(1) \to 0,$$

since the remainder $r_k = u_k - S_k$ converges $\mathcal{G}^{\frac{N-mp}{p}}$-weakly to 0, so that

$$\int_{\mathbb{R}^N} |\nabla^m u_k|^p dx \geq \int_{\mathbb{R}^N} |\nabla^m S_k|^p dx + o(1) = \sum_n \int_{\mathbb{R}^N} |\nabla^m w^{(n)}|^p dx + o(1), \tag{4.19}$$

where the last evaluation is again based on the asymptotically disjoint supports of individual blowups. Taking into account (4.17) and Lemma 4.6.2, we have the following profile decomposition (noting that the remainder r_k vanishes in $L^{p_m^*}(\mathbb{R}^N)$ since the embedding $\dot{H}^{m,p}(\mathbb{R}^N) \hookrightarrow L^{p_m^*}(\mathbb{R}^N)$, $N > mp$, is $\mathcal{G}^{\frac{N-mp}{p}}$-cocompact).

Theorem 4.6.4 (Sergio Solimini). *Let (u_k) be a bounded sequence in $\dot{H}^{m,p}(\mathbb{R}^N)$, $m \in \mathbb{N}$, $1 < p < N/m$. Then it has a renamed subsequence and a concentration family $(w^{(n)}, (g_k^{(n)})_{k\in\mathbb{N}})_{n\in\mathbb{N}}$, $w^{(n)} \in \dot{H}^{m,p}(\mathbb{R}^N)$, $g_k^{(n)} u \stackrel{\text{def}}{=} 2^{\frac{N-mp}{p} j_k^{(n)}} u(2^{j_k^{(n)}}(\cdot - y_k^{(n)}))$, $j_k^{(n)} \in \mathbb{Z}$, $y_k^{(n)} \in \mathbb{R}^N$, with*

$$j_k^{(1)} = 0, \quad y_k^{(1)} = 0, \quad |j_k^{(n)} - j_k^{(m)}| + (2^{j_k^{(n)}} + 2^{j_k^{(m)}})|y_k^{(n)} - y_k^{(m)}| \to \infty, \quad m \neq n, \tag{4.20}$$

such that $[g_k^{(n)}]^{-1} u_k \rightharpoonup w^{(n)}$ in $\dot{H}^{m,p}(\mathbb{R}^N)$,

$$u_k - \sum_{n\in\mathbb{N}} g_k^{(n)} w^{(n)} \to 0 \quad \text{in } L^{p_m^*}(\mathbb{R}^N), \tag{4.21}$$

the series $\sum_{n\in\mathbb{N}} g_k^{(n)} w^{(n)}$ converges in $\dot{H}^{m,p}(\mathbb{R}^N)$ unconditionally and uniformly in k, and

$$\sum_{n\in\mathbb{N}} \int_{\mathbb{R}^N} |\nabla^m w^{(n)}|^p dx \leq \liminf \int_{\mathbb{R}^N} |\nabla^m u_k|^p dx. \tag{4.22}$$

For sequences bounded in the space $H^{m,p}(\mathbb{R}^N)$, $1 < p < \infty$, we similarly have the following profile decomposition (noting that the remainder r_k vanishes in $L^q(\mathbb{R}^N)$, $p < q < p_m^*$ since the embedding $H^{m,p}(\mathbb{R}^N) \hookrightarrow L^q(\mathbb{R}^N)$ is cocompact relative to the group of integer shifts).

Theorem 4.6.5. *Let (u_k) be a bounded sequence in $H^{m,p}(\mathbb{R}^N)$, $m \in \mathbb{N}$, $1 < p < \infty$. Let $q \in (p, \infty)$ if $N \leq mp$ and $q \in (p, p_m^*)$ if $N > mp$. Then (u_k) has a renamed subsequence and a concentration family $(w^{(n)}, (u \mapsto u(\cdot - y_k^{(n)}))_{k\in\mathbb{N}})_{n\in\mathbb{N}}$, $w^{(n)} \in \dot{H}^{m,p}(\mathbb{R}^N)$, $y_k^{(n)} \in \mathbb{Z}^N$, with*

$$y_k^{(1)} = 0, \quad |y_k^{(n)} - y_k^{(m)}| \to \infty, \quad m \neq n, \tag{4.23}$$

such that $u_k(\cdot + y_k^{(n)}) \rightharpoonup w^{(n)}$ *in* $H^{m,p}(\mathbb{R}^N)$,

$$u_k - \sum_{n\in\mathbb{N}} w^{(n)}(\cdot - y_k^{(n)}) \to 0 \quad \text{in } L^q(\mathbb{R}^N), \tag{4.24}$$

the series $\sum_{n\in\mathbb{N}} w^{(n)}(\cdot - y_k^{(n)})$ *converges in* $H^{m,p}(\mathbb{R}^N)$ *unconditionally and uniformly in k, and*

$$\sum_{n\in\mathbb{N}} \int_{\mathbb{R}^N} |\nabla^m w^{(n)}|^p \mathrm{d}x \le \liminf \int_{\mathbb{R}^N} |\nabla^m u_k|^p \mathrm{d}x. \tag{4.25}$$

Note that in addition to (4.22) and (4.25) one has an analogous relation for the L^q-norms, which is a particular case of the "iterated Brezis–Lieb lemma" below. Conditions of this statement are satisfied in the case of Theorem 4.6.4 for $q = p_m^*$, and in the case of Theorem 4.6.5 for $p \le q < p_m^*$.

4.7 Decoupling of nonlinear functionals

In this section, we characterize behavior of the nonlinear mapping, for example, $u \mapsto \int_{\mathbb{R}^N} F(u)\mathrm{d}x$, in relation to profile decompositions in Sobolev spaces.

Theorem 4.7.1 ("Iterated Brezis-Lieb Lemma"). *Let E be a Banach space of functions on* $\mathbb{R}^N$ *cocompactly embedded into* $L^q(\mathbb{R}^N)$ *for some* $q \in [1,\infty)$ *relative to a subgroup* $\mathcal{G}$ *of the rescaling group* $\mathcal{G}^{q/N}$. *Assume that weak convergence in E implies convergence almost everywhere on* $\mathbb{R}^N$. *Let* (u_k) *be a sequence in E that has a profile decomposition relative to* $\mathcal{G}$. *Then*

$$\int_{\mathbb{R}^N} |u_k|^q \mathrm{d}x \to \sum_{n\in\mathbb{N}} \int_{\mathbb{R}^N} |w^{(n)}|^q \mathrm{d}x. \tag{4.26}$$

Proof. Without loss of generality, we may assume that the sum (4.3) is finite and contains $M \in \mathbb{N}$ terms. Let us prove by induction that for any $m \le M$,

$$\int_{\mathbb{R}^N} |u_k|^q \mathrm{d}x = \int_{\mathbb{R}^N} \left| u_k - \sum_{n=1}^{m} g_k^{(n)} w^{(n)} \right|^q \mathrm{d}x + \sum_{n=1}^{m} \int_{\mathbb{R}^N} |w^{(n)}|^q \mathrm{d}x + o(1). \tag{4.27}$$

Since weak convergence in E implies convergence almost everywhere, we have by the Brezis–Lieb lemma

$$\int_{\mathbb{R}^N} |u_k|^q \mathrm{d}x = \int_{\mathbb{R}^N} |u_k - w^{(1)}|^q \mathrm{d}x + \int_{\mathbb{R}^N} |w^{(1)}|^q \mathrm{d}x + o(1),$$

which is (4.27) for $M = 1$. Assume now that (4.27) holds for some m and let us prove it for $m+1$. Indeed, let us apply the Brezis–Lieb lemma to $[g_k^{(m+1)}]^{-1}[u_k - \sum_{n=1}^{m} g_k^{(n)} w^{(n)}] \rightharpoonup$

$w^{(m+1)}$:

$$\int_{\mathbb{R}^N} \left| [g_k^{(m+1)}]^{-1} \left[u_k - \sum_{n=1}^{m} g_k^{(n)} w^{(n)} \right] \right|^q dx$$

$$= \int_{\mathbb{R}^N} \left| [g_k^{(m+1)}]^{-1} \left[u_k - \sum_{n=1}^{m} g_k^{(n)} w^{(n)} \right] - w^{(m+1)} \right|^q dx + \int_{\mathbb{R}^N} |w^{(m+1)}|^q dx + o(1). \tag{4.28}$$

Taking into account that scalings $g_k^{(m+1)}$ preserve the L^q-norm, this can be rewritten as

$$\int_{\mathbb{R}^N} \left| u_k - \sum_{n=1}^{m} g_k^{(n)} w^{(n)} \right|^q dx = \int_{\mathbb{R}^N} \left| u_k - \sum_{n=1}^{m+1} g_k^{(n)} w^{(n)} \right|^q dx + \int_{\mathbb{R}^N} |w^{(m+1)}|^q dx + o(1).$$

Substituting the value of the left-hand side from (4.27) into the relation above, we get (4.27) for $m+1$ and, therefore, for all $m \le M$. Setting $m = M$ in (4.27), we have

$$\int_{\mathbb{R}^N} |u_k|^q dx \to \sum_{n=1}^{M} \int_{\mathbb{R}^N} |w^{(n)}|^q dx,$$

which concludes the proof. □

Lemma 4.7.2. *Let $q \in (1, \infty)$ and let $F : \mathbb{R} \to \mathbb{R}$ be a continuous function satisfying $|F(s)| \le C|s|^q$, $s \in \mathbb{R}$. Let (Ω, μ) be a measure space. Then the map $\varphi(u) = \int_\Omega F(u) d\mu$ is continuous in $L^q(\Omega, \mu)$.*

Proof. Let $u_k \to u$ in $L^q(\Omega, \mu)$ and assume that there is a renamed subsequence such that $\lim \varphi(u_k) \not\to \varphi(u)$. Then it will have a further renamed subsequence (see, e. g., Theorem 4.9 in [23]) such that for some $u_0 \in L^q(\Omega, \mu)$, $|u_k| \le u_0$ and $u_k \to u$ a. e. in Ω. Then $F(u_k) \to F(u)$ a. e. in Ω and $|F(u_k)| \le C|u_0|^q \in L^1(\Omega, \mu)$. Then by Lebesgue dominated convergence $F(u_k) \to F(u)$ in $L^1(\Omega, \mu)$, and thus $\varphi(u_k) \to \varphi(u)$, a contradiction that proves the lemma. □

Theorem 4.7.3. *Let $q \in (1, \infty)$ and let $F : \mathbb{R} \to \mathbb{R}$ be a continuous function satisfying $|F(s)| \le C|s|^q$, $s \in \mathbb{R}$. Assume that (u_k) is a sequence in L^q that has defect of compactness relative to the group $\mathcal{G}_{\mathbb{Z}^N}$ of the form (4.24), with the series $\sum_{n\in\mathbb{N}} w^{(n)}(\cdot - y_k^{(n)})$ convergent in $L^q(\mathbb{R}^N)$ unconditionally and uniformly with respect to k, $|y_k^{(n)} - y_k^{(m)}| \to \infty$ whenever $m \ne n$, and $u_k(\cdot + y_k^{(n)}) \to w^{(n)}$ almost everywhere. Then*

$$\int_{\mathbb{R}^N} F(u_k) dx \to \sum_{n\in\mathbb{N}} \int_{\mathbb{R}^N} F(w^{(n)}) dx. \tag{4.29}$$

Proof. By Lemma 4.7.2, $\varphi(u) = \int_{\mathbb{R}^N} F(u) dx$ is continuous in $L^q(\mathbb{R}^N)$ and, therefore, it suffices to prove the theorem when the profile decomposition has finitely many, say

m, terms and no remainder, that is,

$$u_k = \sum_{n=1}^{m} w^{(n)}(\cdot - y_k^{(n)}).$$

Moreover, by continuity of φ and by density of $C_0^\infty(\mathbb{R}^N)$ in $L^q(\mathbb{R}^N)$ we may assume that all profiles $w^{(n)}$ have compact support. Then there exists $k_0 \in \mathbb{N}$ large enough so that $w^{(n)}(\cdot - y_k^{(n)})$, $n = 1, \dots, m$ have pairwise disjoint supports for all $k \geq k_0$. Then for all $k \geq k_0$

$$\int_{\mathbb{R}^N} F(u_k)\mathrm{d}x = \sum_{n=1}^{m} \int_{\mathbb{R}^N} F(w^{(n)})\mathrm{d}x. \qquad \square$$

Corollary 4.7.4. *Let (u_k) be a sequence in $H^{m,p}(\mathbb{R}^N)$, $m \in \mathbb{N}$, $p \in (1, \infty)$, and assume that $u_k \rightharpoonup u$ in $H^{m,p}(\mathbb{R}^N)$. Let $q \in (p, \infty)$ if $N \leq mp$ and $q \in (p, p_m^*)$ if $N > mp$ and let $F : \mathbb{R} \to \mathbb{R}$ be a continuous function satisfying $|F(s)| \leq C|s|^q$, $s \in \mathbb{R}$. Then*

$$\int_{\mathbb{R}^N} F(u_k)\mathrm{d}x = \int_{\mathbb{R}^N} F(u)\mathrm{d}x + \int_{\mathbb{R}^N} F(u_k - u)\mathrm{d}x + o(1). \tag{4.30}$$

Proof. Consider a renamed subsequence of (u_k) where (4.30) does not hold. Consider a further renamed subsequence such that (u_k) has a profile decomposition given by Theorem 4.6.5 with $w^{(1)} = u$. Then, by continuity of the Sobolev embedding $H^{m,p}(\mathbb{R}^N) \hookrightarrow L^q(\mathbb{R}^N)$, we may apply Theorem 4.7.3, so that

$$\int_{\mathbb{R}^N} F(u_k)\mathrm{d}x \to \sum_{n=1}^{\infty} \int_{\mathbb{R}^N} F(w^{(n)})\mathrm{d}x$$

and

$$\int_{\mathbb{R}^N} F(u_k - u)\mathrm{d}x \to \sum_{n=2}^{\infty} \int_{\mathbb{R}^N} F(w^{(n)})\mathrm{d}x.$$

Taking the difference of the two relations, we have

$$\int_{\mathbb{R}^N} F(u_k)\mathrm{d}x - \int_{\mathbb{R}^N} F(u_k - u)\mathrm{d}x \to \int_{\mathbb{R}^N} F(u)\mathrm{d}x,$$

which is a contradiction proving the corollary. $\square$

Corollary 4.7.5. *Assume all conditions of Corollary 4.7.4 except the condition on $F(s)$, and instead let $F(x, s)$ be a continuous function on $\mathbb{R}^N \times \mathbb{R}$ such that $\lim_{|x|\to\infty} F(x, s) = F_\infty(s)$ and $|F(x, s)| \leq C|s|^q$, $s \in \mathbb{R}$, $x \in \mathbb{R}^N$. Then*

$$\int_{\mathbb{R}^N} F(x, u_k)\mathrm{d}x \to \int_{\mathbb{R}^N} F(x, w^{(1)})\mathrm{d}x + \sum_{n=2}^{\infty} \int_{\mathbb{R}^N} F_\infty(w^{(n)})\mathrm{d}x \tag{4.31}$$

and

$$\int_{\mathbb{R}^N} F(x, u_k)dx = \int_{\mathbb{R}^N} F(x, u)dx + \int_{\mathbb{R}^N} F_\infty(u_k - u)dx + o(1). \tag{4.32}$$

Proof. Note that the functional $\int_{\mathbb{R}^N}(F(x,u) - F_\infty(u))dx$ is weakly continuous in $H^{m,p}(\mathbb{R}^N)$ because for any $\varepsilon > 0$ there exists $R > 0$ such that

$$\int_{\mathbb{R}^N\setminus B_R} |F(x,u) - F_\infty(u)|dx \le \varepsilon \int_{\mathbb{R}^N} |u|^q dx \le \varepsilon \|u\|^q_{H^{m,p}},$$

while $\int_{B_R}(F(x,u) - F_\infty(u))dx$ is weakly continuous. Then

$$\int_{\mathbb{R}^N} F(x, u_k)dx - \int_{\mathbb{R}^N} F_\infty(x, u_k)dx \tag{4.33}$$

$$= \int_{B_R} (F(x, u_k) - F_\infty(u_k))dx \to \int_{B_R} (F(x, u) - F_\infty(u))dx. \tag{4.34}$$

It remains to apply conclusions of Theorem 4.7.3 and Corollary 4.7.4 with $F = F_\infty$. □

We now consider effects of decoupling on nonlinear functional of nonlocal character from a family that includes an expression involved in the Hartree-Fock equation.

Let $N \ge 3$, $\mu \in (0, N)$, $\alpha \in (\frac{2N-\mu}{N}, \frac{2N-\mu}{N-2})$, and consider

$$\Phi(u, v) = \int_{\mathbb{R}^N}\int_{\mathbb{R}^N} \frac{|u(x)|^\alpha |v(y)|^\alpha}{|x-y|^\mu} dx\, dy, \tag{4.35}$$

Lemma 4.7.6. *Let $q_1 > \frac{2N}{2N-\mu} > q_2 \ge 1$. There exists $C > 0$ such that whenever $u \in L^{\alpha q_1}(\mathbb{R}^N) \cap L^{\alpha q_2}(\mathbb{R}^N)$,*

$$\Phi(u, u) \le C(\|u\|^{2\alpha}_{\alpha q_1} + \|u\|^{2\alpha}_{\alpha q_2}). \tag{4.36}$$

Proof. Changing the integration variables (x, y) to $(x, z) = (x, x - y)$, we represent $\Phi(u, u)$ as $\Phi_1(u, u) + \Phi_2(u, u)$, where

$$\Phi_1(u, u) = \int_{|z|<1} \left(\int_{\mathbb{R}^N} |u(x)|^\alpha |u(x - z)|^\alpha dx\right)|z|^{-\mu} dz$$

and

$$\Phi_2(u, u) = \int_{|z|\ge 1} \left(\int_{\mathbb{R}^N} |u(x)|^\alpha |u(x - z)|^\alpha dx\right)|z|^{-\mu} dz.$$

Let p satisfy $\frac{1}{p} + 1 = \frac{2}{q_1}$, which implies $p > \frac{N}{N-\mu}$ and $p' = \frac{1}{1-p^{-1}} < \frac{N}{\mu}$. Then from the Hölder inequality, we obtain

$$\begin{aligned}\Phi_1(u,u) &\le \Big(\int_{B_1(0)} |z|^{-p'\mu} dz\Big)^{\frac{1}{p'}} \\ &\quad \times \int_{\mathbb{R}^N} \Big(\int_{\mathbb{R}^N} |u(x)|^\alpha |u(x-z)|^\alpha \mathrm{d}x\Big)^p \mathrm{d}z \\ &= C \Big\| \int_{\mathbb{R}^N} |u(x)|^\alpha |u(x-z)|^\alpha \mathrm{d}x \Big\|_p^p .\end{aligned}$$

Then, since $\frac{1}{p} + 1 = \frac{2}{q_1}$, by the Young inequality for convolutions ($\|f * g\|_p \le \|f\|_q \|g\|_2$, $\frac{1}{p} + 1 = \frac{1}{q} + \frac{1}{r}$, $q, r \ge 1$) we have

$$\Phi_1(u,u) \le C \big\| |u|^\alpha \big\|_{q_1}^2 = C \|u\|_{\alpha q_1}^{2\alpha}.$$

The same argument applies to Φ_2: the only modification is that the choice of $q_2 < \frac{2N}{2N-\mu}$ yields $p < \frac{N}{N-\mu}$ in the relation $\frac{1}{p} + 1 \overset{\text{def}}{=} \frac{2}{q_2}$, which assures that $|z|^{-p'\mu}$ is integrable in the exterior of the ball. Consequently,

$$\Phi_2(u,u) \le C \|u\|_{\alpha q_2}^{2\alpha}.$$ □

Corollary 4.7.7. *There exist $p_1, p_2 \in (2, 2^*)$ such that the map (4.35) is continuous in $L^{p_1}(\mathbb{R}^N) \cap L^{p_2}(\mathbb{R}^N) \times L^{p_1}(\mathbb{R}^N) \cap L^{p_2}(\mathbb{R}^N)$.*

Proof. Obviously, one can choose q_i in (4.36) so that $p_i \overset{\text{def}}{=} \alpha q_i \in (2, 2^*)$, $i = 1, 2$. The proof of continuity of Φ is then analogous to the argument in Lemma 4.7.2 by means of replacing Lebesgue convergence with dominated convergence. In particular, if $|u_k| \le u \in L^{p_1}(\mathbb{R}^N) \cap L^{p_2}(\mathbb{R}^N)$ and $|v_k| \le v \in L^{p_1}(\mathbb{R}^N) \cap L^{p_2}(\mathbb{R}^N)$, then

$$\frac{|u_k(x)|^\alpha |v_k(y)|^\alpha}{|x-y|^\mu} \le \frac{|u(x)|^\alpha |v(y)|^\alpha}{|x-y|^\mu} \in L^1(\mathbb{R}^N \times \mathbb{R}^N)$$

by Cauchy inequality and (4.36). □

Theorem 4.7.8. *Let (u_k, v_k) be a bounded sequence in $H^{1,2}(\mathbb{R}^N) \times H^{1,2}(\mathbb{R}^N)$ that has a profile decomposition relative to the diagonal group of integer shifts*

$$^2\mathcal{G}_{\mathbb{Z}^N} = \{g_y : (u,v) \mapsto (u(\cdot - y), v(\cdot - y))\}_{y \in \mathbb{Z}^N},$$

and let $((w^{(n)}, \tilde{w}^{(n)}), (y_k^{(n)}))$ be the complete concentration family for (u_k, v_k). Then

$$\Phi(u_k, v_k) \to \sum_n \Phi(w^{(n)}, \tilde{w}^{(n)}). \tag{4.37}$$

Proof. By Corollary 4.7.7 and continuity of the series in the profile decomposition (see Section 4.5), it suffices to prove (4.37) when $u_k = \sum_{n=1}^{\ell} w^{(n)}(\cdot - y_k^{(n)})$, $v_k = \sum_{n=1}^{\ell} \tilde{w}^{(n)}(\cdot - y_k^{(n)})$, $\ell \in \mathbb{N}$, and $w^{(n)}, \tilde{w}^{(n)} \in C_0^\infty(\mathbb{R}^N)$. Then for all k large enough

$$\begin{aligned}\Phi(u_k, v_k) &= \sum_{m,n=1,\dots,\ell} \int\int \frac{|w^{(n)}(x - y_k^{(n)})|^\alpha |\tilde{w}^{(m)}(y - y_k^{(m)})|^\alpha}{|x-y|^\mu} \mathrm{d}x\mathrm{d}y \\ &= \sum_{n=1}^{\ell} \Phi(w^{(n)}, \tilde{w}^{(n)}) \\ &\quad + \sum_{m \neq n} \int\int \frac{|w^{(n)}(x)|^\alpha |\tilde{w}^{(m)}(y)|^\alpha}{|x - y + y_k^{(n)} - y_k^{(m)}|^\mu} \mathrm{d}x\mathrm{d}y.\end{aligned}$$

Since $|y_k^{(n)} - y_k^{(m)}| \to \infty$ when $m \neq n$, the second sum vanishes:

$$\begin{aligned}&\int\int \frac{|w^{(n)}(x)|^\alpha |\tilde{w}^{(m)}(y)|^\alpha}{|x - y + y_k^{(n)} - y_k^{(m)}|^\mu} \mathrm{d}x\mathrm{d}y \\ &\quad \leq \|w^{(n)}\|_\infty \|\tilde{w}^{(m)}\|_\infty \sup_{x \in \operatorname{supp} w^{(n)}, y \in \operatorname{supp} \tilde{w}^{(m)}} \frac{1}{|x - y + y_k^{(n)} - y_k^{(m)}|^\mu} \to 0.\end{aligned}$$

□

Corollary 4.7.9. *Let $u_k \rightharpoonup u$, $v_k \rightharpoonup v$ in $H^{1,2}(\mathbb{R}^N)$. Then, on a renumbered subsequence,*

$$\Phi(u_k, v_k) - \Phi(u, v) - \Phi(u_k - u, v_k - v) \to 0. \tag{4.38}$$

Proof. Relation (4.37) applied to the sequences $(u_k - w^{(1)})$, $(v_k - \tilde{w}^{(1)})$, gives

$$\Phi(u_k - w^{(1)}, v_k - \tilde{w}^{(1)}) \to \sum_{n=2}^{\infty} \Phi(w^{(n)}, \tilde{w}^{(n)}). \tag{4.39}$$

Substitution of the right-hand side of (4.39) into (4.37) gives (4.38). □

4.8 Profile decomposition for the Moser–Trudinger inequality

Let $B \subset \mathbb{R}^2$ be the open unit disk, centered at the origin. In this section, we study profile decompositions for the radial subspace $H_{0,\mathrm{rad}}^{1,2}(B)$, relative to the group (3.66).

It is easy to see that group (3.66) satisfies (4.5). Indeed, we have

$$g_{s_k} \in \mathcal{G}, g_{s_k} \rightharpoonup 0 \Leftrightarrow |\log s_k| \to \infty. \tag{4.40}$$

If $s_k \to 0$, then for any $v \in C_0^\infty(B \setminus \{0\})$, $g_{s_k} v = 0$ for k sufficiently large since $|x|^{s_k} \to 1$ uniformly on $\operatorname{supp} v$. If $s_k \to \infty$, then

$$\left| \int u(x) g_{s_k} v(x) \mathrm{d}x \right| \leq C s_k^{-1/2} \to 0.$$

Consequently, $(u, g_{s_k} v) \to 0$ in both cases, and by density this extends to all $v \in H_{0,\mathrm{rad}}^{1,2}(B)$. Then (4.5) follows from compactness of closed intervals on $\mathbb{R}$.

Theorem 4.8.1. *Let $u_k \rightharpoonup 0$ in $H^{1,2}_{0,\mathrm{rad}}(B)$. There exist $s_k^{(n)} \in (0,\infty)$, $k \in \mathbb{N}$, $n \in \mathbb{N}$, such that for a renumbered subsequence,*

$$w^{(n)} = \text{w-lim}\, s_k^{(n)^{-1/2}} u_k(r^{-s_k^{(n)}}), \tag{4.41}$$

$$|\log(s_k^{(m)}/s_k^{(n)})| \to \infty \quad \text{for } n \neq m, \tag{4.42}$$

$$\sum_{n\in\mathbb{N}} \int_B |\nabla w^{(n)}|^2 dx \leq \limsup \int_B |\nabla u_k|^2 dx, \tag{4.43}$$

$$r_k = u_k - \sum_{n\in\mathbb{N}} s_k^{(n)^{-1/2}} w^{(n)}(r^{s_k^{(n)}}) \to 0 \tag{4.44}$$

in $L^\infty(B, (\log \frac{1}{r})^{-1/2})$ and in $\exp L^2(B)$, and the series $\sum_{n\in\mathbb{N}} s_k^{(n)^{-1/2}} w^{(n)}(r^{s_k^{(n)}})$ converges in $H_0^{1,2}(B)$ unconditionally and uniformly in k.

Proof. The theorem is a particular case of Theorem 4.1.6, with the decoupling relation (4.2) realized as (4.42) as a result of (4.40) and the multiplicative character of the group: $g_s^{-1} = g_{1/s}$ and $g_s g_t = g_{st}$. Relation (4.43) follows from (1.15). Relation (4.44) follows from cocompactness of embeddings of $H^{1,2}_{0,\mathrm{rad}}(B)$ into $L^\infty(B, (\log \frac{1}{r})^{-1/2})$ and into $\exp L^2(B)$. □

4.9 Bibliographic notes

Theorem 4.1.6 is proved in [113]. It generalizes both the Sobolev space version of [112] and the Hilbert space version of [104]. The earliest profile decomposition that we found in literature is by Struwe [119], for Palais–Smale sequences for semilinear elliptic functionals. Profile decompositions, proved independently afterwards for particular classes of sequences are too numerous to be quoted here. We refer the reader to the next chapter for more bibliographic references concerning specific profile decompositions.

The condition in Theorem 4.1.6 that the set $\mathcal{G}$ consists of bijective isometries can be relaxed. We have not pursued generalization of Theorem 4.1.6 in this direction, but Theorem 3.1 in [127] gives an analog of Theorem 4.1.6 for the Hilbert space with a group of quasi-isometries, namely, a group of linear bijective operators satisfying

$$\inf_{g\in\mathcal{G}} \|g\| > 0. \tag{4.45}$$

Theorem 4.6.4 is a minor generalization of the result of [112]. Theorem 4.4.1 was conjectured by Michael Cwikel (personal communication). Verification in Section 4.2 of Opial condition for Besov and Triebel–Lizorkin spaces with respective norms (3.23) and (3.24) is based on an unpublished paper [34].

Profile decomposition for Besov and Triebel–Lizorkin spaces in Section 4.6 is a corollary of Theorem 4.1.6, combined with cocompactness results of Chapter 3. Its pioneering version in [13] has a weaker remainder, similar to the one in [68], and is based on a stronger property than cocompactness, [13, Assumption 1.1]. As it was observed in [13, Remark 3.1], not all embedding pairs that we use in Theorem 3.5.1 satisfy [13, Assumption 1.1]. Since Theorem 4.1.6 is based on cocompactness, verified in Chapter 3, rather than on a stronger [13, Assumption 1.1], it yields profile decomposition for the full range of known embeddings except the endpoint values $p, q = 1, \infty$.

Decoupling in the nonlocal nonlinearity in Section 4.7 is a corrected and expanded version of Section 10.4 in [127], that gives proper attention to the diagonal action of the scaling group (Section 4.5).

Theorem 4.8.1 was proved in [4]. A nonradial counterpart of it was first provided in [7]. A related profile decomposition for the Adams inequality in $H^{m,2}(\mathbb{R}^{2m})$, but with a different form of elementary concentrations was obtained in [14].

5 More cocompact embeddings

In this chapter, we prove cocompactness of several embeddings relative to respective groups other than the rescaling group $\mathcal{G}^r$ or the group of shifts. A natural analog of the group of shifts for functional spaces of a Riemannian manifold M is the action group of isometries $u \mapsto u \circ \eta$, $\eta \in \mathrm{Iso}(M)$. Similarly, action of the conformal group on M gives rise to an analog of the rescaling group in the Euclidean case. In Chapter 7, we extend the notion of profile decomposition to Sobolev spaces of manifolds that do not necessarily have a nontrivial group of isometries, outside of the functional-analytic framework of Chapter 4. We consider cocompactness of Sobolev embeddings on manifolds as a particular case of the energy form of Laplace-Beltrami operator with magnetic shifts, as well as cocompactness of Sobolev embeddings on Lie groups. We continue with cocompactness in the Strichartz inequality for time-dependent nonlinear Schrödinger equation, followed by a study of cocompactness related to the affine Laplacian.

5.1 Sobolev spaces with periodic magnetic field

This section considers the modification of Sobolev spaces associated with the Schrödinger operator in presence of external magnetic field on periodic (also called cocompact) manifolds. As a particular case of zero magnetic field, it also gives results on cocompactness and profile decompositions for standard Sobolev spaces of a periodic manifold.

Let (M,g) be a complete smooth connected N-dimensional Riemannian manifold and let α be a smooth differential 1-form on M, to be denoted $\alpha \in \Lambda^1$. We consider a space $H^{1,2}_\alpha(M)$ defined as a closure of $C_0^\infty(M;\mathbb{C})$ with respect to the norm given by

$$\|u\|^2 \stackrel{\text{def}}{=} \int_M (|d_x u - iu\alpha(x)|_g^2 + |u|^2)\mathrm{d}v_g. \tag{5.1}$$

Throughout this section, d denotes external derivative of differential forms (including covariant derivative of scalar functions) and d_x indicates the value taken at a point $x \in M$, $|\cdot|_g^2$ is evaluated by the Riemannian complex scalar product $g_x(\cdot,\cdot)$, and v_g denotes the Riemannian measure on M. When $N = 3$, quadratic form (5.1) is the energy functional for a charged particle in presence of an external magnetic field. The linear form $\alpha \in \Lambda^1$ is called the magnetic potential, associated with the magnetic field $\beta = d\alpha \in \Lambda^2$ given by the external differential on M. An elementary calculation shows that the norm (5.1) is invariant under the gauge transformation $(\alpha, u) \mapsto (\alpha + d\varphi, e^{i\varphi}u)$ with an arbitrary smooth φ on M. Magnetic potential $\alpha \in \Lambda^1$ for a given magnetic field $\beta \in \Lambda^2$ is nonunique: the form $\alpha + d\varphi$ with any smooth φ is also a magnetic potential for β.

https://doi.org/10.1515/9783110532432-005

By the well-known (see [77, Theorem 7.21]) diamagnetic inequality,

$$|d_x u - iu\alpha|_g \geq |d_x(|u|)|_g \quad \text{a. e. in } M, u \in H^{1,2}_\alpha(M), \tag{5.2}$$

the space $H^{1,2}_\alpha(M)$ is continuously embedded into $H^{1,2}(M)$.

Let G be a discrete subgroup of $\mathrm{Iso}(M)$. One calls the magnetic field $\beta \in \Lambda^2(M)$ *G-periodic* if (using the pullback action of $\mathrm{Iso}(M)$, $\Lambda^2(M) \to \Lambda^2(M)$) $\eta\beta = \beta$ for all $\eta \in G$ or, in terms of the magnetic potential $\alpha \in \Lambda^1$, if $d(\eta\alpha - \alpha) = 0$. We require a somewhat stronger condition, noting that if M is simply connected then the form $\eta\alpha - \alpha$ is a differential of a function. That is, we assume that for every $\eta \in G$ there exists a C^∞-function $\psi_\eta(\cdot) : M \to \mathbb{R}$, such that

$$\eta\alpha - \alpha = d\psi_\eta. \tag{5.3}$$

From (5.3), it follows that the magnetic field $\beta = d\alpha$ is $\mathrm{Iso}(M)$-periodic. Moreover, (5.3) implies that

$$d\psi_{\eta^{-1}} = \eta^{-1}\alpha - \alpha = -\eta^{-1} d\psi_\eta = -d(\psi_\eta \circ \eta^{-1}),$$

and it is easy to see that for each $\eta \in G$ we may choose an additive constant for ψ_η such that $\psi_{\eta^{-1}} = -\psi_\eta \circ \eta^{-1}$. In particular, this gives

$$\psi_{\mathrm{id}}(x) = 0, \quad x \in M. \tag{5.4}$$

Similar to shifts $u \mapsto u \circ \eta$, $\eta \in G$, that define a group of isometries on $H^{1,2}(M)$, one can use the function ψ_η to define an isometric action of G on $H^{1,2}_\alpha(M)$, known as magnetic shifts:

$$\mathcal{G}'_{\alpha,G} = \{g_\eta : u \mapsto e^{i\psi_\eta} u \circ \eta, u \in C_0^\infty(M)\}_{\eta \in G}. \tag{5.5}$$

Note that if M is the Euclidean space and $G = \{x \mapsto x + \eta\}_{\eta \in M}$, then every G-periodic (here, constant) magnetic field corresponds to the magnetic potential with the component vector Bx where B is a constant antisymmetric matrix, and the magnetic shifts corresponding to the field α use $\psi_\eta = B\eta \cdot x$.

Magnetic shifts do not generally form a group. Let us look at that in more detail. Note first that, with some constant $\gamma(\eta_1, \eta_2) \in \mathbb{R}$,

$$\psi_{\eta_1\eta_2} = \psi_{\eta_1} \circ \eta_2 + \psi_{\eta_2} + \gamma(\eta_1, \eta_2), \tag{5.6}$$

since the derivatives of the left- and the right-hand side coincide by (5.3).

Then for every $\eta_1, \eta_2 \in G$, $\theta_1, \theta_2 \in \mathbb{R}$, using (5.6), we have

$$\begin{aligned} g_{\eta_2,\theta_2} g_{\eta_1,\theta_1} u &= e^{i(\theta_1+\theta_2)} e^{i\psi_{\eta_2}} e^{i\psi_{\eta_1}\circ\eta_2} u \circ (\eta_1\eta_2) \\ &= e^{i(\theta_1+\theta_2-\gamma(\eta_1,\eta_2))} e^{i\psi_{\eta_1\eta_2}} u \circ (\eta_1\eta_2) = g_{\eta_1\eta_2,\theta_1+\theta_2-\gamma(\eta_1,\eta_2)}. \end{aligned} \tag{5.7}$$

Therefore, the set $\mathcal{G}_{\alpha,G}$, defined by

$$\mathcal{G}_{\alpha,G} \stackrel{\text{def}}{=} \{g_{\eta,\theta} : u \mapsto e^{i\theta} e^{i\psi_\eta} u \circ \eta, u \in C_0^\infty(M)\}_{\eta \in G, \theta \in \mathbb{R}}, \tag{5.8}$$

is closed with respect to the operator multiplication. We see below that it is a scaling group.

Lemma 5.1.1. *The set of isometries $\mathcal{G}_{\alpha,G}$ is a scaling group on $H_\alpha^{1,2}(M)$ and*

$$g_{\eta,\theta}^{-1} = g_{\eta^{-1},-\theta}, \quad \eta \in G, \theta \in \mathbb{R}. \tag{5.9}$$

Proof. 1. Let us show first (5.9), which in turn is obviously true once we prove it for $\theta = 0$. As we have already shown above,

$$\psi_\eta = -\psi_{\eta^{-1}} \circ \eta \tag{5.10}$$

which corresponds to $\gamma(\eta, \eta^{-1}) = 0$. Then solving the equation $g_{\eta,0} u = v$, one has $v = e^{-i\psi_\eta \circ \eta^{-1}} u \circ \eta^{-1} = e^{i\psi_{\eta^{-1}}} u \circ \eta^{-1}$.

2. Now we show that the set $\mathcal{G}_{\alpha,G}$ consists of isometries on $H_\alpha^{1,2}(M)$. Without loss of generality, we may consider only elements $g_{\eta,0}$, $\eta \in G$:

$$\begin{aligned}
&(u, g_{\eta,0} v)_{H_\alpha^{1,2}(M)} \\
&= \int_M e^{-i\psi_\eta} g_x(du + iu\alpha, d(v \circ \eta) - id\psi_\eta v \circ \eta + i(v \circ \eta)\alpha) dv_g + \int_M e^{-i\psi_\eta} u \, \bar{v} \circ \eta \, dv_g \\
&= \int_M e^{-i\psi_\eta \circ \eta^{-1}} g_x((du) \circ \eta^{-1} + i(u \circ \eta^{-1})\eta^{-1}\alpha, dv + iv\alpha) dv_g + \int_M e^{-i\psi_\eta} u \, \bar{v} \circ \eta \, dv_g \\
&= \int_M e^{i\psi_{\eta^{-1}}} g_x(d(u \circ \eta^{-1}) + i(u \circ \eta^{-1})(\alpha + d\psi_{\eta^{-1}}), dv + iv\alpha) dv_g + \int_M e^{i\psi_{\eta^{-1}}} u \, \bar{v} \, dv_g \\
&= (g_{\eta^{-1},0} u, v)_{H_\alpha^{1,2}(M)}, \quad u, v \in C_0^\infty(M),
\end{aligned}$$

which proves that $g_{\eta,0}^* = g_{\eta^{-1},0}$, $\eta \in G$, and thus $g_{\eta,\theta}^* = g_{\eta^{-1},-\theta}$. By (5.9), we have $g_{\eta,\theta}^* = g_{\eta,\theta}^{-1}$.

3. Let us verify now the four axiomatic properties that define a group.

(i) By (5.7), the set is closed with respect to operator multiplication.
(ii) By (5.4), the identity element is g_{id}.
(iii) Associativity follows from the identity

$$(g_{\eta_3} g_{\eta_2}) g_{\eta_1} u = g_{\eta_3} (g_{\eta_2} g_{\eta_1}) u = e^{i[\psi_{\eta_3} + \psi_{\eta_2} \circ \eta_3 + \psi_{\eta_1} \circ (\eta_2 \eta_3)]} u \circ (\eta_3 \eta_2 \eta_1)$$

that can be obtained by direct computation.
(iv) Existence of the inverse is immediate from (5.9).

4. It remains to show (4.5), while (4.6) is trivially true in Hilbert spaces. Note that (g_{η_k,θ_k}) does not converge weakly to zero if and only if (η_k) has a constant subsequence. Thus, on a renamed subsequence we have $\eta_k = \eta_1$, while θ_k, taken modulo 2π, converges to some $\theta_0 \in [0, 2\pi)$. Then $g_{\eta_k,\theta_k} = e^{i\theta_k} g_{\eta_1,0} \to e^{i\theta_0} g_{\eta_1,0}$ and $g_{\eta_k,\theta_k} = e^{-i\theta_k} g_{\eta_1^{-1},0} \to e^{i\theta_0} g_{\eta_1,0}$ in the strong operator sense of $H^{1,2}_\alpha(M)$, and the lemma is proved. □

Definition 5.1.2. A Riemannian manifold M is called periodic (or cocompact) relative to a subgroup G of its isometries if for some open geodesic ball $V \subset M$,

$$\bigcup_{\eta\in G} \eta V = M.$$

Lemma 5.1.3. *If the group G is discrete, then the covering in Definition 5.1.2, $\{\eta V\}_{\eta\in G}$, is of uniformly finite multiplicity.*

Proof. Let $m(x)$ be the number of $\eta \in G$ such that $x \in \eta V$. If the covering does not have uniformly finite multiplicity, there exists a sequence (x_k) in M such that $m_k \stackrel{\text{def}}{=} m(x_k) \to \infty$. For each k, there exists $\eta_k \in G$ such that $y_k \stackrel{\text{def}}{=} \eta_k x_k \in V$. Note that $m(y_k) = m_k$ by isometry. Thus there exist distinct elements $\zeta_1^{(k)}, \dots, \zeta_{m_k}^{(k)} \in G$ such that $\zeta_j^{(k)} y_k \in V$, $j = 1, \dots, m_k$. Since V is bounded, $\inf_{j\neq\ell} d(\zeta_j^{(k)} y_k, \zeta_\ell^{(k)} y_k) \to 0$ as $k \to 0$ (otherwise V would contain infinitely many disjoint geodesic balls of fixed radius). Thus there exist j_k and n_k, $j_k \neq n_k$, such that $d(\zeta_{j_k}^{(k)} y_k, \zeta_{n_k}^{(k)} y_k) \to 0$. Passing to a renamed subsequence, we have $y_k \to y \in \bar{V}$. Then $d(\zeta_{j_k}^{(k)} y, \zeta_{n_k}^{(k)} y) \to 0$ which, since the elements $\zeta_{j_k}^{(k)}$, $\zeta_{n_k}^{(k)}$ are distinct, contradicts the assumption that G is discrete. □

Remark 5.1.4. It is easy to see that when M is periodic, the norm $H^{1,2}_\alpha(M)$ is equivalent to the Sobolev norm $H^{1,2}(M)$.

Theorem 5.1.5. *Let G be a discrete subgroup of* Iso(M) *and assume that M is a complete G-periodic Riemannian N-manifold. Then for any $p \in (2, 2^*)$ the embedding $H^{1,2}_\alpha(M) \hookrightarrow L^p(M)$ is $\mathcal{G}_{\alpha,G}$-cocompact.*

Proof. Let V be as in Definition 5.1.2. From the Sobolev inequality on a bounded domain, and since the usual Sobolev norm is dominated by the $H^{1,2}_\alpha$-norm by the diamagnetic inequality we have

$$\int_{\eta(V)} |u|^p \mathrm{d}v_g \le C\|u\|^2_{H^{1,2}_\alpha(\eta V)} \Big(\int_{\eta(V)} |u|^p \mathrm{d}v_g \Big)^{1-2/p}, \quad \eta \in G. \tag{5.11}$$

By adding terms in (5.11) over $\eta \in G$, taking into account Lemma 5.1.3 we obtain

$$\int_M |u|^p \mathrm{d}v_g \le C\|u\|^2_{H^{1,2}_\alpha(M)} \sup_{\eta\in G} \Big(\int_V |g_{\eta^{-1},0} u|^p \mathrm{d}v_g \Big)^{1-2/p}. \tag{5.12}$$

Replacing V with M in the right-hand side of (5.12) and taking into account that $g_{\eta^{-1},0}$ is an isometry on $L^p(M)$, we have that the embedding $H^{1,2}_\alpha(M) \hookrightarrow L^p(M)$ is continuous.

Let (u_k) be a sequence in $H^{1,2}_\alpha(M)$ and assume that $u_k \overset{\mathcal{G}_{\alpha,G}}{\rightharpoonup} 0$. Applying (5.12) to (u_k), we have

$$\begin{aligned}\int_M |u_k|^p \mathrm{d}v_g &\le C\|u_k\|^2_{H^{1,2}(M)} \sup_{\eta\in G}\Big(\int_V |g_{\eta^{-1},0}u_k|^p \mathrm{d}v_g\Big)^{1-2/p} \\ &\le C\Big(\int_V |g_{\eta_k,0}u_k|^p \mathrm{d}v_g\Big)^{1-2/p}\end{aligned} \tag{5.13}$$

for an appropriately chosen "near-supremum" sequence (η_k) in G. It remains to note that by compactness of the Sobolev embedding for a ball in M, $g_{\eta_k,0}u_k \to 0$ in $L^p(V)$, so that the assertion of the lemma follows from (5.13). □

As a consequence of the cocompact embedding, we have the following profile decomposition.

Theorem 5.1.6. *Let M be a periodic manifold with respect to a discrete group of isometries G and let $\alpha \in \Lambda^1$ be a magnetic potential of a G-periodic magnetic field. Any bounded sequence in $H^{1,2}_\alpha(M)$ equipped with the group $\mathcal{G}_{\alpha,G}$ and a has a subsequence that admits a profile decomposition relative to the subset*

$$\{u \mapsto e^{i\psi_\eta} u \circ \eta\}_{\eta\in G} \subset \mathcal{G}_{\alpha,G}, \tag{5.14}$$

such that

$$u_k - \sum_{n\in\mathbb{N}} \exp(i\psi_{\eta_k^{(n)}}) w^{(n)} \circ \eta_k^{(n)} \to 0 \quad \text{in } L^p(M),\ p \in (2, 2^*), \tag{5.15}$$

and sequences $(\eta_k^{(m)^{-1}} \circ \eta_k^{(n)})$ are discrete whenever $m \neq n$.

Proof. Since $\mathcal{G}_{\alpha,G}$ is a scaling group by Lemma 5.1.1, and embedding $H^{1,2}_\alpha(M) \hookrightarrow L^p(M)$ is $\mathcal{G}_{\alpha,G}$-cocompact by Theorem 5.1.5, we may apply Theorem 4.1.6 and Corollary 4.1.9, getting a profile decomposition relative to the whole group $\mathcal{G}_{\alpha,G}$.

We reduce it now to a profile decomposition relative to the set (5.14), that is, to the one with $\theta_k^{(n)} = 0$ by using compactness of the sequences $(e^{i\theta_k^{(n)}})_{k\in\mathbb{N}}$. By extraction of convergent subsequences and standard diagonalization, we may assume that $\theta_k^{(n)} \to \theta_n \in [0, 2\pi]$, and rename $e^{i\theta_n} w^{(n)}$ as $w^{(n)}$.

Finally, we interpret the asymptotic decoupling relation (4.2). If a sequence $(\zeta_k^{-1}\eta_k)_{k\in\mathbb{N}}$ is discrete and $v, w \in C_0^\infty(M)$, then $(g_{\zeta_k,0}v, g_{\eta_k,0}w)_{H^{1,2}(M)} = 0$ for all k sufficiently large. By density, this implies

$$g^*_{\zeta_k,0} g_{\eta_k,0} = g^{-1}_{\zeta_k,0} g_{\eta_k,0} \rightharpoonup 0.$$

On the other hand, if $(\zeta_k^{-1}\eta_k)$ is not discrete, since G is discrete, $(\zeta_k^{-1}\eta_k)$ has a renamed constant subsequence and $\eta_k = \zeta_k\eta$ with some $\eta \in G$. Then

$$|g_{\zeta_k,0}^{-1} g_{\eta_k,0} w| = |w \circ \eta^{-1}|,$$

which implies that $g_{\zeta_k,0}^{-1} g_{\eta_k,0}$ does not weakly converge to zero in $H_\alpha^{1,2}(M)$. Consequently, the decoupling property (4.2) is equivalent to discreteness of $(\eta_k^{(m)^{-1}} \circ \eta_k^{(n)})_{k\in\mathbb{N}}$ whenever $m \neq n$. □

Theorem 5.1.6 in the case of zero magnetic field takes the following form.

Corollary 5.1.7. *Let M be a smooth Riemannian manifold, periodic relative to a discrete group G of its isometries. Any bounded sequence in $H^{1,2}(M)$ has a subsequence that admits a profile decomposition relative to the group $\mathcal{G}(G) = \{u \mapsto u \circ \eta, \eta \in G\}$:*

$$u_k - \sum_{n\in\mathbb{N}} w^{(n)} \circ \eta_k^{(n)} \to 0 \quad \text{in } L^p(M),\ p \in (2, 2^*), \tag{5.16}$$

with

$$w^{(n)} = \text{w-lim}\, u_k \circ \eta_k^{(n)^{-1}}, \tag{5.17}$$

and the sequence $(\eta_k^{(m)^{-1}} \circ \eta_k^{(n)})$ is discrete whenever $m \neq n$.

5.2 Cocompactness of subelliptic Sobolev embeddings

Let G be a Carnot group, that is, a connected and simply connected Lie group associated with a nilpotent Lie algebra $\mathbf{G}$, generated (as a Lie algebra) by a subspace $V_1 \subset \mathbf{G}$, and endowed with a stratification $\mathbf{G} = V_1 \oplus \cdots \oplus V_\ell$ such that $[V_i, V_j] \subset V_{i+j}$. We denote a basis for V_1 as $Y_1, \dots, Y_m$.

Let us fix on G exponential coordinates, which allows to use the same notation for an element Y of $\mathbf{G}$, the left invariant vector field on G defined by Y and the first-order differential operator $Yu = u \mapsto du(Y)$ associated with this vector field. In these notation, an element of $\eta \in G$ is represented by a point $y \in \mathbb{R}^N$.

An example of a Carnot group is the Heisenberg group $\mathbb{H}_n$. In exponential coordinates $(x_1, \dots, x_n, y_1, \dots, y_n, z)$, it has a stratified basis consisting of $Y_i = \partial x_i + 2y_i \frac{\partial}{\partial z}$, $i = 1, \dots, n$, $Y_{i+n} = \frac{\partial}{\partial y_i} - 2x_i \frac{\partial}{\partial z}$, $i = 1, \dots, n$, spanning V_1, and $\frac{\partial}{\partial z}$, spanning V_2.

Using exponential coordinates, one defines *anisotropic dilations* $\delta_t : G \to G, t > 0$, by means of a mapping $y \mapsto t^j y$ on V_j. Note that the Jacobian of δ_t in the exponential coordinates is t^Q, where $Q \overset{\text{def}}{=} \sum_{j=1}^p j \dim V_j$ is called the homogeneous dimension. For example, the homogeneous dimension of $\mathbb{R}^N$ is N, and the homogeneous dimension of the $N = 2n+1$-dimensional Heisenberg group $\mathbb{H}_n$ is $Q = 1 \cdot 2n + 2 \cdot 1 = 2n + 2 = N + 1$. It is known that the left-shift invariant Haar measure on Carnot groups coincides with the Lebesgue measure. We endow the group G with a left-invariant metric tensor by fixing its value at the origin as an inner product on **G** where the basis is orthonormal, and extending it to all points of G by the pullback action of the left shifts.

Definition 5.2.1. The homogeneous subelliptic Sobolev space $\dot{H}^{1,p}(G)$, $p \in [1,\infty)$, is a completion of $C_0^\infty(G)$ in the norm

$$\|u\| = \left(\int_{\mathbb{R}^N} \sum_{i=1}^{\ell} |Y_i u|^p \mathrm{d}y\right)^{\frac{1}{p}}.$$

Let $\Omega \subset G$ be an open set. The inhomogeneous subelliptic Sobolev space $H^{1,p}(\Omega)$ is a completion of $C^\infty(\Omega)$ in the norm

$$\|u\| = \left(\int_{\Omega} \left(\sum_{i=1}^{\ell} |Y_i u|^p + |u|^p\right) \mathrm{d}y\right)^{\frac{1}{p}}.$$

As in the Euclidean case, $\dot{H}^{1,p}(G)$ is not necessarily continuously embedded into a Lebesgue space, that is, it cannot be identified as a space of measurable functions. Similar to the Euclidean case, there is a continuous Sobolev embedding when $Q > p$, where Q is the homogeneous dimension of the Carnot group G. In this case, $\dot{H}^{1,p}(G) \hookrightarrow L^{p_Q^*}(G)$, where $p_Q^* = \frac{pQ}{Q-p}$. When Ω is a domain with a piecewise smooth boundary, there exists a continuous embedding $H^{1,p}(\Omega) \hookrightarrow L^q(\Omega)$ for all $q \in (p, p_Q^*]$ when $p < Q$ and for all $q > p$ when $p \geq Q$, as well as $H^{1,p}(\Omega) \hookrightarrow C(\Omega)$ for $p > Q$. If, furthermore, Ω is a bounded domain, the embedding is compact. For $p = Q$ and bounded Ω, there is also an embedding of Moser–Trudinger type, $H^{1,Q}(\Omega) \hookrightarrow \exp L^{\frac{Q}{Q-1}}(\Omega)$.

For the Heisenberg group $\mathbb{H}_n$, we have

$$\|u\|^p_{\dot{H}^{1,p}} = \int_{\mathbb{R}^{2n+1}} \left(\sum_{i=1}^{n} \left|\frac{\partial u}{\partial x_i} + 2y_i \frac{\partial u}{\partial z}\right|^2 + \sum_{i=1}^{n} \left|\frac{\partial u}{\partial y_i} u - 2x_i \frac{\partial u}{\partial z}\right|^2\right)^{\frac{p}{2}} \mathrm{d}x\mathrm{d}y\mathrm{d}z.$$

Note that $\dot{H}^{1,p}(G)$-norms and $L^q(G)$-norms are invariant with respect to the group of left shifts

$$\mathcal{G} = \{g_\eta : u \mapsto u \circ \eta, \eta \in G\}. \tag{5.18}$$

Furthermore, for $Q > p$, the $\dot{H}^{1,p}(G)$-norm and the $L^{p_Q^*}(G)$-norm are invariant with respect to the action of anisotropic dilations

$$h_s(u) \stackrel{\text{def}}{=} 2^{rs} u \circ \delta_{2^s}, \quad r = \frac{Q-p}{p}, \quad s \in \mathbb{R}. \tag{5.19}$$

We will equip $\dot{H}^{1,p}(G)$ with a group of linear bijective isometries that is a product group of discrete anisotropic dilations and left shifts:

$$\mathcal{G}_G^r \stackrel{\text{def}}{=} \{u \mapsto 2^{rj} u \circ \delta_{2^j}, j \in \mathbb{Z}\} \times \{g_\eta : u \mapsto u \circ \eta, \eta \in G\}. \tag{5.20}$$

Note that every element in $\mathcal{G}_G^r$ can be written in the form $u \mapsto 2^{rj}u(\delta_j\eta\cdot)$ as well in the form $u \mapsto 2^{rj}u(\zeta(\delta_j\cdot))$, $j \in \mathbb{Z}$, $\eta, \zeta \in G$.

Let G be a Carnot group and let $\mathcal{G}$ be the group of left shifts (5.18). Since Carnot groups can be identified as groups of nilpotent matrices, a Carnot group always contains a discrete subgroup G_0, corresponding to the matrices with integer components, such that there exists an open bounded neighborhood $V \subset G$ of the zero element of G satisfying

$$\bigcup_{\eta\in G_0} \eta V = G. \tag{5.21}$$

For example, Heisenberg group $\mathbb{H}_N$ has a discrete subgroup consisting of elements, whose canonic coordinates (x, y, z) take integer values. The group of left shifts by G_0, $\{u \mapsto u \circ \eta\}_{\eta\in G_0}$ will be denoted $\mathcal{G}_0$.

Remark 5.2.2. The covering $\{\eta V\}_{\eta\in G_0}$ of G has uniformly finite multiplicity. The argument is analogous to that for Lemma 5.1.3 and can be omitted.

Theorem 5.2.3. *The embedding $H^{1,p}(G) \hookrightarrow L^q(G)$, $p \in [1,\infty)$, $q \in (p, p_Q^*)$, is $\mathcal{G}_0$-cocompact.*

Proof. Consider the embedding $H^{1,p}(V) \hookrightarrow L^q(V)$ with V as in (5.21). We have, using the change of variables $x \mapsto \eta x$, we have

$$\int_{\eta V} |u_k|^q dy \le C\|u_k\|^p_{H^{1,p}(\eta V)}\left(\int_{\eta V} |u_k|^q dy\right)^{1-p/q}, \quad \eta \in G_0. \tag{5.22}$$

Then adding the terms in (5.22) over $\eta \in G_0$ we obtain

$$\begin{aligned}\int_G |u_k|^q dy &\le C\|u_k\|^p_{H^{1,p}(G)} \sup_{\eta\in G_0}\left(\int_V |u_k \circ \eta^{-1}|^q dy\right)^{1-p/q} \\ &\le C\left(\int_V |u_k \circ \eta_k^{-1}|^q dy\right)^{1-p/q}\end{aligned} \tag{5.23}$$

where $\eta_k \in G_0$ is any sequence satisfying

$$\int_V |u_k \circ \eta_k^{-1}|^q dy \ge \frac{1}{2} \sup_{\eta\in G_0} \int_V |u_k \circ \eta^{-1}|^q dy.$$

Since $u_k \circ \eta_k^{-1} \rightharpoonup 0$, by compactness of the Sobolev embedding for bounded domains, $u_k \circ \eta_k^{-1} \to 0$ in $L^q(V)$. By (5.23), this implies $u_k \to 0$ in $L^q(G)$. □

Proposition 5.2.4. *Group $\mathcal{G}_0$ satisfies* (4.5). *Relation* (4.2) *is satisfied if and only if the sequence $(\eta_k^{(m)^{-1}}\eta_k^{(n)})_{k\in\mathbb{N}}$ for $m \ne n$ is discrete.*

Proof. Let $g_k u = u \circ \eta_k$, $\eta_k \in G$. Since group G_0 is discrete, either the sequence $(\eta_k)_{k\in\mathbb{N}}$ is discrete or has a constant subsequence. If the sequence is discrete, considering without loss of generality $u \in C_0^\infty(G)$, support of $u \circ \eta_k$ will be disjoint from the support of any test function from $C_0^\infty(G)$, provided that k is sufficiently large. Thus $g_k \rightharpoonup 0$; otherwise, $g_k u = u \circ \eta \neq 0$ unless $u = 0$ and (4.5) is verified.

The second assertion of the proposition follows once we note that $(u\circ\eta)\circ\zeta = u\circ(\eta\zeta)$ for all $\eta, \zeta \in G_0$. □

Theorem 5.2.5. *Let G be a Carnot group and let (u_k) be a bounded sequence in $H^{1,2}(G)$. Then (u_k) has a renamed subsequence that has a profile decomposition relative to the group $\mathcal{G}_0$, of the form*

$$u_k - \sum_{n\in\mathbb{N}} w^{(n)} \circ \eta_k^{(n)} \to 0 \quad \text{in } L^q(G), q \in (2, 2_Q^*),$$

with $\eta_k^{(n)} \in G_0$. Elementary concentrations $w^{(\ell)} \circ \eta_k^{(\ell)}$ are asymptotically decoupled in the sense that

$$(\eta_k^{(m)^{-1}} \eta_k^{(n)})_{k\in\mathbb{N}} \text{ is discrete whenever } m \neq n, \tag{5.24}$$

and (1.15) *holds for respective $H^{1,2}(G)$-norms.*

Proof. Since $H^{1,2}(G)$ is Hilbert space, it satisfies Opial condition, and thus (4.6) is also satisfied, so Theorem 10.4.4, and Corollary 4.1.9 apply and the decoupling relation (4.2) and takes the form (5.24) by Proposition 5.2.4. □

We now consider the homogeneous space $\dot{H}^{1,p}(G)$, $p < Q$, equipped with the group of anisotropic rescalings (5.20).

Theorem 5.2.6. *Let G be a Carnot group, let $1 \le p < Q$, and let $\mathcal{G}_G^r$ be the group* (5.20). *The embedding $\dot{H}^{1,p}(G) \hookrightarrow L^{p_Q^*}(G)$ is $\mathcal{G}_G^r$-cocompact.*

Proof. Let $u_k \overset{\mathcal{G}_G^r}{\rightharpoonup} 0$. Let $\chi \in C_0^\infty((\frac{1}{2}, 4), [0, 3])$, such that $\chi(t) = t$ whenever $t \in [1, 2]$ and $|\chi'| \le 2$. Let V be as in (5.21). By the local Sobolev embedding,

$$\left(\int_{\eta V} \chi(|u_k|)^{p_Q^*} dx\right)^{p/p_Q^*} \le C \int_{\eta V} \left(\sum |Y_i u_k|^2 + \chi(u_k)^2\right)^{\frac{p}{2}} dx,$$

from which it follows, if we take into account that $\chi(t)^{p_Q^*} \le C t^p$,

$$\begin{aligned}\int_{\eta V} \chi(|u_k|)^{p_Q^*} dx &\le C \int_{\eta V} \left(\sum |Y_i u_k|^2 + \chi(u_k)^2\right)^{\frac{p}{2}} dx \left(\int_{\eta V} \chi(|u_k|)^{p_Q^*} dx\right)^{1-p/p_Q^*} \\ &\le C \int_{\eta V} \left(\sum |Y_i u_k|^2 + \chi(u_k)^2\right)^{\frac{p}{2}} dx \left(\int_{\eta V} |u_k|^p dx\right)^{1-p/p_Q^*}.\end{aligned}$$

Since $\chi(t)^p \le Ct^{p_Q^*}$, we have

$$\int_G \chi(|u_k|)^p \mathrm{d}x \le C\|u_k\|_{\dot{H}^{1,p}(G)}^{p_Q^*} \le C. \tag{5.25}$$

By Remark 5.2.2, the covering $\{\eta V\}_{\eta\in\mathcal{G}_0}$ has uniformly finite multiplicity, so adding the above inequalities over $\eta \in \mathcal{G}_0$ and using (5.25), we obtain

$$\int_G \chi(|u_k|)^{p_Q^*}\mathrm{d}x \le C \sup_{\eta\in\mathcal{G}_0}\Big(\int_{\eta V} |u_k|^p \mathrm{d}x\Big)^{1-p/p_Q^*}. \tag{5.26}$$

Let $\eta_k \in \mathcal{G}_0$ be such that

$$\sup_{\eta\in\mathcal{G}_0}\Big(\int_{\eta V} |u_k|^p \mathrm{d}x\Big)^{1-p/p_Q^*} \le 2\Big(\int_{\eta_k V} |u_k|^p \mathrm{d}x\Big)^{1-p/p_Q^*}.$$

Since $u_k \overset{\mathcal{G}_G^r}{\rightharpoonup} 0$, $u_k \circ \eta_k^{-1} \rightharpoonup 0$ in $\dot{H}^{1,p}(G)$, and by compactness of the local Sobolev embedding,

$$\int_{\eta_k V} |u_k|^p \mathrm{d}x = \int_V |u_k \circ \eta_k^{-1}|^p \mathrm{d}x \to 0.$$

Substituting this into (5.26), we obtain

$$\int_G \chi(|u_k|)^{p_Q^*}\mathrm{d}x \to 0.$$

Let

$$\chi_j(t) = 2^{rj}\chi(2^{-rj}t)), \quad j \in \mathbb{Z}.$$

Since for any sequence $j_k \in \mathbb{Z}$, recalling that anisotropic dilations are defined in (5.19), $h_{j_k} u_k \overset{\mathcal{G}_G^r}{\rightharpoonup} 0$, we have also, with arbitrary $j_k \in \mathbb{Z}$,

$$\int_G \chi_{j_k}(|u_k|)^{p_Q^*}\mathrm{d}y \to 0. \tag{5.27}$$

Note now that, with $j \in \mathbb{Z}$, we have

$$\Big(\int_G \chi_j(|u_k|)^{p_Q^*}\mathrm{d}x\Big)^{p/p_Q^*} \le C \int_{2^{r(j-1)}\le|u_k|\le 2^{r(j+2)}} \Big(\sum |Y_i u_k|^2\Big)^{\frac{p}{2}}\mathrm{d}x,$$

which can be rewritten as

$$\int_G \chi_j(|u_k|)^{p_Q^*}\mathrm{d}x \le C \int_{2^{r(j-1)}\le|u_k|\le 2^{r(j+2)}} \Big(\sum |Y_i u_k|^2\Big)^{\frac{p}{2}}\mathrm{d}x\Big(\int_G \chi_j(|u_k|)^{p_Q^*}\mathrm{d}x\Big)^{1-\frac{p}{p_Q^*}}. \tag{5.28}$$

Adding the inequalities (5.28) over $j \in \mathbb{Z}$ and taking into account that the sets $2^{r(j-1)} \le |u_k| \le 2^{r(j+2)}$ cover $\mathbb{R}^N$ with uniformly finite multiplicity, we obtain

$$\int_G |u_k|^{p_Q^*}dx \le C\|u_k\|^p_{\dot H^{1,p}} \sup_{j\in\mathbb{Z}}\Big(\int_G \chi_j(|u_k|)^{p_Q^*}\mathrm{d}x\Big)^{1-p/p_Q^*}. \tag{5.29}$$

Let j_k be such that

$$\sup_{j\in\mathbb{Z}}\Big(\int_G \chi_j(|u_k|)^{p_Q^*}\mathrm{d}x\Big)^{1-p/p_Q^*} \le 2\Big(\int_G \chi_{j_k}(|u_k|)^{p_Q^*}\mathrm{d}x\Big)^{1-p/p_Q^*},$$

and note that the right-hand side converges to zero due to (5.27). Then from (5.29) follows that $u_k \to 0$ in $L^{p_Q^*}$, which proves the theorem. □

Theorem 5.2.7. *Let G be a Carnot group and let (u_k) be a bounded sequence in $\dot H^{1,2}(G)$. Then (u_k) has a renamed subsequence that has a profile decomposition relative to the group $\mathcal{G}_G^r$, of the form*

$$u_k - \sum_{n\in\mathbb{N}} 2^{rj_k^{(n)}} w^{(n)}(\delta_{2^{j_k^{(n)}}}\eta_k^{(n)}\cdot) \to 0 \quad \text{in } L^{2_Q^*}(G),$$

with $\eta_k^{(n)} \in G$ and $j_k^{(n)} \in \mathbb{Z}$. Elementary concentrations $2^{rj_k^{(\ell)}} w^{(\ell)}(\delta_{2^{j_k^{(\ell)}}}\eta_k^{(\ell)}\cdot)$ are asymptotically decoupled in the sense that

$$\begin{aligned}&(\delta_{2^{j_k^{(n)}}}\eta_k^{(m)^{-1}}\eta_k^{(n)})_{k\in\mathbb{N}} \text{ is discrete}\\ &\text{on any subsequence where } (j_k^{(n)} - j_k^{(m)})_{k\in\mathbb{N}} \text{ is bounded.}\end{aligned} \tag{5.30}$$

Proof. The proof of relation (4.5) is analogous to that in the beginning of Section 4.6 and can be omitted. Since $\dot H^{1,2}(G)$ is Hilbert space, it satisfies Opial condition and thus (4.6) is also satisfied. Thus Theorem 10.4.4 and Corollary 4.1.9 apply. It is easy to see that any element of the group $\mathcal{G}_G^r$ can be written in the form $u \mapsto 2^{rj}u(\delta_{2_j}\eta\cdot)$, $j \in \mathbb{Z}$, $\eta \in G$. Proof of the decoupling relation (5.30) is analogous to the proof of Lemma 4.6.2. □

5.3 Cocompactness of a Strichartz embedding

Let $N \ge 1$. We will use notation $v = e^{it\Delta}u$ for the solution $v(t,x)$ of the initial value problem

$$\begin{cases}\frac{1}{i}v_t(t,x) = \Delta v(t,x), & (t,x) \in \mathbb{R}^{1+N},\\ v(0,x) = u(x), & x \in \mathbb{R}^N.\end{cases} \tag{5.31}$$

Here and throughout this section, all functions on $\mathbb{R}^{1+N}$ are complex-valued. We consider the following Strichartz inequality for the nonlinear Schrödinger equation (see [61, 118]):

$$\|e^{it\Delta}u\|_{L_{t,x}^{2+\frac{4}{N}}(\mathbb{R}^{1+N})} \leq C\|u\|_{L_x^2(\mathbb{R}^N)}. \tag{5.32}$$

We will denote the completion of $L_x^2(\mathbb{R}^N)$ in the norm $\|e^{it\Delta}u\|_{L_{t,x}^{2+\frac{4}{N}}(\mathbb{R}^{1+N})}$ as $E(\mathbb{R}^N)$, so that (5.32) expresses, in short notation, a continuous embedding $L^2(\mathbb{R}^N) \hookrightarrow E(\mathbb{R}^N)$.

Let $\mathcal{G}_0$ be the product of the following groups, all acting isometrically on $L_x^2(\mathbb{R}^N)$:

$$\begin{aligned}
\mathcal{G}_1 &= \{u \mapsto e^{i\theta}u,\ \theta \in \mathbb{R}^N\},\\
\mathcal{G}_2 &= \{u \mapsto u(\cdot - y),\ y \in \mathbb{R}^N\},\\
\mathcal{G}_3 &= \{u \mapsto 2^{\frac{N}{2}j}u(2^j\cdot), j \in \mathbb{Z}\},\\
\mathcal{G}_4 &= \{u \mapsto e^{i\xi\cdot x}u(x),\ \xi \in \mathbb{R}^N\}.
\end{aligned}$$

An arbitrary element of $\mathcal{G}_0$ can be always written as

$$g_{[\theta,y,j,\eta]}u(x) = 2^{jN/2}e^{i\theta}e^{i\eta\cdot x}u(2^j(x-y)). \tag{5.33}$$

The propagation group

$$\mathcal{G}_{i\Delta} = \{e^{it\Delta}u,\ t \in \mathbb{R}^N\},$$

also acts isometrically on $L_x^2(\mathbb{R}^N)$. Moreover,

$$e^{it\Delta}g_{[\theta,y,j,\eta]}u(x) = 2^{jN/2}e^{i\theta}e^{i\eta\cdot x}e^{-it|\eta|^2}[e^{i2^{2j}t\Delta}u](2^j(x-y-2\eta t)), \tag{5.34}$$

or, equivalently,

$$e^{it\Delta}g_{[\theta,y,j,\eta]} = g_{[\theta-t|\eta|^2,y+2\eta t,j,\eta]}e^{i2^{2j}t\Delta}.$$

This implies that any element in $\mathcal{G}_0 \times \mathcal{G}_{i\Delta}$ can be written as a product $g'g''$ with $g' \in \mathcal{G}_0$ and $g'' \in \mathcal{G}_{i\Delta}$ or vice versa.

Definition 5.3.1. The set $\mathcal{D}$ of dyadic cubes in $\mathbb{R}^N$ is the union

$$\mathcal{D} \stackrel{\text{def}}{=} \{2^j([0,1)^N + y)\}_{j\in\mathbb{Z},y\in\mathbb{Z}^N}.$$

For any function $u \in L^2(\mathbb{R}^N)$ and $Q \in \mathcal{D}$, we define u^Q by

$$\mathcal{F}[u^Q](\xi) \stackrel{\text{def}}{=} \mathbb{1}_Q(\xi)\mathcal{F}u(\xi), \quad \xi \in \mathbb{R}^N.$$

We cite the following refinement of the inequality (5.32).

Proposition 5.3.2 (Begout and Vargas, [15]). *Let $q = \frac{2(N^2+3N+1)}{N^2}$. Then there exist $C > 0$ such that*

$$\|u\|_E \le C\|u\|_{L^2_x(\mathbb{R}^N)}^{\frac{N+1}{N+2}}\Big(\sup_{Q\in\mathcal{D}}|Q|^{\frac{N+2}{Nq}-\frac{1}{2}}\|e^{it\Delta}u^Q\|_{L^q_{t,x}(\mathbb{R}^{1+N})}\Big)^{\frac{1}{N+2}}. \tag{5.35}$$

Corollary 5.3.3. *Let $\alpha = \frac{1}{(N+1)(N+2)}$. There exist $C > 0$ such that*

$$\|u\|_E \le C\|u\|_{L^2_x(\mathbb{R}^N)}^{1-\alpha}\sup_{Q\in\mathcal{D}}|Q|^{-\alpha/2}\|e^{it\Delta}u^Q\|_{L^\infty_{t,x}(\mathbb{R}^{1+N})}^{\alpha}. \tag{5.36}$$

Proof. Using Hölder inequality and taking into account that $\|u^Q\|_E \le \|u\|_E$, for any $Q \in \mathcal{D}$, we have from (5.35):

$$\begin{aligned}\|u\|_E^{N+2} &\le C\|u\|_{L^2(\mathbb{R}^N)}^{N+1}\sup_{Q\in\mathcal{D}}|Q|^{\frac{N+2}{Nq}-\frac{1}{2}}\|u^Q\|_E^{\frac{N(N+2)}{N^2+3N+1}}\|e^{it\Delta}u^Q\|_{L^\infty_{t,x}(\mathbb{R}^{1+N})}^{\frac{N+1}{N^2+3N+1}} \\ &\le C\|u\|_{L^2}^{N+1}\|u\|_E^{\frac{N(N+2)}{N^2+3N+1}}\sup_{Q\in\mathcal{D}}|Q|^{\frac{N+2}{Nq}-\frac{1}{2}}\|e^{it\Delta}u^Q\|_{L^\infty_{t,x}(\mathbb{R}^{1+N})}^{\frac{N+1}{N^2+3N+1}}.\end{aligned} \tag{5.37}$$

Collecting the powers of $\|u\|_E$ in the left-hand side and raising the left- and the right-hand side to appropriate power we arrive at (5.36). □

Theorem 5.3.4. *Strichartz embedding $L^2(\mathbb{R}^N) \hookrightarrow E(\mathbb{R}^N)$, expressed by (5.32), is cocompact relative to the group $\mathcal{G}_0 \times \mathcal{G}_{i\Delta}$.*

Proof. Let (u_k) be a sequence in $L^2_x(\mathbb{R}^N)$, $\mathcal{G}$-weakly convergent to zero. Then, in particular, for any sequences (t_k, y_k) in $\mathbb{R}^{1+N}$, (j_k) in $\mathbb{Z}$ and (ξ_k) in $\mathbb{R}^N$, we have

$$2^{\frac{N}{2}j_k}e^{\xi_k\cdot(2^{j_k}x+y_k)}[e^{it_k\Delta}u_k](2^{j_k}x+y_k) \rightharpoonup 0 \tag{5.38}$$

in $L^2(\mathbb{R}^N)$. By (5.36),

$$\|u_k\|_E \le C\|u_k\|_{L^2(\mathbb{R}^N)}^{1-\alpha}\sup_{Q\in\mathcal{D}}|Q|^{-\alpha/2}\|e^{it\Delta}u_k^Q\|_{L^\infty_{t,x}(\mathbb{R}^{1+N})}^{\alpha}. \tag{5.39}$$

Since (u_k) is bounded in $L^2(\mathbb{R}^N)$, it suffices to show that

$$\sup_{Q\in\mathcal{D}}|Q|^{-1/2}\|e^{it\Delta}u_k^Q\|_{L^\infty_{t,x}(\mathbb{R}^{1+N})} \to 0. \tag{5.40}$$

Let $Q_k \in \mathcal{D}$ and $(t_k, y_k) \in \mathbb{R}^{1+N}$, $k \in \mathbb{N}$, be such that

$$|Q_k|^{-1/2}|e^{it_k\Delta}u_k^{Q_k}|(y_k) \ge \frac{1}{2}\sup_{Q\in\mathcal{D}}|Q|^{-1/2}\|e^{it\Delta}u_k^Q\|_{L^\infty_{t,x}(\mathbb{R}^{1+N})}. \tag{5.41}$$

Let Q_k have side length 2^{-j_k}, $j_k \in \mathbb{Z}$, so that $|Q_k| = 2^{-Nj_k}$. Then, combining (5.40), (5.41) and (5.39) we have

$$\|u_k\|_E^{1/\alpha} \leq C2^{\frac{N}{2}j_k}|e^{it_k\Delta}u_k^{Q_k}|(y_k). \tag{5.42}$$

Let ξ_k be the center of the cube Q_k, $k \in \mathbb{N}$. Define a function h by as a Fourier transform of a characteristic function of the cube $[-\frac{1}{2}, \frac{1}{2})^N$:

$$h = \mathcal{F}^{-1}\mathbb{1}_{[-\frac{1}{2},\frac{1}{2})^N}.$$

Then from (5.42), using Definition 5.3.1, we have

$$\begin{aligned}\|u_k\|_E^{1/\alpha} &\leq C2^{\frac{N}{2}j_k}|e^{it_k\Delta}u_k^{Q_k}|(y_k)\\ &\leq C\left|\int_{\mathbb{R}^N} \bar{h}(x)2^{-\frac{N}{2}j_k}e^{-i\xi_k\cdot(2^{-j_k}x+y_k)}[e^{it_k\Delta}u_k](2^{-j_k}x+y_k)dx\right|.\end{aligned}$$

The integral in the last expression is a scalar product $(h, g_k u_k)$ in $L^2(\mathbb{R}^N)$ with certain $g_k \in \mathcal{G}\times\mathcal{G}_{i\Delta}$. By (5.38), this scalar product converges to zero, which proves the theorem. □

As a consequence of the cocompactness, we have a profile decomposition for the Strichartz embedding. Before we formulate the statement, we give an analytic characterization of the decoupling relation (4.2).

Proposition 5.3.5. *Let $g_k^{(m)}, g_k^{(n)} \in \mathcal{G} \times \mathcal{G}_{i\Delta}$, $k, m, n \in \mathbb{N}$, be given by the expression*

$$g_k^{(\ell)}u = 2^{j_k^{(\ell)}N/2}e^{i\theta_k^{(\ell)}}e^{i\eta_k^{(\ell)}\cdot x}[e^{it_k^{(\ell)}\Delta}u](x - y_k^{(\ell)}). \tag{5.43}$$

Then $g_k^{(m)^{-1}}g_k^{(n)} \rightharpoonup 0$ if and only if

$$\begin{aligned}&|j_k^{(m)} - j_k^{(n)}| + 2^{j_k^{(m)}+j_k^{(n)}}|\xi_k^{(m)} - \xi_k^{(n)}|^2\\ &\quad + 2^{-(j_k^{(m)}+j_k^{(n)})}|y_k^{(n)} - y_k^{(m)} - 2t_k^{(n)}2^{2j_k^{(n)}}(\xi_k^{(m)} - \xi_k^{(n)})|^2\\ &\quad + 2^{-(j_k^{(m)}+j_k^{(n)})}|t_k^{(n)}2^{2j_k^{(n)}} - t_k^{(m)}2^{2j_k^{(m)}}| \to \infty.\end{aligned} \tag{5.44}$$

Proof. We give only an outline. See [15] for details. Note that parameters $\theta_k^{(m)}, \theta_k^{(n)}$ can be set to zero without loss of generality. Show that $g_{[\theta_k,y_k,j_k,\eta_k]}e^{it_k\Delta} \rightharpoonup 0$ if and only if on every subsequence where $|j_k|$ is bounded,

$$|\eta_k| + |t_k| + |y_k| \to \infty,$$

and use the commutation formula (5.34) to express $g_k^{(m)^{-1}}g_k^{(n)}$ in the form $g_{[0,y_k,j_k,\eta_k]}e^{it_k\Delta}$. □

Theorem 5.3.6. *Let (u_k) be a bounded sequence in $L^2_x(\mathbb{R}^N)$. A renamed subsequence of (u_k) has a profile decomposition relative to the group $\mathcal{G}_0 \times \mathcal{G}_{i\Delta}$, namely, there exist $w^{(n)} \in L^2_x(\mathbb{R}^N)$ and sequences $(t_k^{(n)})$ in $\mathbb{R}$, $(y_k^{(n)})$ in $\mathbb{R}^N$, $(j_k^{(n)})$ in $\mathbb{Z}$, and $(\eta_k^{(n)})$ in $\mathbb{R}^N$, $n \in \mathbb{N}$, such that $g_k^{(1)} = \mathrm{id}$ and sequences $(g_k^{(n)})_k \in \mathbb{N} \stackrel{\text{def}}{=} (g_{[0,y_k^{(n)},j_k^{(n)},\eta_k^{(n)}]})_{k\in\mathbb{N}}$ (cf. (5.33)), $n \in \mathbb{N}$, are asymptotically decoupled in the sense of Proposition 5.3.5;*

$$e^{-it_k^{(n)}\Delta} g_k^{(n)^{-1}} u_k \rightharpoonup w^{(n)} \quad \text{in } L^2(\mathbb{R}^N),\ n \in \mathbb{N}; \tag{5.45}$$

$$\text{the series } S_k \stackrel{\text{def}}{=} \sum_{n\in\mathbb{N}} g_k^{(n)} e^{it_k^{(n)}\Delta} w^{(n)} \tag{5.46}$$

converges in $E(\mathbb{R}^N)$ unconditionally and uniformly in k; and

$$\|e^{it\Delta}(u_k - S_k)\|_{L_{t,x}^{2+\frac{4}{N}}(\mathbb{R}^{1+N})} \to 0. \tag{5.47}$$

Furthermore,

$$\sum_{n\in\mathbb{N}} \|w^{(n)}\|^2_{L^2} + \|u_k - S_k\|^2_{L^2} \le \|u_k\|^2_{L^2} + o(1). \tag{5.48}$$

Proof. Note that one can without loss of generality set $\theta_k^{(n)} = 0$. Indeed, passing to renamed subsequence for each $n \in \mathbb{N}$, and using standard diagonalization, one can have $\theta_k^{(n)} - 2\pi\ell_k^{(n)} \to \theta_n$ with suitable $\ell_k^{(n)} \in \mathbb{Z}$. Then one can set θ_n by renaming $e^{i\theta_n} w^{(n)}$ as $w^{(n)}$. Taking this into account, the assertion of the theorem follows immediately from Corollary 4.1.9 of Theorem 4.1.6, once we note that by Theorem 5.3.4, $\mathcal{G}_0 \times \mathcal{G}_{i\Delta}$-convergence of the remainder to zero implies its vanishing in the Strichartz norm. □

Profile decomposition above is applied to nonlinear Schrödinger equations with mass-critical nonlinearity (see [73, 122]). In the case of energy-critical nonlinearity, one studies sequences bounded in the gradient norm. We set $\|u\|_F \stackrel{\text{def}}{=} \|e^{it\Delta}u\|_{L_{t,x}^{\frac{2(N+2)}{N-2}}}$.

Theorem 5.3.7. *Let (u_k) be a bounded sequence in $\dot H^{1,2}_x(\mathbb{R}^N)$, $N > 2$. A renamed subsequence of (u_k) has a profile decomposition relative to the group $\mathcal{G}^{\frac{N-2}{2}} \times \mathcal{G}_{i\Delta}$, namely, there exist $w^{(n)} \in \dot H^{1,2}(\mathbb{R}^N)$ and sequences $(t_k^{(n)})$ in $\mathbb{R}$, $(y_k^{(n)})$ in $\mathbb{R}^N$, and $(j_k^{(n)})$ in $\mathbb{Z}$, $n \in \mathbb{N}$, such that $g_k^{(1)} = \mathrm{id}$ and sequences*

$$(g_k^{(\ell)})_{k\in\mathbb{N}},\ \ell \in \mathbb{N}, g_k^{(\ell)} \stackrel{\text{def}}{=} 2^{j_k^{(\ell)}\frac{N-2}{2}}[e^{it_k^{(\ell)}\Delta}u](x - y_k^{(\ell)}), \tag{5.49}$$

are asymptotically decoupled in the sense of (4.2), equivalently,

$$\begin{aligned} &|j_k^{(m)} - j_k^{(n)}| + 2^{-(j_k^{(m)}+j_k^{(n)})}|y_k^{(n)} - y_k^{(m)}|^2 \\ &\quad + 2^{-(j_k^{(m)}+j_k^{(n)})}|t_k^{(n)}2^{2j_k^{(n)}} - t_k^{(m)}2^{2j_k^{(m)}}| \to \infty; \end{aligned} \tag{5.50}$$

$$\text{for each } n \in \mathbb{N}, \quad e^{-it_k^{(n)}\Delta} g_k^{(n)^{-1}} u_k \rightharpoonup w^{(n)} \quad \text{in } \dot H^{1,2}(\mathbb{R}^N); \tag{5.51}$$

$$\textit{the series } S_k \stackrel{\text{def}}{=} \sum_{n\in\mathbb{N}} g_k^{(n)} e^{it_k^{(n)}} w^{(n)} \tag{5.52}$$

converges in F unconditionally and uniformly in k; and

$$\|e^{it\Delta}(u_k - S_k)\|_{L_{t,x}^{\frac{2(N+2)}{N-2}}} \to 0. \tag{5.53}$$

Furthermore,

$$\sum_{n\in\mathbb{N}} \|\nabla w^{(n)}\|_2^2 + \|\nabla u_k - \nabla S_k\|_2^2 \le \|\nabla u_k\|_2^2 + o(1). \tag{5.54}$$

Proof. We only sketch the proof. One applies Theorem 5.3.6 to ∇u_k, synchronizing profile decompositions of components with help of Section 4.5. Note that the curl of profile vectors of ∇u_k is zero, so they are L_x^2-integrable derivatives of distributions and, therefore, elements of $\dot{H}_x^{1,2}$. It remains to observe that if a sequence of parameters $(\xi_k^{(n)})_{k\in\mathbb{N}}$ in the profile decomposition of Theorem 5.3.6 has an unbounded subsequence for some $n \in \mathbb{N}$, then necessarily $w^{(n)} = 0$. Thus $(\xi_k^{(n)})_{k\in\mathbb{N}}$ are necessarily bounded. Passing to a renamed subsequence for each n and diagonalizing, we get $\xi_k^{(n)} \to \xi_n \in \mathbb{R}^N$, and renaming $e^{i\xi_n\cdot x}w^{(n)}(x)$ as $w^{(n)}(x)$, we have the profile decomposition of the form (5.52). Relation (5.53) follows from (5.47) for $\nabla u_k - \nabla S_k$ and the Sobolev inequality. □

5.4 Affine Sobolev inequality and affine Laplacian

In this section, we study the case $p = 2$ of the functional from the affine Sobolev inequality, [86, 136]:

$$J_p(u) \stackrel{\text{def}}{=} \left(\int_{S_1^{N-1}} \frac{1}{\|\omega \cdot \nabla u\|_p^N} d\omega \right)^{-1/N} \ge C\|u\|_{p^*}, \quad 1 \le p < N, \tag{5.55}$$

where S_1^{N-1} denotes a unit sphere in $\mathbb{R}^N$ centered at the origin. Unlike the limiting Sobolev inequality, the affine Sobolev inequality is invariant not only with respect to rescalings $\mathcal{G}^r$, but also with respect to action of the group SL(N), the group of all matrices with determinant 1. This also suggests that defect of compactness for sequences with a bounded J_p should involve, in addition of dilations and translations, actions of SL(N)-matrices. It is easy to calculate that $J_p(u)$ is a multiple of $\|\nabla u\|_p$ when u is a radial function, and that in general $J_p(u)$ is dominated by $\|\nabla u\|_p$. From the right inequality in the relation (5.66) below, one can easily conclude that $J_p(u)$ does not dominate $\|\nabla u\|_p$, therefore, (5.55) is a nontrivial refinement of the standard Sobolev inequality.

Functional J_p allows the following representation:

$$J_p(u) = \left(\frac{1}{(N-1)!} \int_{\mathbb{R}^N} e^{-\|\xi\cdot\nabla u\|_p} d\xi \right)^{-1/N}. \tag{5.56}$$

Indeed, the integral in (5.55) is obtained by radial integration in the integral in (5.56).

In what follows, we always assume $p = 2$ and $N > 2$. If we set

$$\mathcal{A}_{i,j}[u](x) \overset{\text{def}}{=} \frac{\partial u}{\partial x_i}\frac{\partial u}{\partial x_j}, \tag{5.57}$$

we can represent the L^2-norm in (5.55) as

$$\|\xi \cdot \nabla u\|_2^2 = \int_{\mathbb{R}^N} \mathcal{A}[u](x)\xi \cdot \xi \, \mathrm{d}x, \quad \xi \in \mathbb{R}^N. \tag{5.58}$$

Let now

$$A_{i,j}[u] \overset{\text{def}}{=} \int_{\mathbb{R}^N} \mathcal{A}_{i,j}[u](x)\mathrm{d}x. \tag{5.59}$$

Substituting (5.57) into (5.56) and taking $\eta = A[u]^{1/2}\xi$, we have

$$\begin{aligned}
\int_{\mathbb{R}^N} e^{-\|\xi\cdot\nabla u\|_2}\mathrm{d}\xi &= \int_{\mathbb{R}^N} e^{-(\int_{\mathbb{R}^N}\mathcal{A}[u](x)\xi\cdot\xi\,\mathrm{d}x)^{1/2}}\mathrm{d}\xi \\
&= \int_{\mathbb{R}^N} e^{-(A[u]\xi\cdot\xi)^{\frac{1}{2}}}\mathrm{d}\xi = \int_{\mathbb{R}^N} e^{-|\eta|}(\det A[u])^{-1/2}\mathrm{d}\eta \\
&= \omega_N(N-1)!(\det A[u])^{-1/2},
\end{aligned}$$

where ω_N is the area of a unit sphere in $\mathbb{R}^N$. We conclude that

$$J_2(u) = \omega_N^{-1/N}(\det A[u])^{1/2N}. \tag{5.60}$$

Note that this expression presumes that the matrix $A[u]$ is well-defined, which is the case if and only if $\nabla u \in L^2(\mathbb{R}^N)$. In what follows we will fix the domain of J_2 as $\dot{H}^{1,2}(\mathbb{R}^N)$.

We will also consider later a functional

$$J_{2,\Omega}(u) \overset{\text{def}}{=} \omega_N^{-1/N}(\det A_\Omega[u])^{1/2N}$$

where $A_\Omega[u] = \int_\Omega \mathcal{A}_{i,j}[u](x)\mathrm{d}x$, $\Omega \subset \mathbb{R}^N$ is an open set. Note that if $J_{2,\Omega}(u) = 0$, and Ω is convex, then there is a family of parallel hyperplanes, such that u is constant on their intersections with Ω.

We would like to characterize the behavior of the matrix (5.59) relative to the action of $\mathrm{SL}(N)$.

Lemma 5.4.1. *Let $T \in \mathrm{SL}(N)$ and let $u \in \dot{H}^{1,2}(\mathbb{R}^N)$. Then*

$$A[u \circ T] = T^* A[u]\, T. \tag{5.61}$$

In particular, for every $u \in \dot{H}^{1,2}(\mathbb{R}^N)$, there is a $T_0 \in O(N)$ such that $A[u \circ T_0]$ is diagonal, and a $T \in \mathrm{SL}(N)$ such that $A[u \circ T] = \det(A[u])\mathrm{id}$ and

$$\det A[u]^{1/2N} = \det A[u \circ T]^{1/2N} = \frac{1}{\sqrt{N}}\|\nabla(u \circ T)\|_2. \tag{5.62}$$

Proof. Equation (5.61) follows by elementary computation from the change of variable $Tx = y$, taking into account that $\partial_i u(Tx)\partial_j u(Tx) = [T^* A[u](y)T]_{ij}$ and $dx = dy$. A suitable $T_0 \in O(N)$ makes $T_0^* A[u]\, T_0$ a diagonal matrix.

Applying the same transformation once again, with a diagonal unimodular matrix $T' = \det(A[u\circ T_0])^{1/2} A[u\circ T_0]^{-1/2}$, we get $A[u\circ T_0 T'] = \det(A[u\circ T_0])\mathrm{id} = \det(A[u])\mathrm{id}$. The last assertion follows once we note that $\|\nabla u \circ T_0 T'\|_2^2 = N\det(A[u])^{1/N}$, since the latter expression is the trace of the diagonal matrix $A[u\circ T_0 T']$ with N equal eigenvalues. □

Corollary 5.4.2. *If $u \in \dot{H}^{1,2}(\mathbb{R}^N)$, then*

$$J_2(u) = \frac{\omega_N^{-1/N}}{\sqrt{N}} \min_{T\in \mathrm{SL}(N)} \|\nabla(u \circ T)\|_2. \tag{5.63}$$

Proof. Since for any $v \in \dot{H}^{1,2}(\mathbb{R}^N)$, $\|\nabla v\|_2^2 = \mathrm{tr} A[v]$, it follows from the inequality between the arithmetic and geometric mean that $\det A[u]^{\frac{1}{N}} \le \frac{1}{N}\|\nabla(u \circ T)\|_2^2$ for any $v \in \dot{H}^{1,2}(\mathbb{R}^N)$ and $T \in \mathrm{SL}(N)$. By Lemma 5.4.1, the minimum is attained. □

In view of (5.63), it is convenient to change the scalar multiple in the definition of the energy functional associated with J_2. Namely, we introduce

$$E_2(u) \stackrel{\mathrm{def}}{=} N \det A[u]^{1/N} = N\omega_N^{2/N} J_2(u)^2, \tag{5.64}$$

which allows to rewrite (5.63) as

$$E_2(u) = \min_{T\in \mathrm{SL}(N)} \|\nabla(u \circ T)\|_2^2, \tag{5.65}$$

while for any radial function $u \in \dot{H}^{1,2}_{\mathrm{rad}}(\mathbb{R}^N)$ we have $E_2(u) = \|\nabla u\|_2^2$. We will also use later an analogous functional $E_{2,\Omega}$.

Remark 5.4.3. By [67, Theorem 1.2], the gradient norm and the functional J_p for general $p \ge 1$ are connected by an inequality:

$$C' \min_{T\in \mathrm{SL}(N)} \|\nabla(u \circ T)\|_p \le J_p(u) \le C \min_{T\in \mathrm{SL}(N)} \|\nabla(u \circ T)\|_p. \tag{5.66}$$

The affine Sobolev inequality (5.55) can be now easily derived from the usual Sobolev inequality and (5.66):

$$\|u\|_{p^*} = \inf_{T\in \mathrm{SL}(N)} \|u \circ T\|_{p^*} \le C \inf_{T\in \mathrm{SL}(N)} \|\nabla(u \circ T)\|_p \le CJ_p(u). \tag{5.67}$$

We consider now sequences with an E_2-bound.

Theorem 5.4.4. *Let (u_k) be a sequence in $\dot{H}^{1,2}(\mathbb{R}^N)$ satisfying $\sup_{k\in\mathbb{N}} J_2(u_k) < \infty$. There exist a sequence $(T_k)_{k\in\mathbb{N}}$ in $\mathrm{SL}(N)$, functions $w^{(n)} \in \dot{H}^{1,2}(\mathbb{R}^N)$, and sequences $(y_k^{(n)})_{k\in\mathbb{N}}$*

in $\mathbb{R}^N$, $(j_k^{(n)})_{k\in\mathbb{N}}$ *in* $\mathbb{Z}$ *with* $n \in \mathbb{N}$, *such that* $k_k^{(1)} = 0$, $y_k^{(1)} = 0$, *and for a renumbered subsequence of* (u_k),

$$2^{-\frac{N-2}{2}j_k^{(n)}} u_k(T_k(2^{-j_k^{(n)}} \cdot + y_k^{(n)})) \rightharpoonup w^{(n)}, \quad n \in \mathbb{N}, \tag{5.68}$$

$$|j_k^{(n)} - j_k^{(m)}| + 2^{j_k^{(n)}} |y_k^{(n)} - y_k^{(m)}| \to \infty \quad \text{for } n \neq m, \tag{5.69}$$

$$\sum_{n\in\mathbb{N}} \|\nabla w^{(n)}\|_2^2 \leq \liminf E_2(u_k), \tag{5.70}$$

$$u_k - \left[\sum_{n\in\mathbb{N}} 2^{\frac{N-2}{2}j_k^{(n)}} w^{(n)}(2^{j_k^{(n)}}(\cdot - y_k^{(n)}))\right] \circ T_k^{-1} \to 0 \quad \text{in } L^{2^*}(\mathbb{R}^N), \tag{5.71}$$

and the series in the square brackets above converges in $\dot{H}^{1,2}(\mathbb{R}^N)$ *unconditionally and uniformly with respect to* k.

Proof. Let $T_k \in \mathrm{SL}(N)$ such that, according to Lemma 5.4.1,

$$E_2(u_k) = E_2(u_k \circ T_k) = \|\nabla(u_k \circ T_k)\|_2^2. \tag{5.72}$$

Let $v_k = u_k \circ T_k$ and apply Theorem 4.6.4. To conclude the proof of Theorem 5.4.4, it remains to note that (4.21) gives (5.71) by composing the left- and the right-hand side with T_k^{-1} on the right, and that the right-hand side of (1.15) yields the right-hand side of (5.70) by (5.72). □

A similar decomposition for sequences with bounded $E_2 + \|\cdot\|_2^2$ can be derived in a completely analogous way from Theorem 4.6.5:

Proposition 5.4.5. *Let* $(u_k) \in H^{1,2}(\mathbb{R}^N)$ *be a sequence such that* $E_2(u_k) + \|u_k\|_2^2 \leq C$. *There exist* $w^{(n)} \in H$, (T_k) *in* $\mathrm{SL}(N)$, *and* $(y_k^{(n)})_{k\in\mathbb{N}}$ *in* $\mathbb{Z}^N$, $y_k^{(1)} = 0$, $n \in \mathbb{N}$, *such that, on a renumbered subsequence,*

$$u_k(T_k(\cdot + y_k^{(n)})) \rightharpoonup w^{(n)}, \tag{5.73}$$

$$|y_k^{(n)} - y_k^{(m)}| \to \infty \quad \text{for } n \neq m, \tag{5.74}$$

$$\sum_{n\in\mathbb{N}} \|w^{(n)}\|_{H^{1,2}}^2 \leq \limsup \|u_k\|_{H^{1,2}}^2, \tag{5.75}$$

$$u_k - \left[\sum_{n\in\mathbb{N}} w^{(n)}(\cdot - y_k^{(n)})\right] \circ T_k^{-1} \to 0 \quad \text{in } L^p(\mathbb{R}^N), p \in (2, 2^*), \tag{5.76}$$

and the series in the square brackets above converges in $H^{1,2}(\mathbb{R}^N)$ *unconditionally and uniformly in* k.

5.5 Bibliographic notes

Concentration compactness argument for nonlinear magnetic Schrödinger operator, involving magnetic shifts was developed by Arioli and Szulkin [12], who in turn generalized the work of Esteban and Lions [44] dealing with the case of constant magnetic field. We follow, with corrections concerning the group of magnetic shifts, the presentation in [127]. A recent preprint [37] contains a profile decomposition with generalized magnetic shifts for the case of general bounded (not necessarily periodic) magnetic field. Theorem 5.1.6 can be extended to the case of any Riemannian manifolds cocompact with respect to some group (not necessarily discrete) of its isometries and any magnetic field invariant with respect to this group as it is done in [47] in the non-magnetic case, but this is now a partial case of the profile decomposition in [37].

Section 5.2 on profile decomposition for Carnot groups is based on the paper [105] with some excerpts from [127]. For details on Sobolev spaces on Lie groups see [49, 50, 132]. Earlier studies of noncompact variational problems on Carnot groups can be found, among the rest, in the work of Biagini [20] or Garofalo et al. In particular, the minimizer in the analog of (1.25)

$$\inf_{\|u\|_{\frac{2Q}{Q-2}}=1} \|u\|_{\dot{H}^{1,2}(G)} \tag{5.77}$$

for Carnot groups of rank two, that is, the counterpart of function (1.29) is found in [57, Theorem 1.1], and it equals a scalar multiple of $((1+x^2)^2+16y^2)^{-\frac{Q-2}{4}}$, where x, y are the exponential coordinates corresponding, respectively, to the strata V_1 and V_2.

Theorems 5.2.5 and 5.2.7 most likely extend to any $p > 1$. A possible way to prove it is to use an equivalent $\mathcal{G}_0$-invariant (respectively $\mathcal{G}_G^r$-invariant) Sobolev norm satisfying the Opial condition, provided, similar to the Euclidean case, by the Littlewood–Paley decomposition for Lie groups (see [52]). Alternatively, given the cocompactness, one can reproduce the argument from [112].

Profile decomposition for the Strichartz inequality (5.32) was first proved by Merle and Vega [91] in the case of two space dimensions (see also [22]). The one-dimensional case was treated by Carles and Keraani [27, Theorem 1.4]. The result was obtained for general dimension by Begout and Vargas [15]. All these profile decompositions have a weak form of remainder, similar to [58]. The present version, with a remainder convergent in the Strichartz norm, is due to Tao [123]. Tao's proof, which we follow here, is based on cocompactness of the embedding and profile decomposition in Hilbert spaces of [104] (rendered in this book as Corollary 4.1.9 of Theorem 4.1.6 in this book). Theorems 5.3.4 and 5.3.6 involve only a discrete subgroup of dilations that [15] use, without any change in the argument. An early version of Theorem 5.3.7, in 1 + 3 dimensions and with a weak form of remainder, was proved by Keraani [71]. The idea of deriving Theorem 5.3.7 from Theorem 5.3.6 is found in [73].

Affine Sobolev inequality was introduced by Zhang [136] for the case $p = 1$ and extended to general p in [86]. Representation (5.63) of J_2, definition of the affine Laplace operator and profile decomposition for the affine energy functional are given in [102].

6 Profile decompositions on subsets

In this chapter we study the effect that restriction of sequences to subsets has on profile decompositions. Conditions imposed on a sequence may result in a more specific character of its elementary concentrations, decrease the cardinality of their set, and even make them all disappear, leading to sequential compactness. This is, for example, the case when we look at profile decomposition in $H^{1,p}(\mathbb{R}^N)$, $p \in (1, N)$, relative to the group of shifts $\mathcal{G}_{\mathbb{Z}^N}$, for a bounded sequence (u_k) supported in a bounded domain $\Omega \subset \mathbb{R}^N$. For any sequence (y_k) in $\mathbb{R}^N$, $|y_k| \to \infty$, we have $u_k(\cdot + y_k) \rightharpoonup 0$, so the profile decomposition (4.24) takes the form $u_k - w^{(1)} \to 0$ in L^q, $q \in (p, p^*)$, that is, we get the classical Rellich compactness of embeddings $H_0^{1,p}(\Omega) \hookrightarrow L^q(\Omega)$. An analogous argument will be used below in the case of the affine Sobolev functional.

Compactness of embedding $H_{\mathrm{rad}}^{1,p}(\mathbb{R}^N) \hookrightarrow L^q(\mathbb{R}^N)$, $q \in (p, p^*)$, proved by Strauss [115], can be also derived from (4.24) with an observation that $u_k(\cdot - y_k) \rightharpoonup 0$ whenever $|y_k| \to \infty$ (otherwise by radiality there would be infinitely many concentration profiles equal up to an $O(N)$-transformation). We defer an exposition of this method to the next chapter where we discuss compactness in Sobolev embeddings on noncompact manifolds.

Another example of a consequence of restriction to a subspace is when a profile decomposition relative to some group, such as $\mathcal{G}^r$, has concentrations generated only by a subgroup, such as $\mathcal{G}_{\mathbb{Z}^N}$, as in Theorem 3.8.2. We address this situation in more general terms in Theorem 6.2.2 below.

One more example is the notion of a flask subspace, which is not invariant with respect to the scaling group, but yields concentration profiles in the space nonetheless. In application to Sobolev spaces $H_0^{1,p}(\Omega)$, this can be achieved by Ω having the shape of an infinite flask mirrored at its bottom. We extend the term *flask domain* introduced by del Pino and Felmer [36] to a more general setting.

6.1 Flask subsets. $\mathcal{G}$-compactness

Definition 6.1.1 (Flask subset). Let E be a Banach space endowed with a group of isometries $\mathcal{G}$. A set $A \subset E$ is called a flask subset relative to $\mathcal{G}$, or a $\mathcal{G}$-flask subset, if $\mathcal{G}(A)$ is closed with respect to Delta-convergence, that is, for any sequences $(u_n)_{n\in\mathbb{N}}$ in A and $(g_n)_{n\in\mathbb{N}}$ in $\mathcal{G}$ such that $g_n u_n \rightharpoonup w$, there exists $g \in \mathcal{G}$ such that $gw \in A$.

The theorem below gives a profile decomposition for a flask subset that is generally not an invariant subset of $\mathcal{G}$.

Theorem 6.1.2. *Let E be a uniformly smooth and uniformly convex Banach space with a scaling group of isometries $\mathcal{G}$. Let A be a $\mathcal{G}$-flask subset of E. If (u_n) is a bounded sequence in A, then it has a subsequence with a profile decomposition in E as in Theorem 4.1.6 with $w^{(n)} \in A$, $n \in \mathbb{N}$.*

https://doi.org/10.1515/9783110532432-006

Proof. Consider a profile decomposition provided by Theorem 4.1.6. Since A is a flask subset, for every $n \in \mathbb{N}$ there exists $g_n \in \mathcal{G}$ such that $\tilde{w}^{(n)} \stackrel{\text{def}}{=} g_n w^{(n)} \in A$. Setting $\tilde{g}_k^{(n)} \stackrel{\text{def}}{=} g_k^{(n)} g_n^{-1}$, we get a profile decomposition in E with the scalings $(\tilde{g}_k^{(n)})_{k\in\mathbb{N}}$ and the profiles $\tilde{w}^{(n)} \in A$, once we show that (4.2) holds for the new scalings $(\tilde{g}_k^{(n)})_{k\in\mathbb{N}}$. Indeed,

$$(\tilde{g}_k^{(n)})^{-1}\tilde{g}_k^{(m)} = g_n (g_k^{(n)})^{-1} g_k^{(m)} g_m^{-1} \rightharpoonup 0,$$

since g_n and g_m^{-1} are isometries. □

A stronger property of a subset A relative to a scaling group $\mathcal{G}$ would be to require that Delta-limits of sequences $(g_n w_n)_{n\in\mathbb{N}}$ in E with $w_n \in A$ and $g_n \rightharpoonup 0$ belong to some proper subset of A. In the same spirit one can also introduce a notion of $\mathcal{G}$-compact sets.

Definition 6.1.3. Let E be a Banach space with a group of isometries $\mathcal{G}$. A set $K \subset E$ is called locally $\mathcal{G}$-compact if any bounded sequence in K has a subsequence $\mathcal{G}$-convergent in E.

Proposition 6.1.4. *Let E be a Banach space with a scaling group $\mathcal{G}$. If the set $K \subset E$ is locally $\mathcal{G}$-compact and E is $\mathcal{G}$-cocompactly embedded into a Banach space F, then any bounded subset of K is relatively compact in F.*

Proof. From local $\mathcal{G}$-compactness of K, it follows that every bounded sequence (u_n) satisfies $g_n(u_n - u) \rightharpoonup 0$ with some $u \in E$ for every sequence (g_n) in $\mathcal{G}$, and in particular, $u_n \rightharpoonup u$. Then $u_k \to u$ in F by the definition of a $\mathcal{G}$-cocompact embedding. □

6.2 Profile decompositions for intersection of two spaces

One can use a subgroup of a scaling group to isolate a decoupled portion of a profile decomposition while the rest of the sum is $\mathcal{G}$-weakly vanishing with respect to the subgroup.

Lemma 6.2.1. *Let E be a uniformly convex and uniformly smooth Banach space with a scaling group of isometries $\mathcal{G}$. Let $\mathcal{G}'$ be a subgroup of $\mathcal{G}$. Let (u_k) be a bounded sequence in E that admits a profile decomposition. Then it has a renamed subsequence such that the sum (4.3) is of the form*

$$\sum_{n\in\mathbb{N}} g_k^{(n)} w^{(n)} = \sum_{n\in\mathbb{N}'} h_k^{(n)} v^{(n)} + \sum_{n\notin\mathbb{N}'} g_k^{(n)} w^{(n)}, \tag{6.1}$$

where $\mathbb{N}' \subset \mathbb{N}$, $h_k^{(n)} \in \mathcal{G}'$, $v^{(n)} \in E$, if $n \in \mathbb{N}'$, and for any sequence (h_k) in $\mathcal{G}'$, $h_k^{-1} g_k^{(n)} \rightharpoonup 0$ if $n \notin \mathbb{N}'$.

Proof. By the uniform and unconditional convergence of the sum (4.3), we may assume, without loss of generality, that all but finitely many terms in the sum are zero. Define

$$\mathbb{N}' \stackrel{\text{def}}{=} \{n \in \mathbb{N} : \exists h_k \in \mathcal{G}', h_k^{-1} g_k^{(n)} \not\rightharpoonup 0\}.$$

Let $n \in N'$. Then, by the definition of the scaling group, there exist $(h_k^{(n)})_{k\in\mathbb{N}}$ in $\mathcal{G}'$ and $g_n \in \mathcal{G}$ such that, on a renamed subsequence, $h_k^{(n)^{-1}} g_k^{(n)} \to g_n$, $g_k^{(n)^*} h_k^{(n)^{-1*}} \to g_n^*$, $g_k^{(n)^{-1}} h_k^{(n)} \to g_n^{-1}$ and $h_k^{(n)^*} g_k^{(n)^{-1*}} \to g_n^{-1*}$ in the sense of the strong operator convergence. Then $g_k^{(n)} w^{(n)} - h_k^{(n)} g_n w^{(n)} \to 0$ in E so that $g_k^{(n)} w^{(n)}$ can be replaced in the profile decomposition with $h_k^{(n)} v^{(n)} \to 0$ where $v^{(n)} \stackrel{\text{def}}{=} g_n w^{(n)}$. Observe now that, denoting by $o^E(1)$ any sequence convergent to zero in the norm of E,

$$\begin{aligned} h_k^{(n)^{-1}} u_k - v^{(n)} &= h_k^{(n)^{-1}} u_k - g_n w^{(n)} \\ &= h_k^{(n)^{-1}} (u_k - g_k^{(n)} w^{(n)}) + o^E(1) \rightharpoonup 0, \\ & n \in N', \end{aligned}$$

since for every $\varphi \in E$, using the definition of Delta-convergence,

$$\begin{aligned} \|h_k^{(n)^{-1}} (u_k - g_k^{(n)} w^{(n)})\| &= \|g_k^{(n)^{-1}} u_k - w^{(n)}\| \\ &\le \|g_k^{(n)^{-1}} u_k - w^{(n)} + g_n^{-1}\varphi\| + o(1) = \|h_k^{(n)^{-1}} u_k - g_n w^{(n)} + \varphi\| + o(1) \\ &= \|h_k^{(n)^{-1}} (u_k - g_k^{(n)} w^{(n)}) + \varphi\| + o(1). \end{aligned}$$

Furthermore, whenever $m \neq n$,

$$(h_k^{(n)})^{-1} h_k^{(m)} = g_n (g_k^{(n)})^{-1} g_k^{(m)} g_m^{-1} \rightharpoonup 0$$

and thus, for $m \in \mathbb{N}'$ and $n \notin \mathbb{N}'$,

$$(g_k^{(n)})^{-1} h_k^{(m)} = (g_k^{(n)})^{-1} g_k^{(m)} g_m^{-1} \rightharpoonup 0,$$

(and, similarly, $(h_k^{(m)})^{-1} g_k^{(n)} \rightharpoonup 0$). The uniform and unconditional convergence of the series in (6.1) follows from the uniform and unconditional convergence of the original series. □

Theorem 6.2.2. *Let E and F be two Banach subspaces of some topological vector space. Assume that E is a uniformly smooth and uniformly convex Banach space with a scaling group $\mathcal{G}$, that F is uniformly smooth, and that $E \cap F$ is dense in E. Assume that $\mathcal{G}$ has a subgroup $\mathcal{G}'$ that is a scaling group on $E \cap F$ equipped with the standard intersection norm $\|u\|_E + \|u\|_F$. Asume that both E and F satisfy the Opial condition. If $(u_k)_{k\in\mathbb{N}}$ is a bounded sequence in $E \cap F$, that admits a profile decomposition in E relative to the group $\mathcal{G}$, then* *it has a renamed subsequence that admits a profile decomposition in $E \cap F$ relative to the group $\mathcal{G}'$ of the form*

$$u_k - \sum_{n\in\mathbb{N}'} h_k^{(n)} w^{(n)} \overset{\mathcal{G}'}{\rightharpoonup} 0 \quad \text{in } E \cap F, \tag{6.2}$$

and a profile decomposition in E with the defect of compactness relative to the group $\mathcal{G}$ of the form (6.1) *with the same $(h_k^{(n)})_{k\in\mathbb{N}}$ and $w^{(n)}$, $n \in \mathbb{N}'$, as in* (6.2).

Proof. The second assertion of the Theorem is immediate from Lemma 6.2.1, since under the Opial condition Delta-convergence and weak convergence in E coincide. Note that since E is uniformly convex and both E and F are uniformly smooth, $E \cap F$ is also uniformly convex and uniformly smooth. Furthermore, the Opial condition for $E \cap F$ follows directly from the Opial condition for E and for F.

By Lemma 6.2.1 and the Opial condition, for any sequence (h_k) in $\mathcal{G}'$,

$$h_k^{-1}\Big(u_k - \sum_{n\in\mathbb{N}'} h_k^{(n)} w^{(n)}\Big) \rightharpoonup 0 \quad \text{in } E,$$

and it is easy to see that the expression in brackets coincides, up to passing to a subsequence, with a profile decomposition for (u_k) in $E \cap F$. □

Example 6.2.3. Let $E = \dot{H}^{s,p}(\mathbb{R}^N)$, $F = L^p(\mathbb{R}^N)$, $p \in (1, N/s)$, $s > 0$, let $\mathcal{G}$ be the rescaling group $\mathcal{G}^{\frac{N-ps}{p}}$, and let $\mathcal{G}'$ be the group $\mathcal{G}_{\mathbb{Z}^N}$. We assure that E and F satisfy Opial conditions by equipping them with the equivalent Triebel–Lizorkin norms (3.24) of $F^{s,p,2}$ and $F^{0,p,2}$, respectively. Then, taking into account Corollary 3.5.4, Theorem 6.2.2 implies that every bounded sequence in $H^{s,p}(\mathbb{R}^N)$ has a subsequence with a profile decomposition

$$r_k \overset{\text{def}}{=} u_k - \sum_{n\in\mathbb{N}} w^{(n)}(\cdot - y_k^{(n)}) \overset{\mathcal{G}'}{\rightharpoonup} 0, \tag{6.3}$$

with $|y_k^{(m)} - y_k^{(n)}| \to \infty$ whenever $m \neq n$, $u_k(\cdot + y_k^{(n)}) \rightharpoonup w^{(n)}$, and the remainder r_k converges to zero in $L^q(\mathbb{R}^N)$, $q \in (p, \frac{pN}{N-ps})$, by Theorem 3.7.1 (or by Theorem 3.8.2). We have thus derived profile decomposition of Theorem 4.6.5 from that of Theorem 4.6.4.

Theorem 6.2.4. *Assume that bounded sequences in a Banach space E of functions on $\mathbb{R}^N$ admit profile decompositions relative to the rescaling group $\mathcal{G}^r$ with some $r \in \mathbb{R}$. Let F be a uniformly convex and uniformly smooth Banach space of functions on $\mathbb{R}^N$, let the group $\mathcal{G}^s$, $s \in \mathbb{R}$, act isometrically on F, and assume that for any bounded sequence (v_k) in $E \cap F$,*

$$v_k \rightharpoonup 0 \text{ in } F \Leftrightarrow v_k \rightharpoonup 0 \text{ in } E. \tag{6.4}$$

Let (u_k) be a bounded sequence in $E \cap F$. If $s > r$, then every sequence $(j_k^{(n)})_{k\in\mathbb{N}}$ in the concentration term $2^{rj_k^{(n)}} w^{(n)}(2^{j_k^{(n)}}(\cdot - y_k^{(n)}))$, $n \in \mathbb{N}$, of a profile decomposition of (u_k) in E

relative to the group $\mathcal{G}^r$ is bounded from below. If $s < r$, then every respective sequence $(j_k^{(n)})_{k\in\mathbb{N}}$, $n \in \mathbb{N}$, is bounded from above.

Proof. Assume that (u_k) has a concentration profile $w \neq 0$ such that $2^{-rj_k}u_k(2^{-j_k}\cdot + y_k) \rightharpoonup w$ in E. Then

$$\|2^{-rj_k}u_k(2^{-j_k}\cdot + y_k)\|_F = 2^{(s-r)j_k}\|u_k\|_F \to 0. \tag{6.5}$$

If $s > r$ and $j_k \to -\infty$ or $s < r$ and $j_k \to +\infty$, the right-hand side converges to zero. Then $2^{-rj_k}u_k(2^{j_k}\cdot + y_k) \rightharpoonup 0$ in F. By (6.4), we have $2^{-rj_k}u_k(2^{j_k}\cdot + y_k) \rightharpoonup 0$ in E, so that $w = 0$, a contradiction. □

A particular case of this statement is as follows.

Corollary 6.2.5. *Let $m \in \mathbb{N}$ and $1 < p < N/m$. Let (u_k) be a bounded sequence in the intersection of $\dot{H}^{m,p}(\mathbb{R}^N)$, and $L^q(\mathbb{R}^N)$, $q \neq p_m^*$, equipped with the respective equivalent norm (3.24). If $q < p_m^*$, then every concentration term $2^{rj_k^{(n)}}w^{(n)}(2^{j_k^{(n)}}(\cdot - y_k^{(n)}))$, $r = \frac{N-mp}{p}$, in a profile decomposition of a renamed subsequence of (u_k) in $\dot{H}^{m,p}(\mathbb{R}^N)$ has the sequence $(j_k^{(n)})_{k\in\mathbb{N}}$ bounded from below. If $q > p_m^*$, then every concentration term $2^{rj_k^{(n)}}w^{(n)}(2^{j_k^{(n)}}(\cdot - y_k^{(n)}))$ in a profile decomposition of a renamed subsequence of (u_k) in $\dot{H}^{m,p}(\mathbb{R}^N)$ has the sequence $(j_k^{(n)})_{k\in\mathbb{N}}$ bounded from above.*

We consider now the profile decomposition of Theorem 4.6.4 for sequences of functions with a compact support.

Theorem 6.2.6. *Let (u_k) be a bounded sequence in $\dot{H}^{m,p}(\mathbb{R}^N)$, $m \in \mathbb{N}$, $1 < p < N/m$, and assume that there exists a compact set $K \subset \mathbb{R}^N$ such that $\operatorname{supp} u_k \subset K$, $k \in \mathbb{N}$. Then it has a renamed subsequence with a defect of compactness of the form*

$$S_k = w^{(0)} + \sum_{n\in\mathbb{N}} 2^{\frac{N-mp}{p}j_k^{(n)}}w^{(n)}(2^{j_k^{(n)}}(\cdot - y_k^{(n)})), \tag{6.6}$$

where $w^{(0)}$ is the weak limit of (u_k), $y_k^{(n)} \to y_n \in K$, $j_k^{(n)} \to +\infty$, and $(2^{j_k^{(m)}} + 2^{j_k^{(n)}})|y_k^{(m)} - y_k^{(n)}| \to \infty$ whenever $m \neq n$.

Proof. Consider a weakly convergent subsequence of u_k with a weak limit $w^{(0)}$. It is clear that if the assertion of the theorem holds for the sequence $(u_k - w^{(0)})$; it will hold for (u_k), so we may assume that $u_k \rightharpoonup 0$. Consider the concentration terms $2^{rj_k^{(n)}}w^{(n)}(2^{j_k^{(n)}}(\cdot - y_k^{(n)}))$, $r = \frac{N-mp}{p}$ with $w^{(n)} \neq 0$.

Note that if, on a renamed subsequence, $y_k^{(n)} \to y^{(n)} \notin K$, we have $w^{(n)} = \lim 2^{-rj_k^{(n)}}u_k(2^{-j_k^{(n)}}(\cdot + y_k^{(n)})) = 0$ a. e., since weak convergence in $\dot{H}^{m,p}(\mathbb{R}^N)$ implies convergence almost everywhere and u_k is evaluated in the limit away from the compact set K. The same argument applies when $|y_k^{(n)}| \to \infty$. Applying the standard diagonalization argument, we then may assume, for a renamed subsequence, that $y_k^{(n)} \to y_n \in K$ for every $n \in \mathbb{N}$.

Since (u_k) is supported on the compact set K, it is bounded in L^q, $1 \le q \le p_m^*$. Then, by Corollary 6.2.5, $(j_k^{(n)})_{k\in\mathbb{N}}$ is bounded from below. Moreover, $(j_k^{(n)})_{k\in\mathbb{N}}$ cannot have a bounded subsequence, since this would imply that u_k has a nonzero weak limit, a contradiction. □

Corollary 6.2.7. *Let $\Omega \subset \mathbb{R}^N$ be a bounded domain with a piecewise C^1-boundary and let (u_k) be a bounded sequence in $H^{1,p}(\Omega)$, $1 < p < N$. Then it has a renamed subsequence with a defect of compactness of the form (6.6) with $K = \bar{\Omega}$.*

Proof. By the standard extension theorem, there exists a bounded domain $\Omega' \supset \Omega$ with a smooth boundary and a bounded linear operator $T : H^{1,p}(\Omega) \to H_0^{1,p}(\Omega')$ such that $Tu(x) = u(x)$ whenever $x \in \Omega$. The assertion of the corollary follows from applying Theorem 6.2.6 to the sequence (Tu_k) extended by zero to $\mathbb{R}^N$. □

6.3 Flask domains for Sobolev embeddings

In this section, $\mathcal{G}(G)$ will denote a group of shifts by vectors in $G = \mathbb{Z}^m \times \mathbb{R}^{N-m}$ with some $m \in \{0,\dots,N\}$,

$$\mathcal{G}(G) = \{u \mapsto u(\cdot - y)\}_{y\in G}.$$

Definition 6.3.1. A domain $\Omega \subset \mathbb{R}^N$ is called a G-flask domain (relative to the Sobolev space $H^{1,p}(\mathbb{R}^N)$, $p \in [1,\infty)$) if $H_0^{1,p}(\Omega)$ is a flask subspace of $H^{1,p}(R^N)$ relative to the group $\mathcal{G}(G)$, that is, if $u_k \in H_0^{1,p}(\Omega)$ and $u_k(\cdot - y_k) \rightharpoonup w$ in $H^{1,p}(\mathbb{R}^N)$ for some $y_k \in G$, then there exists a $y \in G$ such that $w(\cdot - y) \in H_0^{1,p}(\Omega)$.

Let us recall the definition of the lower limit of a sequence of abstract sets $(X_k)_{k\in\mathbb{N}}$:

$$\liminf X_k \overset{\text{def}}{=} \bigcup_{n\in\mathbb{N}} \bigcap_{k\ge n} X_k. \tag{6.7}$$

Proposition 6.3.2. *Let $\Omega \subset \mathbb{R}^N$ be a domain with a piecewise-C^1-boundary. If for any sequence (y_k) in G there exists $y \in G$ and a set $Z \subset \mathbb{R}^N$ of measure zero such that*

$$\liminf(\Omega + y_k) \subset (\Omega + y) \cup Z, \tag{6.8}$$

then Ω is a G-flask domain relative to the Sobolev space $H^{1,p}(\mathbb{R}^N)$ for any $p \in [1,\infty)$.

Proof. Let $y_k \in G$ be such that $u_k(\cdot - y_k) \rightharpoonup w$ in $H^{1,p}(\mathbb{R}^N)$. Since weak convergence in $H^{1,p}(\mathbb{R}^N)$ implies convergence almost everywhere, let $Z_0 \subset \mathbb{R}^N$ be a set of zero measure such that $u_k(x-y_k) \to w(x)$ for all $x \in \mathbb{R}^N \setminus Z_0$. Then $w(x) \ne 0$ for $x \notin Z_0$ only if $x-y_k \in \Omega$ for all but finitely many $k \in \mathbb{N}$, that is, only if $x \in \bigcap_{k\ge k(x)}(\Omega + y_k)$ for some $k(x) \in \mathbb{N}$ sufficiently large. In other words, $w = 0$ in the complement of $\liminf(\Omega + y_k)$, except possibly on Z_0. By (6.8), there is a $y \in G$ such that $w(\cdot - y) = 0$ almost everywhere outside of Ω. Since $\partial\Omega$ is piecewise-C^1 and $w \in H^{1,p}(\mathbb{R}^N)$, this implies that $w \in H_0^{1,p}(\Omega)$. Therefore, Ω is a G-flask domain. □

Example 6.3.3. From the definition or from Theorem 6.3.2, we can see that the following domains are flask domains:
1. Any bounded domain is a G-flask domain for any G as above.
2. If $\Omega + y = \Omega$ for any $y \in G$, then Ω is a G-flask domain.
3. If $\Omega = \omega \times \mathbb{R}$, where ω is a bounded domain set with a piecewise-C^1-boundary in $\mathbb{R}^{N-1}$, then Ω is a G-flask domain.
4. If Ω_0 is as in any of two previous cases, $\Omega \supset \Omega_0$ has a piecewise C^1-boundary, $d(\Omega_0, \mathbb{R}^N \setminus (\Omega \cup B_R(0))) \to 0$ as $R \to \infty$, then Ω is a G-flask domain.
5. If $\Omega_1, \dots, \Omega_m$ are G-flask domains whose pairwise intersections are bounded sets, then each connected component of $\bigcup_{i=1}^m \Omega_i$ is a G-flask domain.

Let

$$\Omega_\varepsilon \stackrel{\text{def}}{=} \{x \in \Omega : d(x, \mathbb{R}^N \setminus \Omega) > \varepsilon\}, \quad \varepsilon > 0, \tag{6.9}$$

$$\chi_\varepsilon(x) \stackrel{\text{def}}{=} \min\left\{\frac{1}{\varepsilon} d(x, \mathbb{R}^N \setminus \Omega), 1\right\}, \tag{6.10}$$

and note that for every $p \in [1, N)$ there exists $\eta(\varepsilon) > 0$, $\eta(\varepsilon) \to 0$ as $\varepsilon \to 0$, such that

$$\|\chi_\varepsilon u - u\|_p \le \eta(\varepsilon)\|u\|_{H^{1,p}}. \tag{6.11}$$

Theorem 6.3.4. *Let $p \in [1, N)$. A domain Ω in $\mathbb{R}^N$ with a piecewise C^1-boundary is a G-flask domain (relative to $H^{1,p}(\mathbb{R}^N)$, $p \in [1, N)$), if and only if for every sequence (y_k) in G, $|y_k| \to \infty$, there exists a $y \in G$ such that for every $\varepsilon > 0$,*

$$\liminf(\Omega_\varepsilon + y_k) \subset \Omega + y \text{ up to a set of measure zero.} \tag{6.12}$$

Proof. Let χ_ε be as in (6.10).

Necessity. If Ω is a G-flask domain, then for any (y_k) in G, $|y_k| \to \infty$, there is a $y \in G$ such that on a renamed subsequence $\chi_\varepsilon(\cdot - y_k) \rightharpoonup w_\varepsilon \in H_0^{1,p}(\Omega + y)$, and thus $\chi_\varepsilon(\cdot - y_k) \to 0$ almost everywhere in the complement of $\Omega + y$. On the other hand, for every $x \in \liminf(\Omega_\varepsilon + y_k)$, $\chi_\varepsilon(x - y_k) = 1$ for all k sufficiently large. Therefore, $\liminf(\Omega_\varepsilon + y_k) \subset (\Omega + y)$ up to a set of measure zero.

Sufficiency. Let $u_k \in H_0^{1,p}(\Omega)$, let $w \in H^{1,p}(\mathbb{R}^N)$, and let $y_k \in G$ be such that, on a renamed subsequence, $u_k(\cdot - y_k) \rightharpoonup w$, and thus $u_k(\cdot - y_k) \to w$ almost everywhere. For every $\varepsilon > 0$, we define (on a renumbered subsequence possibly dependent on ε),

$$w_\varepsilon = \text{w-lim}(\chi_\varepsilon u_k)(\cdot - y_k)$$

and note that by the argument of Proposition 6.3.2, $w_\varepsilon = 0$ almost everywhere in $\mathbb{R}^N \setminus (\Omega + y)$. By (6.11)

$$\|w_\varepsilon - w\|_p \le \liminf_{k \in \mathbb{N}} \|(\chi_\varepsilon u_k)(\cdot + y_k) - u_k(\cdot + y_k)\|_p \le \sup_{k \in \mathbb{N}} \|\chi_\varepsilon u_k - u_k\|_p \le \eta(\varepsilon) \sup_{k \in \mathbb{N}} \|u_k\|_{H^{1,p}}.$$

Then $w_\varepsilon \to w$ almost everywhere as on a subsequence of $\varepsilon \to 0$, and thus $w = 0$ almost everywhere in $\mathbb{R}^N \setminus (\Omega + y)$. Therefore, Ω is a G-flask domain. □

Together with the group G, we also consider a subgroup T of $O(N)$ and define (G,T)-flask domains.

Definition 6.3.5. Let G be an additive subgroup of $\mathbb{R}^N$ and let T be a subgroup of $O(N)$. A domain $\Omega \subset \mathbb{R}^N$ is called a (G,T)-flask domain if for every sequence (y_k) in G, $|y_k| \to \infty$, there exist $z \in G$ and $\tau \in T$, such that whenever $u_k \in H_0^{1,p}(\Omega)$ and $u_k(\cdot + y_k) \rightharpoonup w$ in $H^{1,p}(\mathbb{R}^N)$, $w \in H_0^{1,p}(\tau\Omega + z)$.

Note that if $T = \{\mathrm{id}\}$ then a (G,T)-flask domain is a G-flask domain according to Definition 6.3.1. Note that $\mathbb{R}^N$ is a (G,T)-flask domain for any choice of G and T.

The following is a sufficient geometric condition for a domain to be a (G,T)-flask domain.

Proposition 6.3.6. *Let $\Omega \subset \mathbb{R}^N$ be a domain with a piecewise C^1-boundary. It is a (G,T)-flask domain (relative to $H^{1,p}(\mathbb{R}^N)$, $p \in [1,\infty)$) if for any sequence y_k in G there exist $z \in G$, $\tau \in T$, such that, up to a set of measure zero,*

$$\liminf(\Omega + y_k) \subset \tau\Omega + z. \tag{6.13}$$

Proof. The proof is repetitive of the proof of Proposition 6.3.2 and may be omitted. □

Example 6.3.7. The following domains are (G,T)-flask domain:
1. Any G-flask domain.
2. A finite union of intersecting (G,T)-flask domains whose pairwise intersections are bounded;
3. Domain $\Omega = \Omega_0 \cup \Omega_1$, where Ω_0 is a $(\mathbb{R}^N, O(N))$-flask domain, $\Omega_1 \subset \tau\Omega_0$ with some $\tau \in O(N)$, and $\Omega_0 \cap \Omega_1$ is bounded, is a $(\mathbb{R}^N, O(N))$-flask domain.

Proposition 6.3.8. *The following domains are not $(\mathbb{R}^N, O(N))$-flask domains:*
(a) *A domain $\Omega \subset \mathbb{R}^N$, $\overline{\Omega} \neq \mathbb{R}^N$, which for every $R > 0$ contains a ball of radius R (in particular, an open cone) is not a $(\mathbb{R}^N, O(N))$-flask domain;*
(b) *An open cylinder from which one has removed a closed bounded subset with a nonempty interior;*
(c) *A product $\omega \times (0,\infty)$, where $\omega \subset \mathbb{R}^{N-1}$ is a domain.*

Proof. (a). Let $y_k \in \mathbb{R}^N$, $k \in \mathbb{N}$, be such that $B_k(y_k) \subset \Omega$. Let $w \in H^{1,p}(\mathbb{R}^N)$, $\operatorname{supp} w = \mathbb{R}^N$ (e. g., $w(x) = e^{-|x|^2}$), and let $\chi_k \in C_0^\infty(B_k(0),[0,1])$ be equal to 1 on $B_{k-1}(0)$ and satisfy $|\nabla\chi_k| \leq 2$. Clearly, the $H_0^{1,p}(\Omega)$-norm of $u_k \stackrel{\text{def}}{=} \chi_k w(\cdot - y_k)$ is uniformly bounded in $k \in \mathbb{N}$ (in particular, $\operatorname{supp} u_k \subset \operatorname{supp}\chi + y_k \subset B_k(y_k) \subset \Omega$), and the sequence (u_k) is uniformly convergent on compact sets (and, therefore, weakly in $H^{1,p}(\mathbb{R}^N)$) to w. Since $\operatorname{supp} w = \mathbb{R}^N$, $w \notin H_0^1(\tau\Omega + z)$ for any $\tau \in O(N)$ and $z \in \mathbb{R}^N$.

(b) Let $\Omega = \omega \times \mathbb{R} \setminus \overline{U}$, where ω is a domain in $\mathbb{R}^{N-1}$ and U is a nonempty bounded domain contained in $\omega \times \mathbb{R}$. Let $w \in H_0^{1,p}(\omega \times \mathbb{R})$. Let $M = \sup\{x_N : x \in U\}$. $u_k = \chi(x_N)w(\cdot - ke_N)$, where $e_N = (0,\dots,0,1)$ and $\chi \in C^\infty(\mathbb{R},[0,1])$, $\chi(x) = 0$ for $x \leq M$,

$\chi(x) = 1$ for $x \geq M+1$, $|\chi'| \leq 2$. Similar to the argument above, $u_k(\cdot + ke_N) \rightharpoonup w$. It is clear that

$$\inf_{\tau \in O(N), z \in \mathbb{R}^N} \left|(\omega \times \mathbb{R}) \setminus (\tau\Omega + z)\right| > 0$$

which implies that there is a $w \in H_0^1(\omega \times \mathbb{R})$ that is not in $H_0^1(\tau\Omega + z)$ for any $\tau \in O(N)$, $z \in \mathbb{R}^N$.

(c) The proof analogous to the one in the case (b). □

Theorem 6.3.9. *Let $\Omega \subset \mathbb{R}^N$ be a $(\mathbb{Z}^N, T)$-flask domain with a piecewise C^1-boundary. If (u_k) is a bounded sequence in $H_0^{1,p}(\Omega)$, then it has a subsequence that admits a profile decomposition in $H^{1,p}(\mathbb{R}^N)$ relative to the group $\mathcal{G}_{\mathbb{Z}^N}$ as in Theorem 4.6.5 with $w^{(n)} \in H_0^{1,p}(\tau_n\Omega)$, $n \in \mathbb{N}$.*

Proof. Let u_k be a renumbered subsequence with profile decomposition (4.24). By Definition 6.3.5, there exist $\tau_n \in T$ and $z_n \in \mathbb{Z}^N$, such that $w^{(n)} \in H_0^{1,p}(\tau_n\Omega + z_n)$ and $u_k(\cdot + y_k^{(n)} + z_n) \rightharpoonup w^{(n)}(\cdot + z_n) \in H_0^{1,p}(\tau_n\Omega)$. Then it remains to rename $y_k^{(n)} + z_n$ as $y_k^{(n)}$ and $w^{(n)}(\cdot + z_n)$ as $w^{(n)}$. □

Let us give an example of existence of minimizers for a flask domain.

Theorem 6.3.10. *Let $\Omega \subset \mathbb{R}^N$ be a $(\mathbb{Z}^N, O(N))$-flask domain with piecewise-C^1 boundary. Let $p \in (1, N)$ and let $q \in (p, p^*)$. Then the minimum in*

$$\kappa = \inf_{u \in H_0^{1,p}(\Omega), \|u\|_q = 1} \int_\Omega (|\nabla u|^p + |u|^p) dx \tag{6.14}$$

is attained.

Proof. Let (u_k) be a minimizing sequence for (6.14) with a profile decomposition given by Theorem 4.6.5 and refined by Theorem 6.3.9. Then (4.26) will give us

$$1 = \int_\Omega |u_k|^q dx \to \sum_{n \in \mathbb{N}} \int_\Omega |w^{(n)} \circ \tau_n|^q dx. \tag{6.15}$$

Let us define $t_n \stackrel{\text{def}}{=} \int_\Omega |w^{(n)} \circ \tau_n|^q dx$. while from (4.25), (4.26), and we have

$$\begin{aligned}
\kappa &= \int_\Omega (|\nabla u_k|^p + |u_k|^p) dx + o(1) \geq \sum_{n \in \mathbb{N}} \int_{\tau_n\Omega} (|\nabla w^{(n)}|^p + |w^{(n)}|^p) dx + o(1) \\
&= \sum_{n \in \mathbb{N}} \int_\Omega (|\nabla w^{(n)} \circ \tau_n|^p + |w^{(n)} \circ \tau_n|^p) dx + o(1) \geq \kappa \sum_{n \in \mathbb{N}} t_n^{p/q}.
\end{aligned}$$

Since $\sum t_n = 1$ and $p < q$, the relation above is contradictory unless all τ_n except one, say τ_m, are zero, and $\tau_m = 1$. It is easy to see then that $w^{(m)} \circ \tau_m \in H_0^{1,p}(\Omega)$ is a minimizer. □

Remark 6.3.11. If a domain lacks the flask property, a minimizing sequence in an analogous problem may have a profile supported in a larger domain or in a domain with a smaller or equal value of κ in (6.14), which is not conductive for attaining a minimum. Consider, for example, the problem (6.14) for the half-space $\Omega = \mathbb{R}^{N-1} \times (0,\infty)$. If w is the known minimizer for (6.14) when $\Omega = \mathbb{R}^N$, then it is easy to see that $u_k(x) = \chi(x_N)w(x - (0_{N-1}, k))$, where $\chi \in C_0^\infty(0,\infty)$ and $\chi(t) = 1$ for $t \geq 2$, is a minimizing sequence for the problem on the half-space, so that the constant κ for the half-space and for $\mathbb{R}^N$ coincide. Then if the problem on the half-space had a minimizer (vanishing for $x_N \leq 0$), this function would also minimize the problem for the whole space, thus satisfying a corresponding semilinear elliptic equation on $\mathbb{R}^N$ and contradicting the maximum principle.

6.4 Asymptotically null sets, compact Sobolev embeddings

We now consider a sufficient condition for open sets $\Omega \subset \mathbb{R}^N$ such that $H_0^{1,p}(\Omega)$ is $\mathcal{G}_{\mathbb{Z}^N}$-locally compact (see Definition 6.1.3), and thus is compactly embedded into $L^q(\Omega)$, $q \in (p, p^*)$ by Proposition 6.1.4. We shall call such sets asymptotically null.

Definition 6.4.1. An open set $\Omega \subset \mathbb{R}^N$ is called asymptotically null set (relative to $H^{1,p}(\mathbb{R}^N)$ and $\mathcal{G}_{\mathbb{Z}^N}$) if $H_0^{1,p}(\Omega)$ is a $\mathcal{G}_{\mathbb{Z}^N}$-locally compact subspace of $H^{1,p}(\mathbb{R}^N)$.

Proposition 6.4.2. *An open set $\Omega \subset \mathbb{R}^N$ is asymptotically null (relative to $H^{1,p}(\mathbb{R}^N)$ and $\mathcal{G}_{\mathbb{Z}^N}$, $p \in [1,\infty)$) if for any sequence (y_k) in $\mathbb{Z}^N$, $|y_k| \to \infty$, the set $\liminf(\Omega - y_k)$ has measure zero.*

Proof. Let (u_k) be a bounded sequence in $H_0^{1,p}(\Omega)$. Assume without loss of generality that $u_k \rightharpoonup 0$ and let (y_k) in $\mathbb{Z}^N$, $|y_k| \to \infty$, be such that $u_k(\cdot + y_k) \rightharpoonup w$ in $H^{1,p}(\mathbb{R}^N)$. Then $u_k(\cdot + y_k)$ converges almost everywhere. Let $Z \subset \mathbb{R}^N$ be a set of zero measure such that $u_k(x + y_k) \to w(x)$ for all $x \in \mathbb{R}^N \setminus Z$. If $w(x) \neq 0$, $x \notin Z$, then $u_k(x + y_k) \neq 0$ for all k sufficiently large, and thus $x + y_k \in \Omega$ and $x \in \liminf(\Omega - y_k)$. By assumption the latter set has measure zero and thus $w = 0$ a.e. Consequently, $u_k \overset{\mathcal{G}_{\mathbb{Z}^N}}{\rightharpoonup} 0$. We conclude that $H_0^{1,p}(\Omega)$ is $\mathcal{G}_{\mathbb{Z}^N}$-locally compact subspace of $H^{1,p}(\mathbb{R}^N)$ and thus Ω is asymptotically null. □

A necessary and sufficient condition for a set to be asymptotically null, for $p \in [1, N)$, can be formulated in terms of sets (6.9).

Theorem 6.4.3. *An open set Ω in $\mathbb{R}^N$ is asymptotically null (relative to $H^{1,p}(\mathbb{R}^N)$ and $\mathcal{G}_{\mathbb{Z}^N}$, $p \in [1, N)$) if and only if for every sequence $y_k \in \mathbb{Z}^N$ and every $\varepsilon > 0$,*

$$|\liminf(\Omega_\varepsilon - y_k)| = 0. \tag{6.16}$$

Proof. Let $\Omega_\varepsilon, \chi_\varepsilon$ be as in (6.9), respectively (6.10).

Necessity. If Ω is an asymptotically null set, then $\chi_\varepsilon(\cdot+y_k) \rightharpoonup 0$ whenever $|y_k| \to \infty$, and thus $\chi_\varepsilon(\cdot+y_k) \to 0$ almost everywhere. On the other hand, for every $x \in \liminf(\Omega_\varepsilon - y_k)$, $\chi_\varepsilon(x+y_k) = 1$ for all k sufficiently large, and thus $\liminf(\Omega_\varepsilon - y_k)$ has measure zero.

Sufficiency. Let $u_k \in H_0^{1,p}(\Omega)$, let $w \in H^{1,p}(\mathbb{R}^N)$, and let $y_k \in \mathbb{Z}^N$ be such that, on a renamed subsequence, $u_k(\cdot + y_k) \rightharpoonup w$, and thus $u_k(\cdot + y_k) \to w$ almost everywhere. For every $\varepsilon > 0$, we define (on a renumbered subsequence possibly dependent on ε),

$$w_\varepsilon = \text{w-lim}(\chi_\varepsilon u_k)(\cdot + y_k)$$

and note that by the argument of Proposition 6.4.2, $w_\varepsilon = 0$ almost everywhere. Note now, that by (6.11),

$$\|w_\varepsilon - w\|_p \le \liminf_{k\in\mathbb{N}} \|(\chi_\varepsilon u_k)(\cdot + y_k) - u_k(\cdot + y_k)\|_p = \liminf_{k\in\mathbb{N}} \|\chi_\varepsilon u_k - u_k\|_p \le \eta(\varepsilon) \sup_{k\in\mathbb{N}} \|u_k\|_{H^{1,p}}.$$

Then $w_\varepsilon \to w$ almost everywhere as $\varepsilon \to 0$, but $w_\varepsilon = 0$ a.e., so $w = 0$. Therefore, Ω is asymptotically null. □

An "infinitely narrow" flask-shaped set $\{|\bar{x}| < \frac{1}{1+x_N^2}\}$, where $\bar{x} = (x_1, \dots, x_{N-1})$, is an asymptotically null set by Proposition 6.4.2.

6.5 Flask domains and null sets for the affine Sobolev inequality

In this section, we study compactness properties and related isoperimetric problems for the affine Sobolev functional E_2 given by (5.64). We adapt the notions of flask domains and asymptotically null sets to the setting of affine Laplacian.

Let $\Omega \subset \mathbb{R}^N$ be a domain. By analogy with the p-Laplacian which equals the Fréchet derivative of $-\frac{1}{p}\int |\nabla u|^p$, we may also define the affine Laplace operator $\Delta_A(u)$ by differentiation of $-\frac{1}{2}E_2$ in a suitable space, for example, in $\dot{H}_0^{1,2}(\Omega)$ for the Dirichlet affine Laplacian or in $H^{1,2}(\Omega)$ for the Neumann affine Laplacian. Since $A_{i,j}[u]' = (\int_\Omega \nabla_i u \nabla_j u \, dx)' = -2(\nabla_i \nabla_j u)_{ij}$, we have formally,

$$\det A[u]' = \det A[u] \ \operatorname{tr}(A^{-1}[u]A[u]') = -2 \det A \ \operatorname{tr}(A^{-1}[u]u''),$$

where $u''(x)$ is the Hessian of u, that is, the matrix with components $\nabla_i\nabla_j u(x)$. Then

$$\begin{aligned}\Delta_A(u) &= -\frac{N}{2}(\det A[u]^{\frac{1}{N}})' \\ &= -\frac{1}{2}(\det A[u])^{\frac{1}{N}-1}(\det A[u])' \\ &= (\det A[u])^{\frac{1}{N}} \operatorname{tr}(A^{-1}[u]u''). \end{aligned} \tag{6.17}$$

It is easy to see that for any $u \in \dot{H}_0^{1,2}(\Omega)$ this expression is a Fréchet derivative of $-\frac{1}{2}E_2$ and that $E_2 \in C^1(\dot{H}_0^{1,2}(\Omega))$. In what follows, the notation Δ_A will be reserved for the affine Dirichlet Laplacian, that is, for the Fréchet derivative above.

We have the following elementary identity:

$$\Delta_A(u \circ S) = \Delta_A(u) \circ S, \quad S \in \mathrm{SL}(N). \tag{6.18}$$

If $T \in \mathrm{SL}(N)$ is as in the last assertion of Lemma 5.4.1, that is, $A[u \circ T]$ is a multiple of identity, then we have

$$(\Delta_A(u)) \circ T = \Delta(u \circ T). \tag{6.19}$$

Consequently, both the strong and the weak maximum principle apply to classical solutions of $\Delta_A(v) = f$, exactly in the same form as for the classical Laplacian. In what follows, the norm of a matrix T will be denoted as $|T|$. We note that a sequence $(T)_k$ in $\mathrm{SL}(N)$ is either unbounded in norm, or has a subsequence convergent to a matrix in $\mathrm{SL}(N)$.

Definition 6.5.1. We shall say that a function $f \in L^{\frac{2N}{N+2}}(\Omega)$ is of class $L_A(\Omega)$ if for any sequence (T_k) in $\mathrm{SL}(N)$, $|T_k| \to \infty$, one has $f \circ T_k|_\Omega \to 0$ in $L^{\frac{2N}{N+2}}(\Omega)$.

In particular, if Ω is bounded, $L_A(\Omega) = L^{\frac{2N}{N+2}}(\Omega)$, and if $\Omega = \mathbb{R}^N$, $L_A(\Omega) = \{0\}$.

Theorem 6.5.2. *Let $\Omega \subset \mathbb{R}^N$ be a domain with a piecewise-C^1 boundary. If $f \in L_A(\Omega)$, then the infimum*

$$\kappa_f \overset{\text{def}}{=} \inf_{u \in \dot{H}_0^{1,2}(\Omega)} \frac{1}{2} E_2(u) - \int_\Omega f(x)u(x)\mathrm{d}x \tag{6.20}$$

is attained. If, additionally, $f \in L^2(\Omega)$, then this minimizer is a classical solution of

$$\Delta_A(u)(x) + f(x) = 0, \quad x \in \Omega. \tag{6.21}$$

Proof. Note first that $\kappa_f < 0$. Indeed, let $w \in C_0^1(\Omega)$ be such that $\int_\Omega fw\mathrm{d}x < 0$. Then for $t > 0$ sufficiently small the functional in (6.20) will have negative values, since the first term is quadratic in t.

By (5.65), we can rewrite (6.20) as

$$\kappa_f = \inf_{v \in \dot{H}_0^{1,2}(T\Omega),\, T \in \mathrm{SL}(N)} \frac{1}{2} \int_{T\Omega} |\nabla v|^2 \mathrm{d}x - \int_\Omega f(x)\, v(Tx)\mathrm{d}x. \tag{6.22}$$

Let $(v_k, T_k)_{k \in \mathbb{N}}$ in $C_0^\infty(T_k\Omega) \times \mathrm{SL}(N)$ be a minimizing sequence for (6.22). Consider (v_k) as a sequence in $\dot{H}^{1,2}(\mathbb{R}^N)$. Assume first that $|T_k| \to \infty$. Then, since $f \in L_A(\Omega)$, we have $\kappa_f \geq 0$, which is false. Consequently, we have, on a renamed subsequence, $T_k \to T \in \mathrm{SL}(N)$ and $v_k \rightharpoonup v$ in $\dot{H}^{1,2}(\mathbb{R}^N)$ with $v = 0$ outside of $T\Omega$, which means that $u \overset{\text{def}}{=} v \circ T^{-1} \in \dot{H}_0^{1,2}(\Omega)$ is a required minimizer. Equation (6.21) (in the weak sense) follows, and the regularity of the solution is a consequence of the standard elliptic regularity. □

Theorem 6.5.3. *Let J_p, $p \in (1,N)$, be the functional* (5.55). *Let $\Omega \subset \mathbb{R}^N$ be an open set such that for any sequences (T_k) in $\mathrm{SL}(N)$ and (y_k) in $\mathbb{Z}^N$ satisfying $|T_k| + |y_k| \to \infty$, one has*

$$\left|\liminf T_k^{-1}(\Omega - y_k)\right| = 0. \tag{6.23}$$

Then the set $B_1 = \{u \in H_0^{1,p}(\Omega);\ J_p(u) \le 1\}$ is relatively compact in $L^q(\Omega)$, $p < q < p^*$.

Note that, even for bounded Ω the set B_1 is not bounded in $H_0^{1,p}(\Omega)$.

Proof. Let (u_k) be a sequence in B_1 and consider it as a sequence in $H^{1,p}(\mathbb{R}^N)$. Let $T_k \in \mathrm{SL}(N)$ satisfy, in accordance to (5.66),

$$J_p(u_k) \ge \frac{C'}{2}\|\nabla(u_k \circ T_k)\|_p.$$

Let $v_k = u_k \circ T_k$. Then (v_k) is a bounded sequence in $H^{1,p}(\mathbb{R}^N)$. If $|T_k| \to \infty$, then by (6.23), $v_k(\cdot - y_k) \rightharpoonup 0$ in $H^{1,p}(\mathbb{R}^N)$ for any sequence (y_k) in $\mathbb{R}^N$, which implies by cocompactness that $v_k \to 0$ in $L^q(\mathbb{R}^N)$, $p < q < p^*$, and thus $u_k \to 0$ in $L^q(\Omega)$. Otherwise, there is a renamed subsequence of (T_k) convergent to some $T \in \mathrm{SL}(N)$. Passing again to a renamed weakly convergent subsequence, we may assume that $v_k \rightharpoonup v$ in $H^{1,p}(\mathbb{R}^N)$, and thus $u_k \rightharpoonup v \circ T^{-1}$ in $H_0^{1,p}(\Omega)$. On the other hand, from (6.23) we can infer that for any sequence (y_k) in $\mathbb{Z}^N$, $(v_k - v)(\cdot - y_k) \rightharpoonup 0$ in $H^{1,p}(\mathbb{R}^N)$ and thus, setting $u \stackrel{\text{def}}{=} v \circ T^{-1}$, we have $\|u_k - u\|_q \le \|v_k - v\|_q + \|u \circ T - u \circ T_k\|_q \to 0$. □

Note that any bounded set satisfies (6.23). An example of an unbounded set satisfying (6.23) is $\{(x_1, \bar{x}) \in \mathbb{R} \times \mathbb{R}^{N-1} : |\bar{x}| < e^{-x_1^2}\}$. Not every asymptotically null set, considered in the previous section, satisfies (6.23): in particular, such is the set $\{(x_1, \bar{x}) \in \mathbb{R} \times \mathbb{R}^{N-1} : |\bar{x}| < (1 + \log|x_1|)^{-1}\}$.

We consider now an example of a minimization problem. Note that if Ω below were a ball, existence of the minimizer would follow from a rearrangement argument, and the minimizer would coincide with the minimizer for the Sobolev norm.

Theorem 6.5.4. *Let $\Omega \subset \mathbb{R}^N$, $N > 2$, be an open set with a piecewise-C^1-boundary satisfying* (6.23) *[for example, a bounded domain]. Then the minimum in the problem*

$$\kappa_q = \inf_{u \in H_0^{1,2}(\Omega), \|u\|_{p,\Omega}=1} E_2(u), \quad 2 < p < 2^*, \tag{6.24}$$

is attained.

Proof. Let (u_k) be a minimizing sequence and consider it as a sequence in $H^{1,2}(\mathbb{R}^N)$. Let $T_k \in \mathrm{SL}(N)$ be as in (5.72). Repeating the argument in the proof of Theorem 6.5.3, we may assume, for a suitable renamed subsequence, that either $|T_k| \to \infty$ and then $u_k \to 0$ in L^q, or $T_k \to T \in \mathrm{SL}(N)$, and u_k converges weakly in $H_0^{1,2}(\Omega)$ as well as in $L^p(\Omega)$ to some u. The former case is ruled out, since by assumption of minimizing sequence $\|u_k\|_{p,\Omega} = 1$. In the latter case, lower semicontinuity of the norm implies that $\|\nabla u\|_2^2 \le \kappa_p$. Then by (5.65) $E_2(u) \le \kappa_p$, and thus u is necessarily a minimizer. □

Corollary 6.5.5. *Let $\Omega \subset \mathbb{R}^N$ be a bounded domain with a piecewise C^1-boundary. Then (6.24) has a minimizer that, up to a scalar multiple, is a smooth positive classical solution of the boundary problem*

$$-\sum_{i,j=1}^{N}(A^{-1}[u])_{ij}\frac{\partial^2 u}{\partial x_i \partial x_j} = u^{p-1} \quad \text{in } \Omega,\ u|_{\partial\Omega} = 0. \tag{6.25}$$

Proof. Note that if $u \in H_0^1(\Omega)$ is a minimizer for (6.24), then so is $|u|$ by (5.65):

$$\begin{aligned}
&\inf_{u\in H_0^{1,2}(\Omega),\|u\|_{p,\Omega}=1} E_2(u) \\
&= \inf_{u\in H_0^{1,2}(\Omega),\|u\|_{p,\Omega}=1,T\in \mathrm{SL}(N)} \|\nabla(u\circ T)\|_2^2 \\
&= \inf_{u\in H_0^{1,2}(\Omega),\|u\|_{p,\Omega}=1,T\in \mathrm{SL}(N)} \|\nabla|u\circ T|\|_2^2 \\
&= \inf_{u\in H_0^{1,2}(\Omega),\|u\|_{p,\Omega}=1} E_2(|u|),
\end{aligned}$$

so we can without loss of generality assume that $u \geq 0$. Then, for some $\lambda > 0$, the function u satisfies, in the weak sense,

$$-\sum_{i,j=1}^{N}(A^{-1}[u])_{ij}\frac{\partial^2 u}{\partial x_i \partial x_j} = \lambda u^{p-1} \quad \text{in } \Omega. \tag{6.26}$$

Note that $A[u]^{-1}$ is a positive constant matrix, as an inverse of a positive matrix, so the standard elliptic regularity and the bootstrap argument yield the smoothness of the solution. The solution is strictly positive by the maximum principle for uniformly elliptic operators. Finally, note that the left-hand side of (6.26) is of homogeneity $-1 \neq p-1$, so a suitable scalar multiple of u satisfies (6.25). □

We extend here condition (6.8) to actions of the affine group on $\mathbb{R}^N$ to state the following minimization result for the affine Sobolev functional E_2.

Definition 6.5.6. A domain Ω in $\mathbb{R}^N$ is an affine flask domain if for any sequences (T_k) in $\mathrm{SL}(N)$ and (y_k) in $\mathbb{R}^N$ there exist a $y \in \mathbb{R}^N$, a $T \in \mathrm{SL}(N)$ and a set $Z \subset \mathbb{R}^N$ of zero measure such that

$$\liminf T_k^{-1}(\Omega - y_k) \subset (T\Omega + y) \cup Z. \tag{6.27}$$

Example 6.5.7. Obviously, any bounded domain (as well as $\mathbb{R}^N$) is an affine flask domain. A collection of unit balls $B_1(n^4 e_0)$, $n \in \mathbb{N}$, where e_0 is a fixed unit vector, connected consecutively by circular cylinders of corresponding radius e^{-n} that have $\mathbb{R}e_0$ as their common axis, is an affine flask domain. On the other hand, a cylindrical domain with a smooth boundary is an affine flask domain only if it is $\mathbb{R}^N$. Indeed, let $\Omega = \mathbb{R} \times \omega$ and let T_k be a diagonal matrix with diagonal entries $k^{1-N}, k, \dots, k$. Then $\liminf T_k\Omega = \mathbb{R}^N$.

Theorem 6.5.8. *Assume that $\Omega \subset \mathbb{R}^N$, $N > 2$, is an affine flask domain with a piecewise-C^1 boundary. Then the minimum in the problem*

$$\kappa = \inf_{u \in H_0^{1,2}(\Omega):\ \|u\|_p = 1} E_2(u) + \|u\|_2^2, \quad 2 < p < p^*, \tag{6.28}$$

is attained.

Proof. Let $(u_k) \subset H_0^{1,2}(\Omega)$ be a minimizing sequence. Consider it as a sequence in $H^{1,2}(\mathbb{R}^N)$. Let $(T_k) \subset \mathrm{SL}(N)$ and let $w^{(n)}$, $n \in \mathbb{N}$, be as in Theorem 5.4.5, so we have $E_2(u_k \circ T_k) = \|\nabla(u_k \circ T_k)\|_2^2$. From the iterated Brezis–Lieb lemma (Theorem 4.7.1), we have

$$1 = \|u_k\|_p^p = \sum_n \|w^{(n)}\|_p^p. \tag{6.29}$$

Let $t_n = \|w^{(n)}\|_p^p$.

By (5.75)

$$\begin{aligned}\kappa &= \lim E_2(u_k(T_k \cdot - y + y_k^{(n)})) + \|u_k(T_k \cdot - y + y_k^{(n)})\|_2^2 \\ &\geq \sum_{n \in \mathbb{N}} (\|\nabla w^{(n)}\|_2^2 + \|w^{(n)}\|_2^2) \geq \sum_{n \in \mathbb{N}} (E_2(w^{(n)}) + \|w^{(n)}\|_2^2).\end{aligned} \tag{6.30}$$

Equation (6.27) implies that with some $T^{(n)} \in \mathrm{SL}(N)$ and some $y_n \in \mathbb{R}^N$ one has

$$u_k(T_k((T^{(n)})^{-1} \cdot - y_n) + y_k^{(n)}) \rightharpoonup w^{(n)}((T^{(n)})^{-1}(\cdot - y_n)) \in H_0^{1,2}(\Omega).$$

From (6.30), we have

$$\kappa \geq \sum_{n \in \mathbb{N}} \kappa t_n^{2/p}, \tag{6.31}$$

which can hold only if $t_n = 0$ for $n \neq m$ and $t_m = 1$ with some $m \in \mathbb{N}$. Consequently, $w^{(m)}((T^{(m)})^{-1}(\cdot - y_m))$ is a minimizer. □

Remark 6.5.9. Any minimizer for the problem (6.27) is, up to a scalar multiple, a positive smooth solution of the boundary value problem

$$-\det A[u]^{1/N} \sum_{i,j=1}^N (A^{-1}[u])_{ij} \frac{\partial^2 u}{\partial x_i \partial x_j} + u = u^{p-1}, \quad u|_{\partial\Omega} = 0. \tag{6.32}$$

The argument copies that of Corollary 6.5.5 with one modification: in the proof of the corollary we omitted the scalar factor $\det A[u]^{1/N}$ in the Fréchet derivative of the left-hand side. We do not omit it here, and as a consequence the left-hand side is now of homogeneity $1 < p - 1$, which allows to replace u by its scalar multiple while setting the Lagrange multiplier to 1.

Since $\mathbb{R}^N$ is an affine flask domain, there is a minimizer in (6.28). A simple argument shows that the minimum in (6.28) is attained at the minimum of the corresponding Euclidean problem.

Theorem 6.5.10. *The minimal values in the problems*

$$\inf_{u\in\dot{H}^{1,2}(\mathbb{R}^N),\|u\|_{2^*}=1} E_2(u), \tag{6.33}$$

and

$$\inf_{u\in H^{1,2}(\mathbb{R}^N),\|u\|_p=1} E_2(u)+\|u\|_2^2, \quad 2<p<2^*, \tag{6.34}$$

are attained at the minima for respective Euclidean problems (with $E_2(u)$ replaced by $\|\nabla u\|_2^2$).

Proof. By (5.65), for every $u\in\dot{H}^{1,2}(\mathbb{R}^N)$ there is $T\in \mathrm{SL}(N)$ such that $E_2(u)=\|\nabla(u\circ T)\|_2^2$. Therefore,

$$\inf_{u\in\dot{H}^{1,2}(\mathbb{R}^N),\|u\|_{2^*}=1} E_2(u)=\inf_{u\in\dot{H}^{1,2}(\mathbb{R}^N),\|u\|_{2^*}=1}\|\nabla u\|_2^2 \tag{6.35}$$

and

$$\inf_{u\in H^{1,2}(\mathbb{R}^N),\|u\|_p=1} E_2(u)+\|u\|_2^2=\inf_{u\in H^{1,2}(\mathbb{R}^N),\|u\|_p=1}\|\nabla u\|_2^2+\|u\|_2^2. \tag{6.36}$$

□

6.6 Bibliographic notes

Sections 6.1 and 6.2 are written in the general spirit of [113]. Flask domains in a more concrete form of Example 6.3.3, case 4, were introduced by del Pino and Felmer [36]. The notion of asymptotically null set is related to the earlier studies of compact Sobolev embeddings for unbounded domains such as [31]. The functional-analytic notion of a flask subspace of a Hilbert space was introduced in [104, Chapter 4], which also contains characterization of flask domains and asymptotically null sets from Sections 6.3 and 6.4 in the case of $H^{1,2}$.

Counterparts of flask domains and of asymptotically null sets for affine Sobolev spaces were introduced in [102] for $p=2$ and in [103] for $1\le p<N$. Presentation in Section 6.5 follows [102]. The first part of Theorem 6.5.10, for the affine p-Laplacian, is proved in [86].

7 Global compactness on Riemannian manifolds

In this chapter, we study loss of compactness, for Sobolev spaces of manifolds, that is not attributable to an action of a group as it is, for example, in Corollary 5.1.7, where the manifold is assumed to be periodic. An analog of profile decomposition is still possible in absence of an invariant group action, but at the cost of concentration profiles emerging as functions on different manifolds.

An elementary example of profiles arising as functions on a different metric structure, as a consequence of noninvariant transformations, is the space $\ell^p(\mathbb{N})$, $p \in [1, \infty)$, equipped with the set of right shifts. If $w \in \ell^p(\mathbb{Z})$, then $u_k(x) \stackrel{\text{def}}{=} w(x-k)$, $x \in \mathbb{N}$, is a bounded sequence in $\ell^p(\mathbb{N})$, which we extend by zero to $\mathbb{Z}$, setting

$$\bar{u}_k(x) = \begin{cases} u_k(x) = w(x-k), & x \in \mathbb{N}, \\ 0, & x \le 0, \end{cases}$$

then

$$\bar{u}_k(x+k) = \begin{cases} w(x), & x =, -k+1, -k+2, \dots, \\ 0, & x = -k, -k-1, \dots \end{cases} \qquad x \in \mathbb{Z},$$

and $u_k(x+k) \to w(x)$, $x \in \mathbb{Z}$. This implies $u_k(\cdot + k) \rightharpoonup w$, and thus $w \in \ell^p(\mathbb{Z})$ may be regarded as a profile of (u_k), which is a sequence in $\ell^p(\mathbb{N})$.

Note that, of course, that profiles that emerge here are not weak limits in the original spaces, like in Definition 4.1.1, and in general the concentration structures in this chapter correspond to this definition only in the sense of analogy.

A paper of Struwe [120] addressed profile decompositions for limiting Sobolev embeddings on compact manifolds, where formation of dilation bubbles by a “zoom-in” into the manifold yields profiles defined on the tangent space. In [110], dealing with subcritical embeddings, profiles are generated by local isometries of the manifold, which results in profiles defined on different manifolds-at-infinity of the given manifold. The profile decomposition of Theorem 7.9.1 below combines both structures. The conclusionary Section 7.10 presents a related result on compactness of embeddings for subspaces of symmetric functions.

7.1 Bounded geometry. Discretizations and a “spotlight lemma”

Throughout this chapter, we consider be a smooth, complete, connected, N-dimensional Riemannian manifold M, $N \ge 2$, with metric g. By $B(x,r) \subset M$, we will denote in this chapter the geodesic ball of M of radius r, centered at $x \in M$, and by Ω_r we will denote the ball of radius r in the Euclidean space, centered at the origin. For every $x \in M$, there exists a maximal $r(x) > 0$, called injectivity radius, such that

https://doi.org/10.1515/9783110532432-007

the Riemannian exponential map $\exp_x$ is a diffeomorphism of $\{v \in T_xM : g_x(v,v) < r\}$ onto $B(x, r(x))$. For each $x \in M$, we choose an orthonormal basis for T_xM which yields an identification $i_x : \mathbb{R}^N \to T_xM$. Then $e_x : \Omega_{r(x)} \to B(x, r(x))$ will denote a geodesic normal coordinate map at x given by $e_x = \exp_x \circ i_x$. We do not require smoothness of the map i_x with respect to x, since the arguments x will be taken from a discrete subset of M.

For k integer, and $f : M \to \mathbb{R}$, we denote by $d^k f$ the kth covariant derivative of u, by ∂_α partial derivative in local coordinates, $\frac{\partial}{\partial \xi_\alpha}$, and by $|d^k f|$ the norm of $d^k f$ defined by a local chart by

$$|d^k f|^2 = g^{i_1 j_1} \cdots g^{i_k j_k} \partial_{i_1} \dots \partial_{i_k} f \partial_{j_1} \dots \partial_{j_k} f .$$

where g^{ij} are the components of the inverse matrix of the metric matrix $g = [g_{ij}]$. Throughout this chapter, we use the notation $\mathbb{N}_0 \stackrel{\text{def}}{=} \mathbb{N} \cup \{0\}$.

Definition 7.1.1 (Definition A.1.1 from [107]). A smooth Riemannian manifold M is of bounded geometry if the following two conditions are satisfied:
(i) The injectivity radius $r(M) = \inf_{x \in M} r(x)$ of M is positive.
(ii) The Riemann curvature tensor R^M of M has bounded derivatives, namely, $d^k R^M \in L^\infty(M)$ for every $k \in \mathbb{N}_0$.

In particular, all compact manifolds, homogeneous spaces, and periodic manifolds are manifolds of bounded geometry. When M is of bounded geometry and $r < r(M)$, the geodesic normal coordinate map e_x, $x \in M$, is a diffeomorphism $\Omega_r \to B(x, r)$. A Riemannian manifold of bounded geometry is always complete. Further properties of such manifolds are given in the Appendix, Section 10.3.

The Sobolev space $H^{1,p}(M)$, $p \in [1, \infty)$, is a completion of $C_0^\infty(M)$ with respect to the norm

$$\|u\|^p_{H^{1,p}} = \int_M g_x(du, du)^{p/2} dv_g + \int_M |u|^p dv_g .$$

Since M is of bounded geometry, the space $H^{1,p}(M)$ is continuously embedded into $L^q(M)$ for every $p \in (1, N)$ and $q \in [p, p^*]$ and the constant in Sobolev embeddings over balls $B(x, r)$ is independent of $x \in M$ (see, e. g., [65, Theorem 3.2]).

Definition 7.1.2. A subset Y of Riemannian manifold M is called (ε, ρ)-discretization of M, $\rho > \varepsilon > 0$, if the distance between any two distinct points of Y is greater than or equal to ε and

$$M = \bigcup_{y \in Y} B(y, \rho).$$

Any Riemannian manifold M has a (ε, ρ)-discretization for any $\varepsilon > 0$ and a suitable ρ. If M is of bounded geometry, then the covering $\{B(y, r)\}_{y \in Y}$ is uniformly locally finite for any $r \geq \rho$; cf. [64, 107, 108]. It is also well known that for any (ε, ρ)-discretization Y, there exists a smooth partition of unity $\{\chi_y\}_{y \in Y}$ on M, subordinated to the covering $\{B(y, \rho)\}_{y \in Y}$, such that for any $\alpha \in \mathbb{N}_0^N$ there exists a constant $C_\alpha > 0$, such that

$$|D^\alpha \chi_y| \leq C_\alpha \tag{7.1}$$

for all $y \in Y$.

Example 7.1.3. Let M be a noncompact manifold of bounded geometry, let $w \in C_0^1(\Omega_r) \setminus \{0\}$, let (x_k) be a discrete sequence on M, and let $u_k = w \circ e_{x_k}^{-1}$ extended to the rest of the manifold by zero. Then it is easy to see that $u_k \rightharpoonup 0$ while $\|u_k\|_p$ is bounded away from zero by (10.38). In other words, for noncompact manifolds of bounded geometry presence of a *local* concentration profile, w, results in a nontrivial defect of compactness.

Theorem 7.3.5 below is an analog of Corollary 5.1.7, based on local concentration profiles in the spirit of Example 7.1.3. It is natural to expect that once we subtract from the sequence all local “runaway bumps” of the form $w \circ e_{y_k}^{-1}$ suitably patched together, the remainder sequence (v_k) should be left without nonzero local profiles, in other word, it should satisfy $v_k \circ e_{y_k} \rightharpoonup 0$ in $H^{1,2}(\Omega_\rho)$ with some $\rho > 0$. This is a condition related to the one in the cocompactness Lemma 2.6 of [47] for periodic manifolds, and it implies that (v_k) vanishes in $L^p(M)$. In strict terms, we have the following property, similar to cocompactness.

Theorem 7.1.4 (“Spotlight vanishing lemma”). *Let M be an N-dimensional Riemannian manifold of bounded geometry, supplied with a (ε, r)-discretization $Y \subset M$ of M, $r < r(M)$. Let (u_k) be a bounded sequence in $H^{1,p}(M)$, $1 < p < N$. Then $u_k \to 0$ in $L^q(M)$ for any $q \in (p, p^*)$ if and only if $u_k \circ e_{y_k} \rightharpoonup 0$ in $H^{1,p}(\Omega_r)$ for any sequence (y_k) in Y.*

Proof. Let us fix $q \in (p, p^*)$ and assume that $u_k \circ e_{y_k} \rightharpoonup 0$ in $H^{1,p}(\Omega_r)$ for any sequence (y_k), $y_k \in Y$. The local Sobolev embedding theorem and the boundedness of the geometry of the manifold implies that there exists $C > 0$ independent of $y \in M$ such that

$$\int_{B(y,r)} |u_k|^q \mathrm{d}v_g \leq C \int_{B(y,r)} (g_x(du, du)_k^{p|2} + |u_k|^p) \mathrm{d}v_g \Big(\int_{B(y,r)} |u_k|^q \mathrm{d}v_g \Big)^{1-p/q}.$$

Adding the terms in the left- and the right-hand side over $y \in Y$ and taking into account the uniform multiplicity of the covering, we have

$$\int_M |u_k|^q \mathrm{d}v_g \leq C \int_M (g_x(du, du)^{p/2} + |u_k|^p) \mathrm{d}v_g \sup_{y \in Y} \Big(\int_{B(y,r)} |u_n|^q \mathrm{d}v_g \Big)^{1-p/q}. \tag{7.2}$$

Boundedness of the sequence (u_n) in $H^{1,p}(M)$ implies that the supremum of the right-hand side is finite. So for any $k \in \mathbb{N}$, we can find a $y_k \in Y$, such that

$$\sup_{y \in Y} \int_{B(y,r)} |u_k|^q \mathrm{d}v_g \leq 2^{\frac{q}{q-p}} \int_{B(y_k,r)} |u_k|^q \mathrm{d}v_g. \tag{7.3}$$

By compactness of the Sobolev embedding $H^{1,p}(\Omega_r) \hookrightarrow L^q(\Omega_r)$ and weak convergence of the sequence in $H^{1,p}(\Omega_r)$, we have $u_k \circ e_{y_k} \to 0$ in $L^q(\Omega_r)$, and thus, by (10.37), $\int_{B(y_k,r)} |u_k|^q dv_g \to 0$. Combining this with (7.2) and (7.3), we have $u_k \to 0$ in $L^q(M)$.

Assume now that $u_k \to 0$ in $L^q(M)$. By Corollary 10.3.3, this implies convergence $u_k \circ e_{y_k} \to 0$ in $L^q(\Omega_r)$ for any sequence (y_k). On the other hand, boundedness of the sequence u_k in $H^{1,p}(M)$ and (10.37) give us the boundedness of any sequence $(u_k \circ e_{y_k})$ in $H^{1,p}(\Omega_r)$. By continuity of the embedding $H^{1,p}(\Omega_r) \hookrightarrow L^q(\Omega_r)$, we get $u_k \circ e_{y_k} \rightharpoonup 0$ in $H^{1,p}(\Omega_r)$. □

As a consequence, we have the following compactness property for functions supported on sets thin at infinity. For an open set M_0 of a Riemannian manifold M, we denote the closure of the space of Lipschitz functions with compact support on M_0 in the norm of $H^{1,p}(M)$ as $H_0^{1,p}(M_0)$.

Proposition 7.1.5. *Let M be a N-dimensional Riemannian manifold of bounded geometry, let M_0 be an open subset of M, and let $1 < p < N$. Let $Y \subset M$ be a (ε, r)-discretization of M, $r < r(M)$. If for any sequence (y_k) in Y,*

$$v_g(M_0 \cap B(y_k, r)) \to 0, \tag{7.4}$$

then $H_0^{1,p}(M_0)$ is compactly embedded into $L^q(M_0)$, $p < q < p^*$.

Proof. Let (u_k) be a sequence in $H_0^{1,p}(M_0)$, weakly convergent to zero and let (y_k) be an arbitrary sequence in Y. Let $\Omega_k \stackrel{\text{def}}{=} e_{y_k}^{-1}(M_0 \cap B(y_k, r))$. Since M has bounded geometry, condition (7.4) implies that $|\Omega_k| \to 0$. Since $u_k \circ e_{y_k}$ is bounded in $L^{p^*}(\Omega_r)$, by Hölder inequality,

$$\int_{\Omega_r} |u_k \circ e_{y_k}|^q d\xi = \int_{\Omega_k} |u_k \circ e_{y_k}|^q d\xi \le \Big(\int_{\Omega_k} |u_k \circ e_{y_k}|^{p^*} d\xi \Big)^{q/p^*} |\Omega_k|^{1-q/p^*} \to 0.$$

Thus $u_k \circ e_{y_k} \to 0$ in $L^q(\Omega_r)$ and, since $u_k \circ e_{y_k}$ is bounded in $H^{1,p}(\Omega_r)$, we have $u_k \circ e_{y_k} \rightharpoonup 0$ in $H^{1,p}(\Omega_r)$. Then by Theorem 7.1.4 $u_k \to 0$ in $L^q(M)$, which proves the proposition. □

7.2 Manifolds at infinity

In what follows, we consider the radius $\rho < \frac{r(M)}{8}$ and $\hat{\rho}$-discretization Y of M, $\frac{\rho}{2} < \hat{\rho} < \rho$.

Definition 7.2.1. Let $(y_k)_{k \in \mathbb{N}}$ be a sequence in Y that is an enumeration of an infinite subset of Y. We shall call a countable family $\{(y_{k;i})_{k \in \mathbb{N}}\}_{i \in \mathbb{N}_0}$ of sequences on Y *a trailing system* of $(y_k)_{k \in \mathbb{N}}$ if for every $k \in \mathbb{N}$ sequence $(y_{k;i})_{i \in \mathbb{N}_0}$ enumerates Y in the order of the distance from y_k, that is, $d(y_{k;i}, y_k) \le d(y_{k;i+1}, y_k)$ for all $i \in \mathbb{N}_0$. In particular, $y_{k;0} = y_k$.

Note that any enumeration of an infinite subset of Y admits a trailing system: it can be constructed inductively, by starting with $y_{k;0} = y_k$ and, given $i \in \mathbb{N}_0$, choosing $y_{k;i+1}$ as any point $y \in Y \setminus \{y_{k;0}, \ldots, y_{k;i}\}$ with the least value of $d(y, y_k)$, $i \in \mathbb{N}_0$. Since there may exist several points of Y with the same distance from y_k for a given k, the trailing system is generally not uniquely defined.

Lemma 7.2.2. *Let $(y_k)_{k\in\mathbb{N}}$ be a sequence in a discretization Y that is an enumeration of an infinite subset of Y. There exists a renamed subsequence of $(y_k)_{k\in\mathbb{N}}$ with the following property: for any $i \in \mathbb{N}_0$, there exist a finite subset J_i of $\mathbb{N}_0$ such that*

$$B(y_{k;i}, \rho) \cap B(y_{k;j}, \rho) \neq \emptyset \iff j \in J_i\,. \tag{7.5}$$

Proof. Let us fix i. If the ball $B(y_{k;j}, \rho)$ intersects $B(y_{k;i}, \rho)$, then $B(y_{k;\ell}, \rho/2) \subset B(y_k, d(y_k, y_{k;i}) + 3\rho)$ for any $\ell \in \{0, 1, \ldots, j\}$. The geometry of M is bounded so the respective volumes of the balls $(B(y_{k;\ell}, \rho/4))$ are bounded from below by a constant depending on ρ but independent of the balls. Note that these balls are pairwise disjoint. Moreover, the Ricci curvature of M is bounded from below, so by the Bishop–Gromov volume comparison theorem the volume of any ball $B(y_{k;\ell}, r)$ can be estimated from above by the constant depending only on the radius. In consequence,

$$Cj \leq \sum_{\ell=0}^{j} v_g(B(y_{k;\ell}, \rho/4)) \leq v_g(B(y_k, d(y_k, y_{k;i}) + 3\rho)) \leq C_i, \tag{7.6}$$

and the constants C, C_i are independent of k. Let $J_{k;i} = \{j : B(y_{k;i}, \rho) \cap B(y_{k;j}, \rho) \neq \emptyset\}$. Then for any k we have $J_{k;i} \subset [0, C_i/C]$. Therefore, there exists a subsequence $k_1, k_2, \ldots$ such that $J_{k_\ell,i} = J_{k_\nu,i}$ for any ℓ and ν. We put $J_i = J_{k_1,i}$.

The assertion of the lemma follows now from the standard diagonalization argument. □

We will always assume throughout the paper that the sequence we work with satisfies the above property. This can be done since passing to subsequence never spoils our construction.

With a given trailing system $\{(y_{k;i})_{k\in\mathbb{N}}\}_{i\in\mathbb{N}_0}$, we associate a manifold $M_\infty^{(y_{k;i})}$ defined by gluing data that will be constructed below. In the construction, we will use definitions from Section 10.3 of the Appendix.

When we define the manifold $M_\infty^{(y_{k;i})}$, we assume that we work with a sequence satisfying (7.5). The following subset of $\mathbb{N}_0^2$ is essential for the construction:

$$\mathcal{K} = \bigcup_{i=0}^{\infty} \{(i, j) : j \in J_i\}.$$

If $(i, j) \in \mathcal{K}$, then passing to a subsequence for any $\xi, \eta \in \Omega_{2\rho}$ we have

$$d(e_{y_{k;j}}\xi, e_{y_{k;i}}\eta) \leq d(e_{y_{k;j}}\xi, y_{k;j}) + d(y_{k;j}, y_{k;i}) + d(y_{k;i}, e_{y_{k;i}}\eta) < 6\rho < \frac{3r(M)}{4}\,.$$

Therefore, on a subsequence, we may consider a diffeomorphism

$$\psi_{ij,k} \stackrel{\text{def}}{=} e^{-1}_{y_{k;i}} \circ e_{y_{k;j}} : \bar{\Omega}_{2\rho} \to \Omega_a, \quad a = \frac{3}{4} r(M).$$

To each pair $(i,j) \in \mathcal{K}$, we associate a subset Ω_{ji} of $\Omega_{2\rho}$ and a diffeomorphism ψ_{ij} defined on Ω_{ji} whenever the latter is nonempty.

By boundedness of the geometry (cf. Lemma 10.3.2) and the Arzela–Ascoli theorem, there is a renamed subsequence of $(\psi_{ij,k})_{k\in\mathbb{N}}$ that converges in $C^\infty(\bar{\Omega}_{2\rho})$ to some smooth function $\psi_{ij} : \bar{\Omega}_{2\rho} \to \Omega_a$ and, moreover, we may assume that the same extraction of $(\psi_{ji,k})_{k\in\mathbb{N}}$ converges in $C^\infty(\bar{\Omega}_{2\rho})$ as well. Note that Lemma 10.3.2 gives that for any $\alpha \in \mathbb{N}_0^N$ there exists a constant $C_\alpha > 0$, such that

$$|d^\alpha \psi_{ij}(\xi)| \le C_\alpha \quad \text{whenever } i,j \in \mathbb{N}_0,\ \xi \in \Omega_\rho.$$

We define $\Omega_{ij} \stackrel{\text{def}}{=} \psi_{ij}(\Omega_\rho) \cap \Omega_\rho$. This set may generally be empty. Let us define a set that we will invoke in our application of Corollary 10.3.8 that will follow:

$$\mathbb{K} \stackrel{\text{def}}{=} \{(i,j) \in \mathcal{K} :\ \Omega_{ij} \ne \emptyset\}. \tag{7.7}$$

To prove the cocycle condition for the gluing data, condition (v) in Corollary 10.3.8, we should extract subsequences in a more restrictive way. First, we consider a subsequence $\psi^1_{01,k}$ of $\psi_{01,k}$ that converges to ψ_{01} and note that on the same subsequence we have convergence of $\psi^1_{10,k}$ to ψ_{10}. Fix an enumeration $n \mapsto (i_n, j_n)$ of the set of all indices $(i,j) \in \mathbb{K}$, $i < j$, and extract the convergent subsequence $\psi^{n+1}_{i_\ell j_\ell,k}$ of the subsequence $\psi^n_{i_\ell j_\ell,k}$ from the previous extraction step, for $\ell = 0, \dots, n+1$. Then the diagonal sequence $\psi^k_{i_\ell j_\ell,k}$ will converge to $\psi_{i_\ell j_\ell}$ for any $\ell \in \mathbb{N}$.

By the definition of Ω_{ij} and ψ_{ij}, we have $\psi_{ij} \circ \psi_{ji} = \mathrm{id}$ on Ω_{ij} and $\psi_{ji} \circ \psi_{ij} = \mathrm{id}$ on Ω_{ji}. Therefore, $\psi_{ji} = \psi_{ij}^{-1}$ in restriction to Ω_{ij}, and ψ_{ji} is a diffeomorphism between Ω_{ij} and Ω_{ji}. Note that this construction gives that $\psi_{ii} = \mathrm{id}$, $\Omega_{ii} = \Omega_\rho$ for all $i \in \mathbb{N}_0$. Thus conditions (i)–(iii) of Corollary 10.3.8 are satisfied.

Note also that the second step of the constructions implies

$$\begin{aligned}\psi_{\ell i} &= \lim_{k\to\infty} e^{-1}_{y_{k;\ell}} \circ e_{y_{k;i}} = \lim_{k\to\infty} e^{-1}_{y_{k;\ell}} \circ e_{y_{k;j}} \circ e^{-1}_{y_{k;j}} \circ e_{y_{k;i}} \\ &= \lim_{k\to\infty} e^{-1}_{y_{k;\ell}} \circ e_{y_{k;j}} \circ \lim_{k\to\infty} e^{-1}_{y_{k;j}} \circ e_{y_{k;i}} = \psi_{\ell j} \circ \psi_{ji},\end{aligned}$$

and

$$\psi_{ij}(\Omega_{ji} \cap \Omega_{jk}) = \psi_{ij}\big(\psi_{ji}(\Omega_\rho) \cap \Omega_\rho \cap \psi_{jk}(\Omega_\rho) \cap \Omega_\rho\big) = \Omega_{ij} \cap \Omega_{ik},$$

which proves condition (iv) of Corollary 10.3.8.

Let $x \in \partial\Omega_{ij} \cap \Omega_p$. Since $\partial\Omega_{ij} \subset \partial\psi_{ij}(\Omega_p) \cup \partial\Omega_p$ and Ω_p is open we conclude that $x \in \partial\psi_{ij}(\Omega_p) = \psi_{ij}(\partial\Omega_p)$. Thus $\psi_{ji}(x) \in \partial\Omega_p$. This proves the condition (v) of Corollary 10.3.8.

We have thus proved the following proposition.

Proposition 7.2.3. *Let M be a Riemannian manifold with bounded geometry and let Y be its discretization.*

For any trailing system $\{(y_{k;i})_{k\in\mathbb{N}}\}_{i\in\mathbb{N}_0}$ related to the sequence (y_k) in Y, there exists a smooth manifold $M_\infty^{(y_{k;i})}$ with an atlas $\{(U_i, \tau_i)\}_{i\in\mathbb{N}_0}$ such that:

(1) $\tau_i(U_i) = \Omega_p$,

and

(2) *there exists a renamed subsequence of k such that for any two charts (U_i, τ_i) and (U_i, τ_i) with $U_i \cap U_j \neq \emptyset$ the corresponding transition map $\psi_{ij} : \tau_j(U_j \cap U_i) \to \tau_i(U_j \cap U_i)$ is given by the C^∞-limit*

$$\psi_{ij} = \lim_{k\to\infty} e_{y_{k;i}}^{-1} \circ e_{y_{k;j}}.$$

For convenience, we will also widely use the “inverse” charts $\varphi_i = \tau_i^{-1}$ so that $\varphi_j^{-1} \circ \varphi_i = \psi_{ji} : \Omega_{ij} \to \Omega_{ji}$.

We now define the Riemannian metric on $M_\infty^{(y_{k;i})}$ in two steps as follows. First, for any $i \in \mathbb{N}_0$ we define a metric tensor $g^{(i)}$ on Ω_p, and afterwards we pull it back onto $U_i = \varphi_i(\Omega_p) \subset M_\infty^{(y_{k;i})}$ via φ_i^{-1} and prove the compatibility conditions.

Tensor $g^{(i)}$ is defined as a C^∞-limit on a suitable renamed subsequence:

$$\tilde{g}_\xi^{(i)}(v, w) \overset{\text{def}}{=} \lim_{k\to\infty} g_{e_{y_{k;i}}(\xi)}(de_{y_{k;i}}(v), de_{y_{k;i}}(w)), \quad \xi \in \Omega_p \text{ and } v, w \in \mathbb{R}^N, \tag{7.8}$$

Existence of the limit follows from the boundedness of the geometry of the manifold M since the coefficients of the tensors $g_{e_{y_{k;i}}}$ form a bounded family of functions in the spaces $C^\infty(\Omega_p)$. Using the standard diagonalization procedure, we can choose the same subsequence for any i. Furthermore, $\tilde{g}^{(i)}$ is a bilinear symmetric positive-definite form. Since we used in the definition (7.8) normal coordinates, we have $\tilde{g}_0^{(i)}(v, v) = |v|^2$. In consequence, by the boundedness of geometry, $\tilde{g}_\xi^{(i)}(v, v) \geq \frac{1}{2}|v|^2$ in Ω_p for all $i \in \mathbb{N}_0$, provided that ρ is fixed sufficiently small.

Now we can define a metric $\tilde{g}$ on $M_\infty^{(y_{k;i})}$ by the following relation:

$$\tilde{g}_x(v, w) \overset{\text{def}}{=} \tilde{g}^{(i)}_{\varphi_i^{-1}(x)}(d\varphi_i^{-1}(v), d\varphi_i^{-1}(w)), \tag{7.9}$$
$$x \in \varphi_i(\Omega_p) \subset M_\infty^{(y_{k;i})} \text{ and } v, w \in T_x M_\infty^{(y_{k;i})}.$$

To prove that the Riemannian metric is well-defined, we should verify the compatibility relation on overlapping charts, that is,

$$\tilde{g}^{(i)}_{\varphi_i^{-1}(x)}(d\varphi_i^{-1}v, d\varphi_i^{-1}w) = \tilde{g}^{(j)}_{\varphi_j^{-1}(x)}(d\varphi_j^{-1}v, d\varphi_i^{-1}w), \tag{7.10}$$
$$\text{if } x \in \varphi_i(\Omega_\rho) \cap \varphi_j(\Omega_\rho) \text{ and } v, w \in T_x M_\infty^{(y_{k;i})}.$$

But $\varphi_j^{-1} \circ \varphi_i = \psi_{ji}$, so it suffices to prove that

$$\tilde{g}^{(i)}_\xi(v, w) = \tilde{g}^{(j)}_{\psi_{ji}(\xi)}(d\psi_{ji}v, d\psi_{ji}w), \quad \text{with } v, w \in T_\xi \Omega_\rho. \tag{7.11}$$

Let $e_{y_{k;j}}^{-1} \circ e_{y_{k;i}}(\xi) = \eta_k$ then $\psi_{j,i}(\xi) = \lim_{k\to\infty} \eta_k$ and $e_{y_{k;i}}(\xi) = e_{y_{k;j}}(\eta_k)$. In consequence,

$$\begin{aligned}\tilde{g}^{(i)}_\xi(v, w) &= \lim_{k\to\infty} g_{e_{y_{k;i}}(\xi)}(de_{y_{k;i}}v, de_{y_{k;i}}w) \\ &= \lim_{k\to\infty} g_{e_{y_{k;j}}(\eta_k)}(de_{y_{k;j}^{-1}}e_{y_{k;i}}v, de_{y_{k;j}^{-1}} \circ e_{y_{k;i}}w) \\ &= g_{\psi_{j,i}(\xi)}(d\psi_{ji}v, d\psi_{ji}w).\end{aligned} \tag{7.12}$$

Definition 7.2.4. A manifold at infinity $M_\infty^{(y_{k;i})}$ of a manifold M with bounded geometry, generated by a trailing system $\{(y_{k;i})_{k\in\mathbb{N}}\}_{i\in\mathbb{N}_0}$ of a sequence (y_k) in Y, is the differentiable manifold given by Theorem 10.3.8, supplied with a Riemannian metric tensor $\tilde{g}$ defined by (7.9).

For the given chart (Ω_ρ, τ_i), components of the metric tensor $\tilde{g}$ are defined by formula (7.8); cf. (7.9). Let $\xi = 0$. The maps $e_{y_{k;i}}$ are normal coordinates systems, so for any k components $g_{\ell,m}$ of the metric tensor g satisfy $g_{\ell,m}(0) = \delta_{\ell,m}$ and $\partial_n g_{\ell,m}(0) = 0$. So by identity (7.8), we get

$$\tilde{g}_{\ell,m}(0) = \delta_{\ell,m} \quad \text{and} \quad \partial_n \tilde{g}_{\ell,m}(0) = 0\,.$$

Moreover, the components $g_{\ell,m}$ are a bounded set in $C^\infty(\Omega_\rho)$ so all the set of $\tilde{g}_{\ell,m}$ is also bounded in $C^\infty(\Omega_\rho)$.

For any k and i, $(\Omega_\rho, e_{y_{k;i}})$ is a normal coordinate system, so for any unit vector v we have on that ball $\Gamma^n_{m,\ell}(tv)v_\ell v_m = 0$, $0 \le t \le \rho$, where $\Gamma^n_{m,\ell}$ denotes Christoffel symbols of a given Riemannian metric on M. But Christoffel symbols can be expressed in terms of components of Riemannian metric tensor and their derivatives, so the Christoffel symbols $\tilde{\Gamma}^n_{m,\ell}$ of the manifold $M_\infty^{(y_{k;i})}$ are limit values in C^∞ of the Christoffel symbols $\Gamma^n_{m,\ell}$ of the manifold M. Therefore, $t \mapsto tv$, $0 \le t \le \rho$, are geodesic curves also for $M_\infty^{(y_{k;i})}$ in the coordinates (Ω_ρ, φ_i). Thus the injectivity radius of $M_\infty^{(y_{k;i})}$ is not smaller then ρ and (Ω_ρ, φ_i) is a normal system of coordinates.

In terms of Definition 7.2.4, the argument above yields the following conclusion.

Proposition 7.2.5. *Let M be a Riemannian manifold of bounded geometry and let Y be its $\hat{\rho}$-discretization, $\frac{\rho}{2} < \hat{\rho} < \rho < \frac{r(M)}{8}$. Then every discrete sequence (y_k) in Y with a* *given trailing system* $\{(y_{k;i})_{k\in\mathbb{N}}\}_{i\in\mathbb{N}_0}$ *has a renamed subsequence* (y_k) *that generates a Riemannian manifold at infinity* $M_\infty^{(y_{k;i})}$ *of the manifold* M. *Moreover, manifold* $M_\infty^{(y_{k;i})}$ *is of bounded geometry and its injectivity radius is not less than* ρ.

Remark 7.2.6. Let M' be another manifold such that M and M' have respective compact subsets M_0 and M'_0 such that $M \setminus M_0$ is isometric to $M' \setminus M'_0$, that is, let M' and M coincide up to a compact perturbation. Then their respective manifolds at infinity for the same trailing systems coincide. From this, it follows that manifold at infinity of the manifold M is not necessarily diffeomorphic to M.

7.3 Local and global profiles. Profile decomposition

Defect of compactness for bounded sequences in $H^{1,2}(M)$ can be formulated using discretizations, related trailing systems described in Definition 7.2.1 and corresponding manifolds at infinity.

Definition 7.3.1. Assume that manifold M has bounded geometry and let Y be its discretization. Let (u_k) be a bounded sequence in $H^{1,2}(M)$. Let (y_k) be a sequence of points in Y and let $\{(y_{k;i})_{k\in\mathbb{N}}\}_{i\in\mathbb{N}_0}$ be its trailing system. One says that $w_i \in H^{1,2}(\Omega_\rho)$ is *a local profile* of (u_k) relative to a trailing sequence $(y_{k;i})_{k\in\mathbb{N}}$, if, on a renamed subsequence, $u_k \circ e_{y_{k;i}} \rightharpoonup w_i$ in $H^{1,2}(\Omega_\rho)$ as $k \to \infty$. If (y_k) is a renamed (diagonal) subsequence such that $u_k \circ e_{y_{k;i}} \rightharpoonup w_i$ in $H^{1,2}(\Omega_\rho)$ as $k \to \infty$ for all $i \in \mathbb{N}_0$, then the family $\{w_i\}_{i\in\mathbb{N}_0}$ is called an *array of local profiles of* (u_k) *relative to the trailing system* $\{(y_{k;i})_{k\in\mathbb{N}}\}_{i\in\mathbb{N}_0}$ *of the sequence* (y_k).

Proposition 7.3.2. *Let* M *be a manifold of bounded geometry with a discretization* Y. *Let* (u_k) *be a bounded sequence in* $H^{1,2}(M)$. *Let* $\{w_i\}_{i\in\mathbb{N}_0}$ *be an array of local profiles of* (u_k) *associated with a trailing system* $\{(y_{k;i})_{k\in\mathbb{N}}\}_{i\in\mathbb{N}_0}$ *related to the sequence* (y_k) *in* Y. *Then there exists a* $H^{1,2}_{\mathrm{loc}}$*-function* $w : M_\infty^{(y_{k;i})} \to \mathbb{R}$ *such that* $w \circ \varphi_i = w_i$, $i \in \mathbb{N}_0$, *where* $\varphi_i : \Omega_\rho \to M_\infty^{(y_{k;i})}$ *are local coordinate maps of* $M_\infty^{(y_{k;i})}$.

Proof. Functions w_i are defined on Ω_ρ that is a domain of definition of φ_i. Set $w \overset{\mathrm{def}}{=} w_i \circ \varphi_i^{-1}$ on $\varphi_i^{-1}(\Omega_\rho)$ and note that if $x \in \varphi_i^{-1}(\Omega_\rho) \cap \varphi_j^{-1}(\Omega_\rho)$ for some $j \in \mathbb{N}_0$, then $\varphi_i(x) \in \Omega_{ij}$, $\varphi_j(x) \in \Omega_{ji}$, and, using the a. e. convergence of $u_k \circ e_{y_{k;i}}$ and $u_k \circ e_{y_{k;j}}$ to w_i and w_j, respectively, and the uniform convergence of $e_{y_{k;i}}^{-1} e_{y_{k;j}}$ to ψ_{ij}, we have

$$\begin{aligned} w_j \circ \varphi_j^{-1} &= \lim_{k\to\infty} u_k \circ e_{y_{k;j}} \circ \varphi_j^{-1} = \lim_{k\to\infty} u_k \circ e_{y_{k;i}} \circ e_{y_{k;i}}^{-1} \circ e_{y_{k;j}} \circ \varphi_j^{-1} \\ &= w_i \circ \psi_{ij} \circ \varphi_j^{-1} = w_i \circ \varphi_i^{-1} \circ \varphi_j \circ \varphi_j^{-1} = w_i \circ \varphi_i^{-1} \end{aligned}$$

almost everywhere in $\varphi_i^{-1}(\Omega_\rho) \cap \varphi_j^{-1}(\Omega_\rho)$. □

Definition 7.3.3. Let $\{w_i\}_{i\in\mathbb{N}_0}$ be a local profile array of a bounded sequence (u_k) in $H^{1,2}(M)$ relative to a trailing system $\{(y_{k;i})_{k\in\mathbb{N}}\}_{i\in\mathbb{N}_0}$. The function $w : M_\infty^{(y_{k;i})} \to \mathbb{R}$ given by Proposition 7.3.2 is called the *global profile* of the sequence (u_k) relative to $(y_{k;i})$.

Given that the discretization Y of M yields a uniformly finite covering of M by geodesic balls $\{B(y,\rho)\}_{y\in Y}$ satisfying (7.1), consider a smooth partition of unity $\{\chi_y\}_{y\in Y}$ subordinated to this covering.

Definition 7.3.4. Let M be a manifold of bounded geometry with a discretization $Y \subset M$. Let $M_\infty^{(y_{k;i})}$ be a manifold at infinity of M generated by a corresponding trailing system $\{(y_{k;i})_{k\in\mathbb{N}}\}_{i\in\mathbb{N}_0}$. An *elementary concentration* associated with a function $w : M_\infty^{(y_{k;i})} \to \mathbb{R}$ is a sequence $(W_k)_{k\in\mathbb{N}}$ of functions $M \to \mathbb{R}$ given by

$$W_k = \sum_{i\in\mathbb{N}_0} \chi_{y_{k;i}} w \circ \varphi_i \circ e_{y_{k;i}}^{-1}, \quad k \in \mathbb{N}, \tag{7.13}$$

where φ_i are the local coordinate maps of manifold $M_\infty^{(y_{k;i})}$.

In heuristic terms, after we find limits w_i, $i \in \mathbb{N}_0$, of the sequence (u_k) under the "trailing spotlights" $(e_{y_{k;i}})_{k\in\mathbb{N}_0}$ that follow different trailing sequences $(y_{k;i})_{k\in\mathbb{N}}$ of (y_k), we give an approximate reconstruction W_k of u_k "centered" on the moving center y_k of the "core spotlight." We do that by first splitting w into local profiles $w \circ \varphi_i$, $i \in \mathbb{N}_0$, on the set Ω_ρ, casting them onto the manifold M in the vicinity of $y_{k;i}$ by composition with $e_{y_{k;i}}^{-1}$, and patching all such compositions together by the partition of unity on M satisfying (7.1). Such reconstruction approximates u_k on geodesic balls $B(y_k, R)$ with any $R > 0$, but it ignores the values of u_k for k large on the balls $B(y_k', R)$, with $d(y_k, y_k') \to \infty$, where u_k is approximated by a different local concentration. Note that in Corollary 5.1.7 we have for the case of periodic manifold that a global reconstruction of u_k, up to a remainder vanishing in $L^p(M)$, is a sum elementary concentrations associated with all such mutually decoupled sequences.

Similarly, the profile decomposition theorem below says any bounded sequence (u_k) in $H^{1,2}(M)$ has a subsequence that, up to a remainder vanishing in $L^p(M)$, $p \in (2, 2^*)$, equals a sum of spatially elementary concentrations.

In the theorem below and in the subsequent sections, we will work with countable families of discrete sequences of the set Y. To each sequence, we assign a trailing system so in consequence also a manifold at infinity. To simplify the notation, we will index the sequences in Y, the related trailing systems the corresponding manifolds, concentration profiles on these manifolds, etc. by n, that is, we will write $y_k^{(n)}$, $y_{k;i}^{(n)}$, $M_\infty^{(n)}$, $w^{(n)}$, etc.

Theorem 7.3.5. *Let M be a manifold of bounded geometry with a discretization $Y \subset M$. Let (u_k) be a sequence in $H^{1,2}(M)$ weakly convergent to some function $w^{(0)}$ in $H^{1,2}(M)$. There exist a renamed subsequence of (u_k), sequences $(y_k^{(n)})_{k\in\mathbb{N}}$ in Y, and global profiles*

$w^{(n)}$ on the respective manifolds at infinity $M_\infty^{(n)}$, $n \in \mathbb{N}$, associated with $(y_k^{(n)})_{k\in\mathbb{N}}$ and their trailing sequences, such that $d(y_k^{(n)}, y_k^{(m)}) \to \infty$ when $n \neq m$, and

$$u_k - w^{(0)} - \sum_{n\in\mathbb{N}} W_k^{(n)} \to 0 \quad \text{in } L^p(M),\ p \in (2, 2^*), \tag{7.14}$$

where $W_k^{(n)} = \sum_{i\in\mathbb{N}_0} \chi_{y_{k;i}^{(n)}} w^{(n)} \circ \varphi_i^{(n)} \circ e^{-1}_{y_{k;i}^{(n)}}$ are elementary concentrations, $\varphi_i^{(n)}$ are the local coordinates of the manifolds $M_\infty^{(n)}$ and $\{\chi_y\}_{y\in Y}$ is the partition of unity satisfying (7.1). The series $\sum_{n\in\mathbb{N}} W_k^{(n)}$ converges in $H^{1,2}(M)$ unconditionally and uniformly in $k \in \mathbb{N}$. Moreover,

$$\|w^{(0)}\|^2_{H^{1,2}(M)} + \sum_{n=1}^{\infty} \|w^{(n)}\|^2_{H^{1,2}(M_\infty^{(n)})} \le \limsup \|u_k\|^2_{H^{1,2}(M)}, \tag{7.15}$$

and

$$\int_M |u_k|^p \mathrm{d}v_g \to \int_M |w^{(0)}|^p \mathrm{d}v_g + \sum_{n=1}^{\infty} \int_{M_\infty^{(n)}} |w^{(n)}|^p \mathrm{d}v_{g^{(n)}}. \tag{7.16}$$

7.4 Auxiliary statements

In Sections 7.5–7.10, we assume that conditions of Theorem 7.3.5 hold true. First, we prove an inequality for the norms defined by Lemma 10.3.4 in the Appendix.

Lemma 7.4.1. *Let (u_k) be a bounded sequence in $H^{1,2}(M)$, let $M_\infty^{(y_{k;i})}$ be a manifold at infinity of M generated by a trailing system $\{(y_{k;i})_{k\in\mathbb{N}}\}_{i\in\mathbb{N}_0}$, and let $w \in H^{1,2}(M_\infty^{(y_{k;i})})$ be the associated global profile of (u_k). Then*

$$\liminf |||u_k|||^2_{H^{1,2}(M)} \ge |||w|||^2_{H^{1,2}(M_\infty^{(y_{k;i})})}.$$

Proof. Let $\{\chi_y\}_{y\in Y}$ be the partition of satisfying (7.1), and let us enumerate it for each $k \in \mathbb{N}$ according to the enumeration $\{y_{k;i}\}_{i\in\mathbb{N}_0}$ of Y, namely $i \mapsto \chi_{y_{k;i}}$, $i \in \mathbb{N}_0$. In other words, for every k the set $\{\chi_{y_{k;i}}\}_{i\in\mathbb{N}_0}$ equals the set $\{\chi_y\}_{y\in Y}$, and only its enumeration depends on the given trailing system $\{(y_{k;i})_{k\in\mathbb{N}}\}_{i\in\mathbb{N}_0}$. By Arzela–Ascoli theorem, we can define for any i a function η_i on Ω_ρ by the formula

$$\eta_i = \lim_{k\to\infty} \chi_{y_{k;i}} \circ e_{y_{k;i}}. \tag{7.17}$$

The functions η_i are smooth functions compactly supported in Ω_ρ. Moreover, using the diagonalization argument if needed, we get

$$\eta_i = \lim_{k\to\infty} \chi_{y_{k;i}} \circ e_{y_{k;j}} \circ e^{-1}_{y_{k;j}} \circ e_{y_{k;i}} = \eta_j \circ \psi_{ji}.$$

Since $\sum_{i\in\mathbb{N}_0}\chi_{y_{k;i}}\circ e_{y_{k;j}}=1$ on Ω_p for any $j\in\mathbb{N}_0$, we have in the limit $\sum_{i\in\mathbb{N}_0:(i,j)\in\mathbb{K}}\eta_i\circ\psi_{ij}=1$ on Ω_p; cf. Lemma 7.2.2. So the family of functions

$$\chi_i^{(y_{k;i})}\stackrel{\text{def}}{=}\eta_i\circ\varphi_i^{-1},\quad i\in\mathbb{N}_0 \tag{7.18}$$

is a partition of unity for $M_\infty^{(y_{k;i})}$, which is subordinated to the covering $\{\varphi_i(\Omega_p)\}_{i\in\mathbb{N}_0}$ of $M_\infty^{(y_{k;i})}$, and one can easily see that it satisfies (7.1).

Both the manifolds M and $M_\infty^{(y_{k;i})}$ have bounded geometry and, therefore,

$$\begin{aligned}\liminf_{k\to\infty}|||u_k|||^2_{H^{1,2}(M)}&=\liminf_{k\to\infty}\sum_{i\in\mathbb{N}_0}\|(\chi_{y_{k;i}}u_k)\circ e_{y_{k;i}}\|^2_{H^{1,2}(\mathbb{R}^N)}\\&\geq\sum_{i\in\mathbb{N}_0}\liminf_{k\to\infty}\|(\chi_{y_{k;i}}u_k)\circ e_{y_{k;i}}\|^2_{H^{1,2}(\mathbb{R}^N)}\geq\sum_{i\in\mathbb{N}_0}\|\eta_iw_i\|^2_{H^{1,2}(\mathbb{R}^N)}\\&=\lim_{k\to\infty}\sum_{i\in\mathbb{N}_0}\|\chi_i^{(y_{k;i})}w\circ\varphi_i\|^2_{H^{1,2}(\mathbb{R}^N)}\geq|||w|||^2_{H^{1,2}(M_\infty^{(y_{k;i})})}.\end{aligned} \tag{7.19}$$

□

Lemma 7.4.2. *Let $\{(y_{k;i})_{k\in\mathbb{N}}\}_{i\in\mathbb{N}_0}$ be a trailing system for a discrete sequence (y_k) and let $w\in H^{1,2}(M_\infty^{(y_{k;i})})$. Then the elementary concentrations $W_k^{(y_{k;i})}$, $k\in\mathbb{N}$, associated with this system belongs to $H^{1,2}(M)$. Moreover, there is a positive constant C independent of k and i such that*

$$\|W_k^{(y_{k;i})}\|_{H^{1,2}(M)}\leq C\,\|w\|_{H^{1,2}(M_\infty^{(y_{k;i})})}. \tag{7.20}$$

If $(y'_k)_{k\in\mathbb{N}}$ is a discrete sequence on M such that $d(y_k,y'_k)\to\infty$, then the elementary concentration $W_k^{(y_{k;i})}$ satisfies

$$W_k^{(y_{k;i})}\circ e_{y'_k}\to 0$$

in $H^{1,2}(\Omega_p)$.

Proof. We recall that

$$W_k^{(y_{k;i})}=\sum_{i\in\mathbb{N}_0}\chi_{y_{k;i}}\,w\circ\varphi_i\circ e_{y_{k;i}}^{-1}; \tag{7.21}$$

cf. (7.13). The functions $\chi_{y_{k;i}}\circ e_{y_{k;i}}$ are smooth compactly supported functions on Ω_p and the family $\{\chi_{y_{k;i}}\circ e_{y_{k;i}}\}$ is a bounded set in $C^\infty(\Omega_p)$. By the boundedness of the geometry (cf. Lemma 7.2.2 and Lemma 10.3.4), and using (7.18), we have

$$\begin{aligned}\|\chi_{y_{k;i}}\circ e_{y_{k;i}}\,w\circ\varphi_i\|^2_{H^{1,2}(\mathbb{R}^N)}&\leq C\|\chi_{y_{k;i}}\circ e_{y_{k;i}}\circ\tau_iw\|^2_{H^{1,2}(M_\infty^{(y_{k;i})})}\\&\leq C\sum_{j:\,(i,j)\in\mathbb{K}}\|\chi_i^{(y_{k;i})}w\|^2_{H^{1,2}(M_\infty^{(y_{k;i})})}.\end{aligned}$$

So using once more Lemma 10.3.4 we get

$$\|W_k^{(y_{k;i})}\|^2_{H^{1,2}(M)} \le C\sum_i \|\chi_{y_{k;i}} \circ e_{y_{k;i}} w \circ \varphi_i\|^2_{H^{1,2}(\mathbb{R}^N)} \le C\sum_i \sum_{j:\,(i,j)\in\mathbb{K}} \|\chi_j^{(y_{k;i})} w\|^2_{H^{1,2}(M_\infty^{(y_{k;i})})} \le C\|w\|^2_{H^{1,2}(M_\infty^{(y_{k;i})})}. \tag{7.22}$$

This proves (7.20).

Let $\varepsilon > 0$. If follows from (7.22) that there exist $N_\varepsilon \in \mathbb{N}$ independent of k such that

$$\sum_{i\ge N_\varepsilon} \|\chi_{y_{k;i}} \circ e_{y_{k;i}} w \circ \varphi_i\|^2_{H^{1,2}(\mathbb{R}^N)} \le \varepsilon. \tag{7.23}$$

By (7.21), we have

$$W_k^{(y_{k;i})} \circ e_{y'_k} = \sum_{i\in I_k} (\chi_{y_{k;i}} w \circ \varphi_i \circ e^{-1}_{y_{k;i}}) \circ e_{y'_k}, \tag{7.24}$$

where $I_k = \{i : B(y'_k, \rho) \cap B(y_{k;i}, \rho) \ne \emptyset\}$. Since $d(y_k, y'_k) \to \infty$, we have

$$\sup_{i\le N_\varepsilon} d(y_{k;i}, y'_k) \ge d(y_k, y'_k) - 2N_\varepsilon\rho \to \infty$$

as $k \to \infty$, and thus $B(y'_k, \rho) \cap B(y_{k;i}, \rho) = \emptyset$ for all $i \le N_\varepsilon$ if k is sufficiently large. Then $\sum_{i=1}^{N_\varepsilon} (\chi_{y_{k;i}} w \circ \varphi_i) \circ e^{-1}_{y_{k;i}} \circ e_{y'_k} = 0$ for all k large, which together with (7.23) proves the lemma. □

Lemma 7.4.3. *Let w be a profile of the sequence (u_k), given by Proposition 7.3.2 relative to a trailing system $\{(y_{k;i})_{k\in\mathbb{N}}\}_{i\in\mathbb{N}_0}$, and let (W_k) be the associated sequence of elementary concentrations. The following holds true:*

$$\lim_{k\to\infty} \langle u_k, W_k\rangle_{H^{1,2}(M)} = \|w\|^2_{H^{1,2}(M_\infty^{(y_{k;i})})}. \tag{7.25}$$

Proof. We use for each $k \in \mathbb{N}$ an enumeration of the covering $\{B(y,\rho)\}_{y\in Y}$ by the points $y_{k;i}$ from the trailing system $\{(y_{k;i})_{k\in\mathbb{N}}\}_{i\in\mathbb{N}_0}$. Taking into account that, as $k \to \infty$, $u_k \circ y_{k;j} \rightharpoonup w_j$, $e^{-1}_{y_{k;i}} \circ e_{y_{k;j}} \to \psi_{ij}$, and $w_i \circ \psi_{ij} = w_j$, and using the expression $o^w(1)$ for any sequence of functions that converges weakly to zero in $H^{1,2}(\Omega_\rho)$, we have

$$\langle u_k, W_k\rangle_{H^{1,2}(M)} = \sum_{j\in\mathbb{N}_0} \int_{B(y_{k;j},\rho)} \chi_{y_{k;j}}(x) u_k(x) W_k(x) dv_g(x) + \sum_{j\in\mathbb{N}_0} \int_{B(y_{k;j},\rho)} \chi_{y_{k;j}}(x) g_x(du_k(x), dW_k(x)) dv_g(x), \tag{7.26}$$

and

$$\|w\|^2_{H^{1,2}(M_\infty^{(y_{k;i})})} = \sum_{j\in\mathbb{N}_0} \int_{B(y_{k;j},\rho)} \chi_j^{(y_{k;j})}(x)|w(x)|^2 dv_{\tilde g}(x) + \sum_{j\in\mathbb{N}_0} \int_{B(y_{k;j},\rho)} \chi_j^{(y_{k;i})}(x)\tilde g_x(dw(x),dw(x))dv_{\tilde g}(x), \tag{7.27}$$

where the functions $\chi_j^{(y_{k;i})}$ are defined by the formulas (7.17)–(7.18) relative to the trailing system $\{(y_{k;i})_{k\in\mathbb{N}}\}_{i\in\mathbb{N}_0}$.

Both coverings are uniformly locally finite, so it is sufficient to prove local identities

$$\lim_{k\to\infty} \int_{B(y_{k;j},\rho)} \chi_{y_{k;j}}(x)u_k(x)W_k(x)dv_g(x) = \int_{B(y_{k;j},\rho)} \chi_j^{(y_{k;i})}(x)|w(x)|^2 dv_{\tilde g}(x) \tag{7.28}$$

and

$$\lim_{k\to\infty} \int_{B(y_{k;j},\rho)} \chi_{y_{k;j}}(x)g_x(du_k(x),dW_k(x))dv_g(x) = \int_{B(y_{k;j},\rho)} \chi_j^{(y_{k;i})}(x)\tilde g_x(dw(x),dw(x))dv_{\tilde g}(x), \tag{7.29}$$

In the first case, we have

$$\begin{aligned}
&\int_{\Omega_\rho} \chi_{y_{k;j}} \circ e_{y_{k;j}}(\xi)u_k \circ e_{y_{k;j}}(\xi) \\
&\quad\times \sum_{i\in\mathbb{N}_0} [\chi_{y_{k;i}}\, w\circ\varphi_i\circ e_{y_{k;i}}^{-1})]\circ e_{y_{k;j}}(\xi)\sqrt{g(\xi)}\, d\xi \\
&= \int_{\Omega_\rho} \chi_{y_{k;j}} \circ e_{y_{k;j}}(\xi)(w_j + o^w(1))(\xi) \\
&\quad\times \sum_{i\in\mathbb{N}_0} \chi_{y_{k;i}} \circ e_{y_{k;j}}\, w_i \circ (\psi_{ij} + o^w(1))(\xi)\sqrt{g(\xi)}\, d\xi \\
&= \int_{\Omega_\rho} \chi_{y_{k;j}} \circ e_{y_{k;j}}(\xi)(w_j + o^w(1))(\xi)\,(w_j + o^w(1))(\xi) \\
&\quad\times \sqrt{(\tilde g + o^w(1))(\xi)}\, d\xi \\
&\longrightarrow \int_{\Omega_\rho} \chi_j^{(y_{k;j})} \circ \varphi_j(\xi)|w_j|^2\, \sqrt{\tilde g(\xi)}\, d\xi\,,
\end{aligned}$$

where the last inequality follows from the identity $\sum_{i\in\mathbb{N}_0}\chi_{y_{k;i}}\circ e_{y_{k;j}} = 1$ on Ω_p; cf. Lemma 7.2.2. This proves (7.28).

To prove (7.29), we first note that

$$\begin{aligned}
&\sum_{\nu,\mu=1}^{N} g^{\nu,\mu}(\xi)\partial_\nu(u_k\circ e_{y_{k;j}})(\xi)\partial_\mu(W_k\circ e_{y_{k;j}})(\xi)\\
&\quad=\sum_{\nu,\mu=1}^{N} g^{\nu,\mu}(\xi)\partial_\nu(u_k\circ e_{y_{k;j}})(\xi)\\
&\qquad\times\partial_\mu\Big(\sum_{i\in\mathbb{N}_0}[\chi_{y_{k;i}}\, w\circ\varphi_i\circ e^{-1}_{y_{k;i}}]\circ e_{y_{k;j}}\Big)(\xi)\\
&\quad=\sum_{\nu,\mu=1}^{N} g^{\nu,\mu}(\xi)\partial_\nu((w_j+o^W(1))\circ e_{y_{k;j}})(\xi)\\
&\qquad\times\partial_\mu(\chi_{y_{k;i}}\circ e_{y_{k;j}}(\xi)\; w_i\circ(\psi_{ij}+o^W(1)))(\xi)\\
&\quad=\sum_{\nu,\mu=1}^{N} g^{\nu,\mu}(\xi)\partial_\nu((w_j+o^W(1))\circ e_{y_{k;j}})(\xi)\partial_\mu(w_j+o(1)))(\xi).
\end{aligned}$$

In consequence,

$$\begin{aligned}
&\int_{\Omega_p}\chi_{y_{k;j}}\circ e_{y_{k;j}}(\xi)\sum_{\nu,\mu=1}^{N} g^{\nu,\mu}(\xi)\partial_\nu(u_k\circ e_{y_{k;j}})(\xi)\partial_\mu(W_k\circ e_{y_{k;j}})(\xi)\sqrt{g(\xi)}\,d\xi\\
&\quad=\int_{\Omega_p}\chi_{y_{k;j}}\circ e_{y_{k;j}}(\xi)\sum_{\nu,\mu=1}^{N} g^{\nu,\mu}(\xi)\partial_\nu((w_j+o^W(1))\circ e_{y_{k;j}})(\xi)\\
&\quad\partial_\mu((w_j+o^W(1))\circ e_{y_{k;j}})(\xi)\sqrt{\bar{g}(\xi)+o(1)}\,d\xi\\
&\quad\longrightarrow\int_{\Omega_p}\chi_j^{(y_{k;i})}\circ\varphi_j(\xi)\sum_{\nu,\mu=1}^{N}\bar{g}^{\nu,\mu}(\xi)\partial_\nu w\circ\varphi_j(\xi)\partial_\mu w\circ\varphi_j(\xi)\sqrt{\bar{g}(\xi)}\,d\xi
\end{aligned}$$

Combining the last calculations with (7.26)–(7.29), we arrive at (7.25). □

Lemma 7.4.4. *Let w be a profile of the sequence u_k, given by Proposition 7.3.2 relative to a trailing system $\{(y_{k;i})_{k\in\mathbb{N}}\}_{i\in\mathbb{N}_0}$, and let $(W_k)_{k\in\mathbb{N}}$ be the associated sequence of elementary concentrations. The following holds true:*

$$\lim_{k\to\infty}\|W_k\|^2_{H^{1,2}(M)}=\|w\|^2_{H^{1,2}(M_\infty^{(y_{i;k})})}. \tag{7.30}$$

Proof. We can proceed in the similar way as in the proof of Lemma 7.4.3. Once more we can reduce the argumentation to local identities using (7.27) and

$$\|W_k\|^2_{H^{1,2}(M)} = \sum_{j\in\mathbb{N}_0} \int_{B(y_{k;j},\rho)} \chi_{y_{k;j}}(x) g_x(W_k(x), W_k(x)) \mathrm{d}v_g(x) \\ + \sum_{j\in\mathbb{N}_0} \int_{B(y_{k;j},\rho)} \chi_{y_{k;j}}(x) |dW_k(x)|^2 \mathrm{d}v_g(x), \tag{7.31}$$

We have

$$\int_{\Omega_\rho} \chi_{y_{k;j}} \circ e_{y_{k;j}}(\xi) \sum_{\nu,\mu=1}^N g^{\nu,\mu}(\xi)\partial_\nu(W_k \circ e_{y_{k;j}})(\xi)\partial_\mu(W_k \circ e_{y_{k;j}})(\xi)\sqrt{g(\xi)}\,\mathrm{d}\xi$$

$$= \int_{\Omega_\rho} \chi_{y_{k;j}} \circ e_{y_{k;j}}(\xi) \sum_{\nu,\mu=1}^N g^{\nu,\mu}(\xi)\partial_\nu((w_j + o^W(1)))(\xi)$$

$$\times\, \partial_\mu((w_j + o^W(1)))\sqrt{\tilde{g}(\xi) + o(1)}\,\mathrm{d}\xi$$

$$\longrightarrow \int_{\Omega_\rho} \chi_j^{(y_{k;i})} \circ \varphi_j(\xi) \sum_{\nu,\mu=1}^N \tilde{g}^{\nu,\mu}(\xi)\partial_\nu(w\circ\varphi_j)(\xi)\partial_\mu(w\circ\varphi_j)(\xi)\sqrt{\tilde{g}(\xi)}\,\mathrm{d}\xi.$$

Also as above,

$$\int_{\Omega_\rho} \chi_{y_{k;j}} \circ e_{y_{k;j}}(\xi) \left|\sum_{i\in\mathbb{N}_0} [\chi_{y_{k;i}}\, w\circ\varphi_i\circ e^{-1}_{y_{k;i}})] \circ e_{y_{k;j}}(\xi)\right|^2 \sqrt{g}(\xi)\,\mathrm{d}\xi$$

$$= \int_{\Omega_\rho} \chi_{y_{k;j}} \circ e_{y_{k;j}}(\xi) \left|\sum_{i\in\mathbb{N}_0} \chi_{y_{k;i}} \circ e_{y_{k;j}}(\xi)\, w_i \circ (\psi_{ij} + o^W(1))(\xi)\right| \sqrt{g}(\xi)\,\mathrm{d}\xi$$

$$= \int_{\Omega_\rho} \chi_{y_{k;j}} \circ e_{y_{k;j}}(\xi) |(w_j + o^W(1))(\xi)|^2 \sqrt{(\tilde{g} + o^W(1))(\xi)}\,\mathrm{d}\xi$$

$$\longrightarrow \int_{\Omega_\rho} \chi_j^{(y_{k;i})} \circ \varphi_j(\xi)|(w_j(\xi)|^2 \sqrt{\tilde{g}}(\xi)\,\mathrm{d}\xi. \qquad \square$$

Below we consider a countable family of trailing systems $\{(y^{(n)}_{k;j})_{k\in\mathbb{N}}\}_{i\in\mathbb{N}_0}$, $n\in\mathbb{N}$, and will abbreviate the notation of the associated manifolds at infinity, $M_\infty^{(y^{(n)}_{k;j})}$, as $M_\infty^{(n)}$. This convention will also extend to all other objects generated by trailing systems $\{(y^{(n)}_{k;i})_{k\in\mathbb{N}}\}_{i\in\mathbb{N}_0}$, but not to objects indexed by points in Y, such as $\chi_{y^{(n)}_{k;i}}$.

Lemma 7.4.5. *Assume that $u_k \rightharpoonup 0$. Assume that trailing systems $\{(y^{(n)}_{k;i})_{k\in\mathbb{N}}\}_{i\in\mathbb{N}_0}$ of discrete sequences $(y^{(n)}_k)_{k\in\mathbb{N}}$, $n\in\mathbb{N}$, generate local profiles $\{w^{(n)}_i\}_{i\in\mathbb{N}_0}$, such that*

$d(y_k^{(n)}, y_k^{(\ell)}) \to \infty$ when $n \neq \ell$. Then

$$\sum_{n=1}^{m} \|w^{(n)}\|^2_{H^{1,2}(M_\infty^{(n)})} \leq \limsup \|u_k\|^2_{H^{1,2}(M)}. \tag{7.32}$$

Proof. Consider for each $n = 1, \dots, m$ the elementary concentrations $W_k^{(n)} = \sum_{i\in\mathbb{N}_0} \chi_{y_{k;i}^{(n)}} w_i^{(n)} \circ e_{y_{k;i}^{(n)}}^{-1}$, $w_i^{(n)} = w^{(n)} \circ \varphi_i^{(n)}$, where $\{\varphi_i, \Omega_\rho\}_{i\in\mathbb{N}_0}$ is the atlas of the manifold at infinity $M_\infty^{(n)} \overset{\text{def}}{=} M_\infty^{(y_{k;i}^{(n)})}$, and let us expand by bilinearity the trivial inequality

$$\left\| u_k - \sum_{n=1}^{m} W_k^{(n)} \right\|^2_{H^{1,2}(M)} \geq 0.$$

For convenience, the subscript in the Sobolev norm will be omitted for the rest of this proof. We have then

$$2\sum_{n=1}^{m} \langle u_k, W_k^{(n)} \rangle - \sum_{n=1}^{m} \|W_k^{(n)}\|^2 \leq \|u_k\|^2 + \sum_{n\neq\ell} \langle W_k^{(n)}, W_k^{(\ell)} \rangle. \tag{7.33}$$

Applying Lemmas 7.4.3 and 7.4.4, we have

$$\sum_{n=1}^{m} \|w^{(n)}\|^2_{H^{1,2}(M_\infty^{(n)})} \leq \|u_k\|^2 + \sum_{n\neq\ell} \langle W_k^{(n)}, W_k^{(\ell)} \rangle + o(1). \tag{7.34}$$

In order to prove the lemma, it suffices therefore to show that $\langle W_k^{(n)}, W_k^{(\ell)} \rangle \to 0$ whenever $n \neq \ell$.

Since $d(y_k^{(n)}, y_k^{(\ell)}) \to \infty$, we also have $d(y_{k;i}^{(n)}, y_{k;j}^{(\ell)}) \to \infty$ for any $i, j \in \mathbb{N}_0$. Let $\varepsilon > 0$ and let $N_\varepsilon \in \mathbb{N}$ be such that, in view of Lemma 7.4.1,

$$\begin{aligned}\sum_{i\geq N_\varepsilon} \int_{\Omega\rho} \chi_i^{(n)}(\xi) \sum_{\nu,\mu=1}^{N} g^{\nu\mu}(\xi)\partial_n(w_i^{(n)})(\xi)\partial_\mu(w_i^{(n)})(\xi) \\ + |w_i^{(n)}(\xi)|^2 \sqrt{g(\xi)}\mathrm{d}\xi \leq \varepsilon, \quad n = 1, \dots, m.\end{aligned} \tag{7.35}$$

Let $W_k^{(n)} = W_k^{(n)'} + W_k^{(n)''}$ where

$$W_k^{(n)'} = \sum_{i<N_\varepsilon} (\chi_{y_{k;i}^{(n)}} w_i^{(n)} \circ e_{y_{k;i}^{(n)}}^{-1}) \quad \text{and} \quad W_k^{(n)''} = \sum_{i\geq N_\varepsilon} (\chi_{y_{k;i}^{(n)}} w_i^{(n)} \circ e_{y_{k;i}^{(n)}}^{-1})$$

and note that for all k sufficiently large, $W_k^{(n)'}$ and $W_k^{(\ell)'}$ have disjoint supports. Thus

$$|\langle W_k^{(n)}, W_k^{(\ell)} \rangle| \leq 2S_k T_k + T_k^2, \tag{7.36}$$

where $S_k = \max_{n=1,\dots m} \|W_k^{(n)'}\|$ and $T_k = \max_{n=1,\dots m} \|W_k^{(n)''}\|$. The estimate for S_k is readily provided by repeating verbally the argument of Lemma 7.4.4, which gives

$$S_k^2 \leq \max_{n=1,\dots,m} \|w^{(n)}\|^2_{H^{1,2}(M_\infty^{(n)})} + o(1),$$

so S_k is bounded by $C\|u_k\| + o(1)$ due to Lemma 7.4.1, while a similar adaptation of Lemma 7.4.4 to summation for $i \geq N_\varepsilon$ yields that T_k^2 is bounded, up to vanishing terms, by the left-hand side of (7.35), and thus $T_k \leq \sqrt{\varepsilon} + o(1)$. Thus from (7.36), we have

$$|\langle W_k^{(n)}, W_k^{(\ell)}\rangle| \leq C\sqrt{\varepsilon}(\|u_k\| + \sqrt{\varepsilon} + o(1)),$$

which implies, in turn, that $\limsup_{k\to\infty} |\langle W_k^{(n)}, W_k^{(\ell)}\rangle| \leq C\sqrt{\varepsilon}$, and since ε is arbitrary, we have $\langle W_k^{(n)}, W_k^{(\ell)}\rangle \to 0$ for $n \neq \ell$, which completes the proof. □

Before we begin the proof of Theorem 7.3.5, we introduce the following technical definition.

Definition 7.4.6. Let $(u_k)_{k\in\mathbb{N}}$ be a bounded sequence in $H^{1,2}(M)$. Let $(y_k^{(\ell)})_{k\in\mathbb{N}}$, $\ell = 1,\dots,m$, $m \in \mathbb{N}$, be discrete sequences of points in Y, satisfying $d(y_k^{(n)}, y_k^{(\ell)}) \to \infty$ for $n \neq \ell$, and generating global profiles $w_1,\dots,w_m$ of a renamed subsequence of (u_k) in respective Sobolev spaces $H^{1,2}(M_\infty^{(\ell)})$. A modulus $\nu^{(u_k)}((y_k^{(1)}),\dots,(y_k^{m}))$ of this subsequence is the supremum of the set of values $\|w\|^2_{H^{1,2}(M_\infty^{(y_{k;i})})}$ of all global profiles w of the renamed subsequence (u_k) generated by a trailing system $\{(y_{i;k})_{k\in\mathbb{N}}\}_{i\in\mathbb{N}_0}$ in Y satisfying $d(y_{k;0}, y_k^{(\ell)}) \to \infty$, $\ell = 1,\dots,m$. If such set is empty, we set $\nu^{(u_k)}((y_k^{(1)}),\dots,(y_k^{(m)})) \stackrel{\text{def}}{=} 0$. For $m = 0$, $\nu^{(u_k)}(\emptyset)$ is defined as the corresponding unconstrained supremum.

7.5 Proof of Theorem 7.3.5

Step 1. It suffices to prove Theorem 7.3.5 for sequences that weakly converge to zero. Indeed, assume that the theorem is true in this case. A general bounded sequence (u_k) in $H^{1,2}(M)$ has a renamed subsequence weakly convergent to some $w^{(0)}$ in $H^{1,2}(M)$. Consider then conclusions of the theorem for the sequence $(u_k - w^{(0)})$. Since for any discrete sequence (y_k) in Y, $w^{(0)} \circ e_{y_k} \rightharpoonup 0$ in $H^{1,2}(\Omega_\rho)$ by Lemma 7.4.1, sequences (u_k) and $(u_k - w^{(0)})$ have identical local profiles under the same trailing systems $\{(y_{i;k}^{(n)})_{k\in\mathbb{N}}\}_{i\in\mathbb{N}_0}$, identical manifolds at infinity and identical concentration terms $W_k^{(n)}$, which yields (7.14). Relation (7.15) follows from the elementary identity for Hilbert space norms (1.17), and (7.15) for the sequence $(u_k - w^{(0)})$. Relation (7.16) follows from the Brezis–Lieb lemma (Corollary 1.3.3) which gives, in our settings,

$$\int_M |u_k|^p dv_g - \int_M |w^{(0)}|^p dv_g - \int_M |u_k - w^{(0)}|^p dv_g \to 0,$$

combined with (7.16) for the sequence $(u_k - w^{(0)})$.

From now on, we assume that $u_k \rightharpoonup 0$.

Step 2. Let us give an iterative construction of sequences $(v_k^{(n)})_{k\in\mathbb{N}}$ in $H^{1,2}(M)$, $n \in \mathbb{N}_0$. We set $v_k^{(0)} = u_k$ and choose $(y_k^{(1)})_{k\in\mathbb{N}}$ so that $\|w^{(1)}\|_{H^{1,2}(M_\infty^{(1)})} \geq \frac{1}{2}\nu^{(u_k)}(\emptyset)$ (cf. Definition 7.4.6).

Assume that we have defined sequences $(v_k^{(0)})_{k\in\mathbb{N}},\dots,(v_k^{(m)})_{k\in\mathbb{N}}$, with the following properties:

- There exists, for a given m, a renamed subsequence of (u_k), sequences $(y_k^{(1)})_{k\in\mathbb{N}}, \dots, (y_k^{(m)})_{k\in\mathbb{N}}$ of points in Y such that $d(y_k^{(\ell)}, y_k^{(n)}) \to \infty$ whenever $\ell \neq n$, with trailing systems $\{(y_{k;i}^{(n)})_{k\in\mathbb{N}}\}_{i\in\mathbb{N}_0}$, defining on a subsequence for each respective $n = 1,\dots,m$, an array of local profiles $\{w_i^{(n)}\}_{i\in\mathbb{N}_0}$ of (the mth extraction of) (u_k), and, consequently, a Riemannian manifold at infinity $M_\infty^{(n)}$ and a global profile $w^{(n)} \in H^{1,2}(M_\infty^{(n)})$. Assume, furthermore, that $\|w^{(n)}\|^2_{H^{1,2}(M_\infty^{(n)})} \geq \frac{1}{2}\nu^{(u_k)}((y_k^{(1)}),\dots,(y_k^{(n-1)}))$, $n = 2,\dots,m$ (cf. Definition 7.4.6). Let $(W_k^{(n)})_{k\in\mathbb{N}}$, $n = 1,\dots,m$, be corresponding elementary concentrations, and define, with the convention that the sum over an empty set of indices equals zero,

$$v_k^{(n)} \overset{\text{def}}{=} u_k - \sum_{\ell=1}^{n} W_k^{(\ell)}, \quad n = 1,\dots m.$$

Under the above assumptions, we construct now a sequence $v_k^{(m+1)}$ that will also satisfy these assumptions. Consider all sequences (y_k) of points in Y such that $d(y_k, y_k^{(\ell)}) \to \infty$ for all $\ell = 1,\dots,m$. We have three complementary cases:

Case 1: for any such sequence, one has $v_k^{(m)} \circ e_{y_k} \rightharpoonup 0$ in $H^{1,2}(\Omega_\rho)$ on a renamed subsequence;

Case 2: there exists a bounded sequence (y_k) of points in Y (so that $d(y_k, y_k^{(\ell)}) \to \infty$ for all $\ell = 1,\dots,m$) such that, on a renamed subsequence, $v_k^{(m)} \circ e_{y_k} \rightharpoonup w \neq 0$;

Case 3: there exists a discrete sequence (y_k) of points in Y such that $d(y_k, y_k^{(\ell)}) \to \infty$ for all $\ell = 1,\dots,m$, and $v_k^{(m)} \circ e_{y_k} \rightharpoonup w \neq 0$.

Case 2 is, in fact, vacuous. Indeed, in this case (y_k) would have a constant subsequence with some value z and $u_k \circ e_z \rightharpoonup w \neq 0$, which contradicts the assumption $u_k \rightharpoonup 0$.

Consider case 1. We prove that in that case $v_k^{(m)} \circ e_{z_k} \rightharpoonup 0$ for any sequence (z_k) in Y. By assumption, we know that it is true if $d(z_k, y_k^{(\ell)}) \to \infty$ for all $\ell = 1,\dots,m$. So let us assume that on a renamed subsequence, $d(z_k, y_k^{(\ell)})$ is bounded for some $\ell \in \{1,\dots m\}$. Then by the definition of the trailing system there exists $i \in \mathbb{N}_0$ such that $z_k = y_{k;i}^{(\ell)}$ on a renamed subsequence. So if $u_k \circ e_{z_k} \rightharpoonup w \neq 0$ then w coincides with the local profile $w_i^{(\ell)}$. Moreover, $d(z_k, y_k^{(n)}) \to \infty$ if $1 \leq n \leq m$ and $n \neq \ell$. So by Lemma 7.4.2, $W_k^{(n)} \circ e_{z_k} \rightharpoonup 0$ if $n \neq \ell$ and $W_k^{(\ell)} \circ e_{z_k} \rightharpoonup w_i$. In consequence, $v_k^{(m)} \circ e_{z_k} \rightharpoonup 0$ Now by Theorem 7.1.4, $v_k^{(m)} \to 0$ in $L^p(M)$, which means that the asymptotic relation (7.14) is proved with a finite sum of elementary concentrations, and we can take $v_k^{(m+1)} = 0$.

Consider now case 3. Now the modulus $\nu^{(u_k)}((y_k^{(1)}),\dots,(y_k^{m})) > 0$ is positive; cf. Definition 7.4.6. We may choose a sequence $y_k^{(m+1)}$, $d(y_k^{(m+1)}, y_k^{(\ell)}) \to \infty$ for all $\ell = 1,\dots,m$, in such a way that the corresponding global profile $w^{(m+1)}$ of (u_k) satisfies

$$\|w^{(m+1)}\|^2_{H^{1,2}(M_\infty^{(m+1)})} \geq \frac{1}{2}\nu^{(u_k)}((y_k^{(1)}),\dots,(y_k^{(m)})). \tag{7.37}$$

Then using the local profiles $w_i^{(m+1)}$, $i \in \mathbb{N}_0$, we may define, for a renamed subsequence, the associated global profile $w^{(m+1)}$ (cf. Proposition 7.3.2), and the corresponding elementary concentration $W_k^{(m+1)}$, and put

$$v_k^{(m+1)} \stackrel{\text{def}}{=} u_k - \sum_{\ell=1}^{m+1} W_k^{(\ell)}.$$

It is easy to see that the sequence $(v_k^{(m+1)})$ has the same properties as $(v_k^{(n)})$, $n = 0,\dots,m$.

Step 3. By Lemma 7.4.5, we have

$$\sum_{n=1}^{m} \|w^{(n)}\|^2_{H^{1,2}(M_\infty^{(n)})} \leq \limsup \|u_k\|^2_{H^{1,2}(M)}$$

for any m, which proves (7.15).

Step 4. In order to prove convergence of the series $\sum_{n=1}^{\infty} W_k^{(n)}$ note first that we may assume without loss of generality that for each $n \in \mathbb{N}$, there exists $r_n > 0$ such that $\operatorname{supp} W_k^{(n)} \subset B(y_k^{(n)}, r_n)$. Indeed, acting like in the proof of Lemma 7.4.5, from the calculations in the proof of Lemma 7.4.4 one can easily see that one can approximate $W_k^{(n)}$ in the $H^{1,2}$-norm by restricting summation in (7.13) to a finite number of terms, with the norm of the remainder bounded by, say, $\varepsilon 2^{-n}$ with a small $\varepsilon > 0$. Then, for any $m \in \mathbb{N}$ one can extract a subsequence $(k_j^{(m)})_{j\in\mathbb{N}}$ of $(k)_{k\in\mathbb{N}}$ such that $d(y_k^{(n)}, y_k^{(\ell)}) > r_n + r_\ell$ whenever $1 \leq \ell < n \leq m$. Then on a diagonal subsequence $(k_m^{(m)})_{m\in\mathbb{N}}$ the elementary concentrations $(W_k^{(n)})_{k=k_m^{(m)}, m\in\mathbb{N}}$ will have pairwise disjoint supports. Together with (7.15), this proves that the convergence is unconditional and uniform with respect to k.

Step 5. Now we prove that $(u_k - \sum_{\ell=1}^{\infty} W_k^{(\ell)}) \circ e_{y_k} \to 0$ in $L^p(M)$ for any sequence y_k in Y.

Let first (y_k) in Y be a bounded sequence. Since it has finitely many values, on each constant subsequence we have $u_k \circ e_y \rightharpoonup 0$ and $W_k^{(\ell)} \circ e_y \rightharpoonup 0$, and thus $(u_k - \sum_{\ell=1}^{\infty} W_k^{(\ell)}) \circ e_{y_k} \rightharpoonup 0$.

Let now (y_k) be a discrete sequence in Y. If there is $\ell \in \mathbb{N}$ such that on a renamed subsequence we have $d(y_k, y_k^{(\ell)})$ is bounded. Then on a renamed subsequence $y_k = y_{k;i}^{(\ell)}$

for some i; cf. Step 2. But then $u_k \circ e_{y_k} \rightharpoonup w_i^{(\ell)}$, $W_k^{(\ell)} \circ e_{y_k} \rightharpoonup w_i^{(\ell)}$ and $W_k^{(n)} \circ e_{y_k} \rightharpoonup 0$ if $n \neq \ell$; cf. Lemma 7.4.2. Thus $(u_k - \sum_{\ell=1}^{\infty} W_k^{(\ell)}) \circ e_{y_k} \rightharpoonup 0$.

Let (y_k) be a discrete sequence in Y, such that $d(y_k, y_k^{(\ell)}) \to \infty$ for any $\ell \in \mathbb{N}_0$. Assume that on a renamed subsequence $(u_k - \sum_{\ell=1}^{\infty} W_k^{(\ell)}) \circ e_{y_k} \rightharpoonup w_0 \neq 0$. Then (y_k) generates a profile w of (u_k) on some manifold ar infinity M_∞ of M that necessarily satisfies $\|w\|_{H^{1,2}(M_\infty)} \leq \nu^{(u_k)}((y_k^{(1)}), \dots, (y_k^{(m)}))$ for any $m \in \mathbb{N}$. By (7.15) and (7.37), we have $\nu^{(u_k)}((y_k^{(1)}), \dots, (y_k^{(m)})) \to 0$ as $m \to \infty$ and, therefore, $w = 0$, which implies $w_0 = 0$. This gives the contradiction.

We conclude that $(u_k - \sum_{\ell=1}^{\infty} W_k^{(\ell)}) \circ e_{y_k} \rightharpoonup 0$ for any sequence (y_k) in Y, and by Theorem 7.1.4 $(u_k - \sum_{\ell=1}^{\infty} W_k^{(\ell)}) \circ e_{y_k} \to 0$ in $L^p(M)$.

Step 6. It was proved in Step 4 that the series of elementary concentration $W_k^{(n)}$ is convergent in $H^{1,2}(M)$. So for any $\varepsilon > 0$ the sum S_k of the elementary concentrations can be approximated by the finite sum S_k^ε, that is,

$$\begin{aligned} \big|\|u_k\|_p - \|S_k^\varepsilon\|_p\big| &\leq \big|\|u_k\|_p - \|S_k\|_p\big| + \|S_k - S_k^\varepsilon\|_p \\ &\leq o(1) + C\|S_k - S_k^\varepsilon\|_{H^{1,2}(M)} \leq C\varepsilon + o(1). \end{aligned} \tag{7.38}$$

Moreover, similar to Step 4, we may assume without loss of generality all $w^{(n)}$ have compact support. In consequence, we may assume that there exists $m \in \mathbb{N}$ such that $w^{(n)} = 0$ for all $n > m$, and that $w^{(n)}$ have compact support if $n \leq m$.

Let us now evaluate $\|S_k^\varepsilon\|_p$. Let us show first that

$$\int_M |W_k^{(n)}|^p \mathrm{d}v_g \to \int_{M_\infty^{(n)}} |w^{(n)}|^p \mathrm{d}v_{\bar{g}^{(n)}}. \tag{7.39}$$

Indeed, omitting for the sake of simplicity the superscript n and taking into account that $w_i \circ e_{y_{k;i}}^{-1} \circ e_{y_{k;j}} \to w_j$, $e_{y_{k;j}}^{-1} \circ e_{y_{k;i}} \to \psi_{ji}$, and $\chi_{y_{k;j}} \circ e_{y_{k;j}} \to \chi_j$ as in the proof of Lemma 7.4.1, we have

$$\begin{aligned} \int_M |W_k|^p \mathrm{d}v_g &= \int_M \Big|\sum_{i \in \mathbb{N}_0} \chi_{y_{k;i}} w_i \circ e_{y_{k;i}}\Big|^p \mathrm{d}v_g \\ &= \sum_{j \in \mathbb{N}_0} \int_{\Omega_p} \chi_{y_{k;j}} \circ e_{y_{k;j}} \Big|\sum_{i \in \mathbb{N}_0} \chi_{y_{k;i}} w_i \circ e_{y_{k;i}}\Big|^p \circ e_{y_{k;j}}^{-1} \sqrt{\eta_{k;j}} \mathrm{d}\xi \\ &= \sum_{j \in \mathbb{N}_0} \int_{\Omega_p} (\chi_j + o(1)) \Big|\sum_{i \in \mathbb{N}_0} \chi_{y_{k;i}} \circ e_{y_{k;j}}^{-1} (w_j + o(1))\Big|^p \sqrt{\bar{g}_j + o(1)} \mathrm{d}\xi \\ &\to \int_{M_\infty} |w|^p \mathrm{d}v_{\bar{g}}. \end{aligned}$$

Note that the notation $o(1)$ above refers to functions vanishing in the sense of C^∞ and that all infinite sums contain uniformly finitely many nonzero terms.

Now, for all k sufficiently large, all elementary concentrations $W_k^{(n)}$ in the sum S_k^ε have pairwise disjoint supports, and since $\ell^1 \hookrightarrow \ell^{\frac{p}{2}}$, taking into account (7.15), we have

$$\Big(\sum_{n\geq\nu}\int_{M_\infty^{(n)}}|w^{(n)}|^p dv_{\bar{g}^{(n)}}\Big)^{\frac{2}{p}} \leq \sum_{n\geq\nu}\Big(\int_{M_\infty^{(n)}}|w^{(n)}|^p dv_{\bar{g}^{(n)}}\Big)^{\frac{2}{p}}$$
$$\leq \sum_{n\geq\nu} C\|w^{(n)}\|^2_{H^{1,2}(M_\infty^{(n)})} \to 0 \quad \text{as } \nu\to\infty.$$

Then (7.16) follows from (7.14), as we may now consider a finite sum; assume, by density, that $w^{(n)}$ have compact support, and apply (7.39). This completes the proof of Theorem 7.3.5. □

7.6 Local and global profile decompositions on periodic manifolds

Let M be now a smooth connected complete Riemannian manifold, periodic (cocompact) relative to a subgroup G of its isometry group, that is, we assume that there exists an open bounded set $\mathcal{O}$ such that $\bigcup_{\eta\in G}\eta\mathcal{O} = M$. Without loss of generality, we may assume that $\mathcal{O}$ is a geodesic ball. Periodic manifolds are obviously of bounded geometry. It is then natural to ask if Theorem 7.3.5 yields Corollary 5.1.7 with the manifolds $M_\infty^{(n)}$ isometric to M.

Theorem 7.6.1. *Let M be a smooth connected N-dimensional Riemannian manifold, let $\rho \in (0, \frac{r(M)}{8})$ and $z \in M$, and assume that there exists a discrete countable subgroup G of isometries on M such that $\{B(\eta z,\rho)\}_{\eta\in G}$ covers M with a uniformly finite multiplicity. Then:*

(i) *one can choose the construction parameters of manifolds $M_\infty^{(n)}$, so that they will coincide, up to isometry, with M, and*

(ii) *there exist sequences $(\eta_k^{(n)})_{k\in\mathbb{N}}$, of elements in G, and functions $w^{(n)} \in H^{1,2}(M)$, $n \in \mathbb{N}$, such that the sequences $([\eta_k^{(\ell)}]^{-1}\eta_k^{(n)})_{k\in\mathbb{N}}$ are discrete whenever $\ell\neq n$, $u_k\circ\eta_k^{(n)} \rightharpoonup w^{(n)}$ in $H^{1,2}(M)$, $n\in\mathbb{N}$, and*

$$W_k^{(n)} = w^{(n)} \circ [\eta_k^{(n)}]^{-1}.$$

Proof. 1. Let us repeat the construction of the manifold at infinity relative to a sequence (y_k) in $Y = \{\eta z\}_{\eta\in G}$. Fix a sequence of elements $\zeta_i \in G$, $\zeta_0 = \mathrm{id}$, such that $d(\zeta_{i+1}z,z)\geq d(\zeta_i z,z)$, $i\in\mathbb{N}_0$, and define the ith trailing sequence of (y_k) by $y_{k;i} \stackrel{\text{def}}{=} \eta_k\zeta_i z$, $k\in\mathbb{N}$. Recall that the normal coordinates at the points $y\in Y$ were defined as $\exp_y$ up to an arbitrarily fixed isometry on T_yM. For the present construction, we set them specifically as $e_{\eta z} \stackrel{\text{def}}{=} \eta\circ e_z$. Under such choice, the transition maps of $M_\infty^{(y_{k;i})}$ are characterized by elements of the group G:

$$\psi_{ij} = \lim_{k\to\infty} e_{y_{k;i}}^{-1} \circ e_{y_{k;j}} = \lim e_z^{-1} \circ [\eta_k\zeta_i]^{-1}\eta_k\zeta_j \circ e_z = e_z^{-1} \circ \zeta_i^{-1}\zeta_j \circ e_z,$$

and the sequences above are in fact constant with respect to k. Consequently, the transition maps ψ_{ij} of the manifold $M_\infty^{(y_{k;i})}$ are $e_z^{-1} \circ \zeta_i^{-1}\zeta_j \circ e_z$ – same as of M itself. In other words, all the gluing data for $M_\infty^{(y_{k;i})}$ are taken from M, which suggests, since Theorem 10.3.6 is based on a suitable list of properties of charts of a manifold that will allow its reconstruction that $M_\infty^{(y_{k;i})}$ is isometric to M. We will, however, apply Corollary 10.3.8 formally, as follows.

Manifold $M_\infty^{(y_{k;i})}$ has an atlas $\{(\varphi_i(\Omega_\rho), \varphi_i^{-1})\}_{i\in\mathbb{N}_0}$ with transition maps $\varphi_i^{-1}\varphi_j = e_z^{-1} \circ \zeta_i^{-1}\zeta_j \circ e_z$, while manifold M has an atlas, enumerated by $\zeta_i \in G$, $\{(B(\zeta_i(z), \rho), e_z^{-1} \circ \zeta_i^{-1})\}_{i\in\mathbb{N}_0}$ with the same transition maps as $M_\infty^{(y_{k;i})}$. Let $T_i \overset{\text{def}}{=} \zeta_i \circ e_z \circ \varphi_i^{-1} : \varphi_i(\Omega_\rho) \to M$, $i \in \mathbb{N}_0$, and note that this defines a smooth map $T : M_\infty^{(y_{k;i})} \to M$, since the values of T_i are consistent on intersections of sets $\varphi_i(\Omega_\rho)$:

$$\zeta_i \circ e_z \circ \varphi_i^{-1} \circ [\zeta_j \circ e_z \circ \varphi_j^{-1}]^{-1} = \zeta_i \circ e_z \circ \psi_{ij} \circ [\zeta_j \circ e_z]^{-1} \tag{7.40}$$

$$= \zeta_i \circ e_z \circ e_z^{-1} \circ \zeta_i^{-1}\zeta_j \circ e_z \circ e_z^{-1} \circ \zeta_j^{-1} = \text{id}. \tag{7.41}$$

Furthermore, T is a diffeomorphism with $T^{-1} = \varphi_i \circ e_z^{-1} \circ \zeta_i^{-1}$, consistently defined on $B(\eta_i z, \rho)$, $i \in \mathbb{N}_0$. Note that (7.11) on $M_\infty^{(y_{k;i})}$ holds because it holds on M with the same transition map for every k, so the Riemannian metric on $M_\infty^{(y_{k;i})}$ in the normal coordinates coincides with the Riemannian metric on M. In what follows, we will identify $M_\infty^{(y_{k;i})}$ as M.

2. Let now (u_k) be a bounded sequence in $H^{1,2}(M)$ and note that its local profile associated with the sequence $(\eta_k\zeta_i)_{k\in\mathbb{N}}$ is given by

$$w_i = \text{w-lim}\, u_k \circ (\eta_k\zeta_i) \circ e_z,$$

and the global profile is by definition $w = w_i \circ \varphi_i^{-1} = w_i \circ e_z^{-1} \circ \zeta_i^{-1} = \text{w-lim}\, u_k \circ \eta_k$, which coincides with the profile of (u_k) as defined in Corollary 5.1.7, relative to the sequence (η_k). Consider now the local concentration defined by the array $\{w_i\}_{i\in\mathbb{N}_0}$ of local profiles:

$$\begin{aligned} W_k &= \sum_{i\in\mathbb{N}_0} \chi_{\eta_k\zeta_i z} w_i \circ e_{y_{k;i}}^{-1} = \sum_{i\in\mathbb{N}_0} \chi_{\eta_k\zeta_i z} w_i \circ e_z^{-1} \circ \zeta_i^{-1} \circ \eta_k^{-1} \\ &= \sum_{i\in\mathbb{N}_0} \chi_{\eta_k\zeta_i z} w \circ \eta_k^{-1} = w \circ \eta_k^{-1}, \end{aligned}$$

which completes the proof. □

7.7 Cocompactness of the limiting Sobolev embedding

We will now proceed with analysis of defect of compactness for the limiting Sobolev embedding $H^{1,2}(M) \hookrightarrow L^{2^*}$. The following property, like Theorem 7.1.4, is similar to cocompactness.

Theorem 7.7.1 (Vanishing lemma – the critical case). *Assume that M is a smooth N-dimensional Riemannian manifold of bounded geometry. If (u_k) is a bounded sequence in $H^{1,2}(M)$, $u_k \to 0$ in $L^p(M)$ for some $p \in (2, 2^*)$, such that for any sequence (y_k) in M, and any sequence or positive numbers (t_k), $t_k \to 0$,*

$$t_k^{\frac{N-2}{2}} u_k \circ e_{y_k}(t_k \xi) \longrightarrow 0 \quad a.e.\ in\ \mathbb{R}^N. \tag{7.42}$$

Then $u_k \to 0$ in $L^{2^*}(M)$.

Proof. Step 1. For any $u \in H^{1,2}(M)$, the following holds:

$$\|u\|_{2^*}^{2^*} \le C\|u\|_{H^{1,2}}^2 \sup_{j \in 2^{\frac{N-2}{2}\mathbb{Z}}} \Bigg(\int\limits_{j \le |u(x)| \le 2^{\frac{N-2}{2}} j} |u|^{2^*} \mathrm{d}v_g \Bigg)^{\frac{2}{N}}. \tag{7.43}$$

Indeed, let $\chi \in C_0^1(2^{-\frac{N-2}{2}}, 2^{N-2})$, extended by zero to $[0, \infty)$ be such that $\chi(s) \in [0, 1]$ for all s and $\chi(s) = 1$ whenever $s \in [1, 2^{\frac{N-2}{2}}]$. Let $\chi_j(s) = j\chi(j^{-1}s)$, $j \in 2^{\frac{N-2}{2}\mathbb{Z}}$.

Applying Sobolev inequality to $\chi_j(|u|)$, we get

$$\Bigg(\int\limits_{j \le |u(x)| \le 2^{\frac{N-2}{2}} j} |u|^{2^*} \mathrm{d}v_g \Bigg)^{2/2^*} \le C \int\limits_{2^{-\frac{N-2}{2}} j \le |u(x)| \le 2^{N-2} j} (g_x(du, du) + |u|^2) \mathrm{d}v_g,$$

from which we have

$$\begin{aligned} &\int\limits_{j \le |u(x)| \le 2^{\frac{N-2}{2}} j} |u|^{2^*} \mathrm{d}v_g \\ &\quad \le C \int\limits_{2^{-\frac{N-2}{2}} j \le |u(x)| \le 2^{N-2} j} (g_x(du, du) + |u|^2) \mathrm{d}v_g \Bigg(\int\limits_{j \le |u(x)| \le 2^{\frac{N-2}{2}} j} |u|^{2^*} \mathrm{d}v_g \Bigg)^{1 - \frac{2}{2^*}}. \end{aligned}$$

Adding the inequalities – and replacing the last term by its upper bound – over $j \in 2^{\frac{N-2}{2}\mathbb{Z}}$, we get (7.43).

Step 2. Let us now consider (7.43) with $u = u_k$. Choose $j_k \in 2^{\frac{N-2}{2}\mathbb{Z}}$ that satisfy

$$\sup_{j \in 2^{\frac{N-2}{2}\mathbb{Z}}} \int\limits_{j \le |u_k(x)| \le 2^{\frac{N-2}{2}} j} |u|_k^{2^*} \mathrm{d}v_g \le 2 \int\limits_{j_k \le |u_k(x)| \le 2^{\frac{N-2}{2}} j_k} |u|^{2^*} \mathrm{d}v_g. \tag{7.44}$$

Then we have from (7.43),

$$\|u_k\|_{2^*} \le C\Big(\int\limits_{j_k \le |u(_k x)| \le 2^{\frac{N-2}{2}} j_k} |u_k|^{2^*} \,dv_g\Big)^{\frac{2}{2^* N}}. \tag{7.45}$$

Without loss of generality, we may consider two following cases: (a) $j_k \le L$ for all $k, L \in \mathbb{Z}$, and (b) $j_k \to +\infty$. In the case (a) we have from (7.45) with any small $\varepsilon > 0$,

$$\|u_k\|_{2^*} \le C\Big(\int\limits_{j_k \le |u_k(x)| \le 2^{\frac{N-2}{2}} j_k} |u_k|^{2^*} \,dv_g\Big)^{\frac{2}{N2^*}} \le C\Big(L^{2^*-p}\int\limits_M |u_k|^p \,dv_g\Big)^{\frac{2}{N2^*}}$$

which vanishes by assumption.

Step 3. Now consider the case (b), that is, $j_k \to \infty$. Let $Y_k \subset M$, $k \in \mathbb{N}$, be a $(t_k\varepsilon, t_k r)$-discretization of M, $\varepsilon \in (0, r)$, $t_k r < r(M)$, so that the collection of balls $\{B(y, t_k r)\}_{y\in Y_k}$ with $t_k = j_k^{-\frac{2}{N-2}}$, is a uniformly finite covering of M. Note that the multiplicity of this covering is also uniformly finite with respect to $k \in \mathbb{N}$. Let $D_k = \{x \in M : |u_k(x)| \in [j_k, 2^{\frac{N-2}{2}} j_k]\}$ and $D_k' = \{x \in M : |u_k(x)| \in [2^{-\frac{N-2}{2}} j_k, 2^{N-2} j_k]\}$.

Applying scaled Sobolev inequality to $\chi_{j_k}(|u_k|)$ on the geodesic balls $B(y, t_k r)$, we have

$$\Big(\int\limits_{B(y,t_k r)\cap D_k} |u_k|^{2^*} \,dv_g\Big)^{2/2^*} \le C \int\limits_{B(y,t_k r)\cap D_k'} (g_x(du_k, du_k) + t_k^{-2}|u_k|^2)\,dv_g,$$

with some C independent of k.

Since the integration domain in the right-hand side is a subset of D_k', we have $t_k^{-2}|u_k|^2 \le C|u_k|^{2^*}$ uniformly in k, and thus

$$\int\limits_{B(y,t_k r)\cap D_k} |u_k|^{2^*} \,dv_g$$
$$\le \int\limits_{B(y,t_k r)\cap D_k'} (g_x(du_k, du_k) + |u_k|^{2^*})\,dv_g \Big(\int\limits_{B(y,t_k r)\cap D_k} |u_k|^{2^*} \,dv_g\Big)^{\frac{2}{N}}.$$

Let us add these inequalities – while replacing the second term in the right-hand side by its upper bound – over $y \in Y_k$:

$$\int\limits_{D_k} |u_k|^{2^*} \,dv_g \le C \sup_{y\in Y_k}\Big(\int\limits_{B(y,t_k r)\cap D_k} |u_k|^{2^*} \,dv_g\Big)^{\frac{2}{N}}. \tag{7.46}$$

Choosing points $y_k \in Y_k$ so that

$$\sup_{y\in Y_k} \int\limits_{B(y,t_k r)\cap D_k} |u_k|^{2^*} \,dv_g \le 2 \int\limits_{B(y_k,t_k r)\cap D_k} |u_k|^{2^*} \,dv_g,$$

we have from (7.45) and (7.46),

$$\|u_k\|_{2^*} \le C\Big(\int_{B(y_k,t_kr)\cap D_k} |u_k|^{2^*}\,\mathrm{d}v_g\Big)^{\frac{4}{2^*N^2}}.$$

Using the geodesic map e_{y_k} to change the variables from a small ball on M to small ball on $\mathbb{R}^N$, noting that the Jacobian is uniformly bounded with respect to k since M is of bounded geometry, and setting $\Delta_k \overset{\text{def}}{=} \{\xi \in \Omega_r : |u_k \circ \exp_{y_k}(\xi)| \in (j_k, 2^{\frac{N-2}{2}} j_k)\}$, we have, for all k sufficiently large,

$$\|u_k\|_{2^*} \le C\Big(\int_{\Omega_{t_kr}\cap\Delta_k} |u_k \circ e_{y_k}(\xi)|^{2^*}\,\mathrm{d}\xi\Big)^{\frac{4}{2^*N^2}}.$$

Let us change the variables again, $\xi \overset{\text{def}}{=} t_k\eta$, $\eta \in \Omega_r$:

$$\|u_k\|_{2^*} \le C\Big(\int_{\{\eta\in\Omega_r : |u_k\circ e_{y_k}(t_k\eta)|\in[j_k,2^{\frac{N-2}{2}}j_k]\}} |t_k^{\frac{N-2}{2}}\, u_k \circ e_{y_k}(t_k\eta)|^{2^*}\,\mathrm{d}\eta\Big)^{\frac{4}{2^*N^2}},$$

and note that the expression under integral is bounded by the constant 2^N and vanishes almost everywhere by assumption. Therefore, by Lebesgue dominated convergence theorem, the right-hand side above vanishes, and the theorem is proved. □

Remark 7.7.2. It is easy to show that (7.42) holds for all $t_k > 0$ if it holds for all $t_k \in a^{\mathbb{Z}}$, with some $a > 1$.

7.8 Profile decomposition for sequences vanishing in L^p, $p < 2^*$

In this section, we provide a profile decomposition for bounded sequences in $H^{1,2}(M)$ that vanish in $L^p(M)$ for some $p \in (2, 2^*)$. This profile decomposition consists of a sum of concentrating bubbles and a remainder vanishing in $L^{p^*}(M)$. This allows to take a profile decomposition of Theorem 7.3.5 whose remainder vanishes in $L^p(M)$, and further expand this remainder into bubbles (that still vanish in $L^p(M)$, $2 < p < 2^*$) with a sharper remainder that vanishes in $L^p(M)$, $2 < p \le 2^*$. We will start with a characterization of decoupling of bubbles involved in our profile decomposition. We remind that we call two sequences (U_k) and (V_k) in a Hilbert space asymptotically orthogonal, if $\langle U_k, V_k\rangle \to 0$. Let us fix $\chi \in C_0^\infty(\Omega_r)$ such that $\chi(\xi) = 1$ whenever $|\xi| \le \frac{r}{2}$, extended by zero to a function on $\mathbb{R}^N$.

Lemma 7.8.1. *Let:*

$$S_k u \overset{\text{def}}{=} 2^{\frac{N-2}{2}j_k} \chi \circ e_{y_k}^{-1}\, u(2^{j_k} e_{y_k}^{-1}\cdot), \quad k \in \mathbb{N}, u \in \dot{H}^{1,2}(\mathbb{R}^N),$$

$$T_k u \overset{\text{def}}{=} 2^{\frac{N-2}{2}\ell_k} \chi \circ e_{z_k}^{-1} u(2^{\ell_k} e_{z_k}^{-1} \cdot), \quad k \in \mathbb{N}, u \in \dot{H}^{1,2}(\mathbb{R}^N),$$

where $j_k, \ell_k \in \mathbb{N}$, $j_k, \ell_k \to \infty$ and $y_k, z_k \in M$, understanding the expressions in the respective right hand sides as functions in $H^{1,2}(M)$ vanishing outside of $B(y_k, r)$. Then (S_k) and (T_k) are bounded sequences of continuous operators from $\dot{H}^{1,2}(\mathbb{R}^N)$ to $H^{1,2}(M)$.

Sequences $(S_k v)$ and $(T_k w)$ are asymptotically orthogonal in $H^{1,2}(M)$ for every v and w in $\dot{H}^{1,2}(\mathbb{R}^N)$ if and only if the following condition holds:

$$|\ell_k - j_k| + (2^{\ell_k} + 2^{j_k}) d(y_k, z_k) \to \infty \tag{7.47}$$

as $k \to \infty$.

Furthermore, given $w \in \dot{H}^{1,2}(\mathbb{R}^N)$, one has $\langle S_k v, T_k w \rangle \to 0$ for every $v \in \dot{H}^{1,2}(\mathbb{R}^N)$ if and only if

$$2^{-j_k \frac{N-2}{2}} W_k \circ e_{y_k}(2^{-j_k}\xi) \to 0 \quad a.e. \text{ in } \mathbb{R}^N. \tag{7.48}$$

Proof. 1. Boundedness of sequences (S_k) and (T_k) follows from the bounded geometry of M.

2. Note that $\int_M |S_k v)|^2 dv_g \to 0$ and $\int_M |T_k w|^2 dv_g \to 0$, so asymptotic orthogonality of $S_k v$ and $T_k w$ is equivalent to $\int_M g_x(d(S_k v), d(T_k w)) dv_g \to 0$. By density of $C_0^\infty(\mathbb{R}^N)$ in $\dot{H}^{1,2}(\mathbb{R}^N)$, we may assume without loss of generality that $v, w \in C_0^\infty(\mathbb{R}^N)$.

2. *Sufficiency of (7.47):* First, note that supports of $S_k v$ and $T_k w$ are contained in $B(y_k, r) \cap e_{y_k}(2^{-j_k} \operatorname{supp} v)$ and $B(z_k, r) \cap e_{z_k}(2^{-\ell_k} \operatorname{supp} w)$, respectively. Thus, if on some renamed subsequence $\inf d(y_k, z_k) > 0$, then the supports of $S_k v$ and $T_k w$ are disjoint for large k and the asymptotic orthogonality follows. Hence we assume in the rest of the proof that $d(y_k, z_k) \to 0$ as $k \to \infty$.

Support of $g_x(d(S_k v, d(T_k w)$ is contained in $B(z_k, r)$, so we can evaluate the integral under the coordinate map e_{z_k}:

$$\int_M g_x(d(S_k v), d(T_k w)) dv_g = \int_{\Omega_r} \sum_{\alpha,\beta=1}^N g^{\alpha\beta}(e_{z_k}(\xi)) \partial_\alpha(S_k v \circ e_{z_k}) \partial_\beta(T_k w \circ e_{z_k}) \sqrt{g(e_{z_k}(\xi))} d\xi.$$

Set $j_k - \ell_k = m_k$, $e_{y_k}^{-1} \circ e_{z_k} = \psi_k$. Setting the new variable $\eta = 2^{\ell_k}\xi$, and taking into account that $g^{\alpha\beta}(z_k) = \delta_{\alpha\beta}$ and $g(z_k) = 1$, the above expression can be written as

$$\begin{aligned}
&\int_M g_x(d(S_k v), d(T_k w)) dv_g \\
&= \int_{B(z_k,r) \cap B(y_k,r)} g_x(d(S_k v), d(T_k w)) dv_g \\
&= 2^{\frac{N-2}{2} m_k} \int_{D_k} \sum_{\alpha,\beta=1}^N g^{\alpha\beta}(e_{z_k}(\xi)) \partial_\alpha(\chi \circ \psi_k(\xi) v(2^{j_k} \psi_k(\xi))) \partial_\beta(\chi(\xi) w(2^{\ell_k}\xi)) \sqrt{g(e_{z_k}(\xi))} d\xi
\end{aligned}$$

$$= 2^{\frac{N}{2}m_k} \int\limits_{2^{\ell_k}D_k \cap \operatorname{supp} w} [\nabla w(\eta)\psi_k'(2^{-\ell_k}\eta) \cdot \nabla v(2^{j_k}\psi_k(2^{-\ell_k}\eta)) + o(1)]d\eta,$$

where $D_k = e_{z_k}^{-1}(B(z_k, r) \cap B(y_k, r))$, ψ_k' is the $N \times N$-matrix derivative of ψ_k. Given that M is a manifold of bounded geometry, may assume (with reference to the Arzela–Ascoli theorem) without loss of generality that ψ_k and its derivatives will converge locally uniformly in Ω_r to some function $\psi \in C^\infty(\Omega_r \to \overline{\Omega_r})$ and its respective derivatives.

Since (7.47) holds, it suffices to consider two cases:

Case 1: $|\ell_k - j_k| \to \infty$.

Without loss of generality assume that $j_k - \ell_k = m_k \to -\infty$ and taking into account that ψ_k is bounded and $v, w \in C_0^\infty(\mathbb{R}^N)$, we get

$$\left| \int\limits_M g_x(d(S_k v), d(T_k w)) dv_g \right| \le C 2^{\frac{N}{2}m_k} \to 0.$$

Case 2: $|\ell_k - j_k|$ is bounded and $(2^{\ell_k} + 2^{j_k})d(y_k, z_k) \to \infty$.

For $\eta \in 2^{\ell_k}D_k \cap \operatorname{supp} w$, one gets

$$2^{j_k}\psi_k(2^{-\ell_k}\eta) = 2^{j_k}\psi_k(0) + 2^{j_k}\psi_k'(0)2^{-\ell_k}\eta + O(2^{j_k - 2\ell_k}).$$

Taking into account that m_k is bounded and that $|\psi_k(0)| = d(y_k, z_k)$ (since distance from the origin in Ω_r is preserved by the geodesic map), we have

$$|2^{j_k}\psi_k(2^{-\ell_k}\eta)| \to \infty, \quad \text{as } k \to \infty$$

(with uniform convergence), and hence $dv(2^{j_k}\psi_k(2^{-\ell_k}\eta)) = 0$ for all $\eta \in \operatorname{supp} w$ as long as k is sufficiently large, and the asymptotic orthogonality follows.

3. *Necessity of* (7.47): If (7.47) if false, then, on a renamed subsequence, $j_k - \ell_k$ is a constant sequence with some value $m \in \mathbb{Z}$, while $2^{j_k}d(y_k, z_k) = |2^{j_k}\psi_k(0)|$ stays bounded. Hence extracting a further subsequence we may assume $2^{j_k}\psi_k(0) \to \eta_0 \in \mathbb{R}^N$. Repeating calculations in the proof of sufficiency we get

$$\int\limits_M g_x(d(S_k v), d(T_k w)) dv_g \to 2^{\frac{N}{2}m} \int\limits_{\mathbb{R}^N} \nabla w(\eta) \cdot \psi'(0) \nabla v(\eta_0 + 2^m \psi'(0)\eta)\, d\eta.$$

Since $\psi'(0) \neq 0$ by properties of the geodesic map, the above expression will be nonzero with a suitable v and $w(\eta) = v(\eta_0 + 2^m\psi'(0)\eta)$.

4. Finally, representing the scalar product of $H^{1,2}(M)$ under the exponential map at y_k,

$$\langle S_k v, T_k w \rangle = \int\limits_{\mathbb{R}^N} \nabla 2^{-j_k \frac{N-2}{2}} (T_k w) \circ e_{y_k}(2^{-j_k}\xi) \cdot \nabla v(\xi) d\xi + o(1),$$

which proves that (7.48) is equivalent to asymptotic orthogonality of $S_k v$ and $T_k w$ for all v. □

Theorem 7.8.2. *Assume that M is a smooth N-dimensional Riemannian manifold of bounded geometry and let $p \in (2, 2^*)$. For any bounded sequence (u_k) in $H^{1,2}(M)$ that vanishes in $L^p(M)$, there exist sequences $(y_k^{(n)})_{k\in\mathbb{N}}$ in M and $(j_k^{(n)})_{k\in\mathbb{N}}$ in $\mathbb{N}$, $j_k^{(n)} \to +\infty$, as well as functions $w^{(n)} \in \dot{H}^{1,2}(\mathbb{R}^N)$, $n \in \mathbb{N}$, such that, for a renamed subsequence,*

$$2^{-j_k^{(n)}\frac{N-2}{2}} u_k \circ e_{y_k^{(n)}}(2^{-j_k^{(n)}}\xi) \to w^{(n)}(\xi) \quad a.e.\ in\ \mathbb{R}^N; \tag{7.49}$$

(AO) *Condition (7.47) holds with $j_k = j_k^{(m)}$, $y_k = y_k^{(m)}$, $\ell_k = j_k^{(n)}$ and $z_k = y_k^{(n)}$ whenever $m \neq n$;*

The series $\mathbf{S}_k \stackrel{\text{def}}{=} \sum_{n\in\mathbb{N}} W_k^{(n)}$, where

$$W_k^{(n)}(x) = 2^{j_k^{(n)}\frac{N-2}{2}} \chi \circ e_{y_k^{(n)}}^{-1}(x)\, w^{(n)}(2^{j_k^{(n)}} e_{y_k^{(n)}}^{-1}(x)), \quad x \in M, \tag{7.50}$$

converges in $H^{1,2}(M)$ unconditionally and uniformly with respect to k,

$$u_k - \mathbf{S}_k \to 0 \quad in\ L^q(M)\ 2 < q \le 2^*, \tag{7.51}$$

and

$$\sum_{n\in\mathbb{N}} \int_{\mathbb{R}^N} |\nabla w^{(n)}|^2 d\xi \le \liminf \int_M g_x(du_k, du_k) dv_g. \tag{7.52}$$

Proof. The proof is largely repetitive of the proofs for profile decompositions earlier in this book, Theorem 7.3.5 in particular, so we give it in an abbreviated form.

1. Consider arbitrary sequences $(j_k^{(1)})$ in M $j_k^{(1)} \to \infty$, and $(y_k^{(1)})$ in $\mathbb{N}$. By the bounded geometry of M, for any $a \in (0, r(M))$, functions $2^{-j_k^{(1)}\frac{N-2}{2}} u_k \circ e_{y_k^{(1)}}(2^{-j_k^{(1)}}\cdot)$ have a uniformly (in $k \in \mathbb{N}$) bounded $H^{1,2}(\Omega_a)$-norm, and thus their sequence is weakly convergent in $H^{1,2}(\Omega_a)$. We may infer by diagonalization that there exists a renamed subsequence of (u_k) such that $2^{-j_k^{(1)}\frac{N-2}{2}} u_k \circ e_{y_k^{(1)}}(2^{-j_k^{(1)}}\cdot)$ converges to some $w^{(1)}$ almost everywhere on $\mathbb{R}^N$. Since M is of bounded geometry, we have $\|\nabla w^{(1)}\|_2^2 \le \limsup C \int_{B(y_k^{(1)},r)} g_x(du_k, du_k) dv_g < \infty$ with the constant C independent of the sequence $(y_k^{(n)})_{k\in\mathbb{N}}$.

Let Ξ_1 be a set of all $w \in \dot{H}^{1,2}(M)$ such that there exist $(j_k^{(1)})$ in $\mathbb{N}$, $j_k^{(1)} \to \infty$, and $(y_k^{(1)})$ in M, such that

$$2^{-j_k^{(1)}\frac{N-2}{2}} u_k \circ e_{y_k^{(1)}}(2^{-j_k^{(1)}}) \to w \quad \text{a.e. in } \mathbb{R}^N,$$

define

$$\beta_1 \stackrel{\text{def}}{=} \sup_{w\in\Xi_1} \|w\|_{\dot{H}^{1,2}}$$

and fix an element $w^{(1)} \in \Xi_1$ and corresponding sequences $j_k^{(1)}$ and $(y_k^{(1)})$ so that $\|w^{(1)}\|_{\dot H^{1,2}} \geq \frac{1}{2}\beta_1$. If $\beta_1 = 0$, then by Theorem 7.7.1 $u_k \to 0$ in $L^{2^*}(M)$ and the theorem is proved. We consider therefore the case $\beta_1 > 0$.

2. We will now construct by iteration sequences $(j_k^{(n)})$ in $\mathbb{N}$, $j_k^{(n)} \to \infty$ and $(y_k^{(n)})$ in M, functions $w^{(n)}$ and numbers β_n, $n \in \mathbb{N}$. Given $\nu \in \mathbb{N}$, assume that for every $n = 1, \dots, \nu$ we already have constructed the objects above with the following properties:

(i) For each $n = 1, \dots, \nu$, there exist a renamed subsequence of (u_k) such that $w^{(n)} \in \dot H^{1,2}(\mathbb{R}^N)$ satisfies (7.49).

(ii) (AO) holds for all $m, n = 1, \dots, \nu$, $m \neq n$.

(iii) If Ξ_n is a set of all $w \in \dot H^{1,2}(M)$ such that

$$2^{-j_k^{(n)}\frac{N-2}{2}} u_k \circ e_{y_k^{(n)}}(2^{-j_k^{(n)}}) \to w \quad \text{a.e. in } \mathbb{R}^N$$

for some choice of $j_k^{(n)} \to \infty$ and $(y_k^{(n)})$ in M satisfying (AO) for $m < n = 1, \dots, \nu$, and if

$$\beta_n \overset{\text{def}}{=} \sup_{w \in \Xi_n} \|w\|_{\dot H^{1,2}}$$

then $w^{(n)} \in \Xi_n$ and corresponding sequences $j_k^{(n)}$ and $(y_k^{(n)})$ satisfy $\|w^{(n)}\|_{\dot H^{1,2}} \geq \frac{1}{2}\beta_n$.

Let $\mathbf{S}_k^{(\nu)} \overset{\text{def}}{=} \sum_{n=1}^{\nu} W_k^{(n)}$ and let $v_k^{(\nu)} \overset{\text{def}}{=} u_k - \mathbf{S}_k^{(\nu)}$. Similar to Step 1, as in the Step 1, we have a renamed subsequence of (u_k) such that $2^{-j_k^{(\nu+1)}\frac{N-2}{2}} u_k \circ e_{y_k^{(\nu+1)}}(2^{-j_k^{(\nu+1)}} \cdot)$ converges almost everywhere on $\mathbb{R}^N$ to some $w^{(\nu+1)} \in \dot H^{1,2}(\mathbb{R}^N)$, for some sequences $(j_k^{(\nu+1)})$ in $\mathbb{N}$, $j_k^{(\nu+1)} \to \infty$, and $(y_k^{(\nu+1)})$ in M. As in Step 1, we consider the class $\Xi_{\nu+1}$ of all such weak limits, and fix $w^{(\nu+1)}$ and corresponding $j_k^{(\nu+1)}$ and $(y_k^{(\nu+1)})$ so that $\|w^{(\nu+1)}\|_{\dot H^{1,2}} \geq \frac{1}{2}\beta_{\nu+1}$.

3. For every $n \leq \nu$, we have now

$$2^{-j_k^{(n)}\frac{N-2}{2}} W_k^{(n)} \circ e_{y_k^{(n)}}(2^{-j_k^{(n)}}\xi) = \chi(2^{-j_k^{(n)}}\xi) w^{(n)}(\xi) \to w^{(n)}(\xi) \quad \text{a.e. in } \mathbb{R}^N, \tag{7.53}$$

for each $n' \leq \nu$, $n' \neq n$, we have, by (AO) and Lemma 7.8.1,

$$2^{-j_k^{(n)}\frac{N-2}{2}} W_k^{(n')} \circ e_{y_k^{(n)}}(2^{-j_k^{(n)}}\xi) \to 0 \quad \text{a.e. in } \mathbb{R}^N,$$

so for every $n \leq \nu$,

$$2^{-j_k^{(n)}\frac{N-2}{2}} S_k^{(\nu)} \circ e_{y_k^{(n)}}(2^{-j_k^{(n)}}\xi) \to w^{(n)}(\xi) \quad \text{a.e. in } \mathbb{R}^N \tag{7.54}$$

and, therefore,

$$\begin{aligned} &2^{-j_k^{(n)}\frac{N-2}{2}} v_k^{(\nu)} \circ e_{y_k^{(n)}}(2^{-j_k^{(n)}}\xi) \\ &= 2^{-j_k^{(n)}\frac{N-2}{2}} (u_k - S_k^{(\nu)}) \circ e_{y_k^{(n)}}(2^{-j_k^{(n)}}\xi) \to 0 \quad \text{a.e. in } \mathbb{R}^N. \end{aligned} \tag{7.55}$$

4. Let us show that (AO) is satisfied for $m = 1, \ldots, \nu$ and $n = \nu + 1$ (or vice versa). Once we show this, we will have completed the construction of $(y_k^{(n)})_{k\in\mathbb{N}}$ in M, $(j_k^{(n)})_{k\in N}$ in $\mathbb{N}$, $j_k^{(n)} \to +\infty$, and $w^{(n)} \in \dot{H}^{1,2}(\mathbb{R}^N)$, $n \in \mathbb{N}$, such that, on a renamed subsequence of (u_k), condition (AO) is satisfied for all $n \in \mathbb{N}$. If $w^{(\nu+1)} = 0$, then necessarily $\beta_{\nu+1} = 0$, and we are free replace $(y_k^{(\nu+1)})_{k\in\mathbb{N}}$ in M and $(j_k^{(\nu+1)})_{k\in N}$ with any sequence that satisfies (7.47) for respective scaling sequences, namely

$$|j_k^{(\nu+1)} - j_k^{(n)}| + (2^{j_k^{\nu+1}} + 2^{j_k^{(n)}})d(y_k^{(\nu+1)}, y_k^{(n)}) \to \infty, \quad n = 1, \ldots \nu.$$

The renamed $w^{(\nu+1)}$ will be necessarily zero since $\beta_{\nu+1} = 0$.

We now may assume that $w^{(\nu+1)} \neq 0$. If (AO) does not hold with the index $\nu+1$ and some index $\ell \leq \nu$, then there exist $m \in \mathbb{Z}$ and $\lambda \in \mathbb{R}$, such that, on a renamed subsequence, $2^{j_k^{(\ell)}} d(y_k^{(\ell)}, y_k^{(\nu+1)})$ is bounded and $j_k^{(\nu+1)} = j_k^{(\ell)} - m$. Let $\psi_k = e^{-1}_{y_k^{(\ell)}} \circ e_{y_k^{(\nu+1)}}$. Note that $d(y_k^{(\ell)}, y_k^{(\nu+1)}) \to 0$, and since M is of bounded geometry, on a renamed subsequence we have ψ_k convergent uniformly, together with its derivatives of every order, and its limit is the identity map. Also there exists, on a renamed subsequence, an $\eta_0 \in \mathbb{R}^N$ such that $2^{j_k^{(\nu+1)}} \psi_k(0) \to \eta_0$ (since $|\psi_k(0)| = d(y_k^{(\nu+1)}, y_k^{(\ell)})$). Then, uniformly on compact subsets of $\mathbb{R}^N$, one has

$$\begin{aligned}
&2^{j_k^{(\ell)}} \psi_k(2^{-j_k^{(\nu+1)}} \xi) \\
&= 2^m 2^{j_k^{(\nu+1)}} \psi_k(0) + 2^m \psi_k'(0)\xi + 2^{j_k^{(\ell)}} O(2^{-2j_k^{(\nu+1)}} \xi) \to 2^m \eta_0 + 2^m \xi.
\end{aligned}$$

Note also that from (AO) and Lemma 7.8.1, one has for any $n = \ell + 1, \ldots, \nu$,

$$2^{j_k^{(\nu+1)} \frac{N-2}{2}} W_k^{(n)} \circ e_{y_k^{(\nu+1)}}(2^{-2j_k^{(\nu+1)}} \xi) \to 0.$$

Substituting the two last calculations into the expression below, one has

$$\begin{aligned}
&2^{j_k^{(\nu+1)} \frac{N-2}{2}} v_k^{(\nu)} \circ e_{y_k^{(\nu+1)}}(2^{-2j_k^{(\nu+1)}} \xi) \\
&= 2^{j_k^{(\nu+1)} \frac{N-2}{2}} v_k^{(\ell)} \circ e_{y_k^{(\nu+1)}}(2^{-2j_k^{(\nu+1)}} \xi) + o(1) \\
&= 2^{-m\frac{N-2}{2}} 2^{j_k^{(\ell)} \frac{N-2}{2}} v_k^{(\ell)} \circ e_{y_k^{(\ell)}} \circ \psi_k(2^{-2j_k^{(\nu+1)}} \xi) \\
&= 2^{-m\frac{N-2}{2}} 2^{j_k^{(\ell)} \frac{N-2}{2}} v_k^{(\ell)} \circ e_{y_k^{(\ell)}}(2^{j_k^{(-\ell)}} [2^{j_k^{(\ell)}} \psi_k(2^{-2j_k^{(\nu+1)}} \xi)]) \\
&= 2^{-m\frac{N-2}{2}} 2^{j_k^{(\ell)} \frac{N-2}{2}} v_k^{(\ell)} \circ e_{y_k^{(\ell)}}(2^{j_k^{(-\ell)}} [2^m \eta_0 + 2^m \xi + o(1)]) \\
&\to 0,
\end{aligned}$$

by (7.55), which by definition of $w^{(\nu+1)}$ implies $w^{(\nu+1)} = 0$, a contradiction.

6. Expanding a trivial inequality $\int_M g_x(d(u_k - \mathbf{S}_k^{(\nu)}), d(u_k - \mathbf{S}_k^{(\nu)}))\mathrm{d}v_g \geq 0$ by bilinearity we get

$$\int_M g_x(du_k, du_k)\mathrm{d}v_g \geq 2 \int_M g_x(du_k, d\mathbf{S}_k^{(\nu)})\mathrm{d}v_g - I_k - I_k', \tag{7.56}$$

where

$$I_k = \sum_{n=1}^{\nu} \int_M g_x(d(2^{j_k^{(n)} \frac{N-2}{2}} \chi \circ e_{y_k^{(n)}}^{-1}(x)\, w^{(n)}(2^{j_k^{(n)}} e_{y_k^{(n)}}^{-1}(x))),$$
$$d(2^{j_k^{(n)} \frac{N-2}{2}} \chi \circ e_{y_k^{(n)}}^{-1}(x)\, w^{(n)}(2^{j_k^{(n)}} e_{y_k^{(n)}}^{-1}(x))))\mathrm{d}v_g,$$

and

$$I_k' = \sum_{m\neq n,\, m,n=1,\dots \nu} \int_M g_x(d(2^{j_k^{(n)} \frac{N-2}{2}} \chi \circ e_{y_k^{(n)}}^{-1}(x)\, w^{(n)}(2^{j_k^{(n)}} e_{y_k^{(n)}}^{-1}(x))),$$
$$d(2^{j_k^{(m)} \frac{N-2}{2}} \chi \circ e_{y_k^{(m)}}^{-1}(x)\, w^{(m)}(2^{j_k^{(m)}} e_{y_k^{(m)}}^{-1}(x))))\mathrm{d}v_g.$$

We evaluate the first term in (7.56) by integration in rescaled geodesic coordinates $\xi = 2^{j_k^{(n)}} e_{y_k^{(n)}}^{-1}(x)$

$$\begin{aligned}
&\int_M g_x(du_k, d\mathbf{S}_k^{(\nu)})\mathrm{d}v_g \\
&= \sum_{n=1}^{\nu} 2^{j_k^{(n)} \frac{N-2}{2}} \int_{B(y_k^{(n)},r)} g_x(du_k, d(\chi \circ e_{y_k^{(n)}}^{-1}(x)\, w^{(n)}(2^{j_k^{(n)}} e_{y_k^{(n)}}^{-1}(x))))\mathrm{d}v_g \\
&= \sum_{n=1}^{\nu} 2^{-j_k^{(n)} \frac{N-2}{2}} \int_{\Omega_{2^{j_k^{(n)}} r}} \sum_{\alpha,\beta=1}^{N} (\delta_{\alpha\beta} + o(1))\partial_\alpha u_k \circ e_{y_k^{(n)}}(2^{-j_k^{(n)}}\xi) \\
&\quad \times (\chi(2^{-j_k^{(n)}}\xi)\partial_\beta w^{(n)}(\xi) + o(1)\partial_\beta w^{(n)}(\xi))\mathrm{d}\xi \\
&= \sum_{n=1}^{\nu} \int_{\mathbb{R}^N} |\nabla w^{(n)}(\xi)|^2 \mathrm{d}\xi \, + \, o(1).
\end{aligned}$$

where $o(1)$ under the integral denotes a sequence of functions on $\mathbb{R}^N$ uniformly vanishing as $k \to \infty$.

An analogous evaluation gives

$$\begin{aligned}
I_k &= \sum_{n=1}^{\nu} 2^{-j_k^{(n)} \frac{N-2}{2}} \int_{\Omega_{2^{j_k^{(n)}} r}} \sum_{\alpha,\beta=1}^{N} (\chi(2^{-j_k^{(n)}}\xi)\partial_\alpha w^{(n)}(\xi) + o(1)\partial_\alpha w^{(n)}(\xi)) \\
&\quad \times (\chi(2^{-j_k^{(n)}}\xi)\partial_\beta w^{(n)}(\xi) + o(1)\partial_\beta w^{(n)}(\xi))(\delta_{\alpha\beta} + o(1))\mathrm{d}\xi \\
&= \sum_{n=1}^{\nu} \int_{\mathbb{R}^N} |\nabla w^{(n)}(\xi)|^2 \mathrm{d}\xi \, + \, o(1).
\end{aligned}$$

By (AO) and Lemma 7.8.1, we have $I'_k \to 0$. Consequently, (7.56) implies

$$\int_M g_x(du_k, du_k)\mathrm{d}v_g \ge 2\sum_{n=1}^{\nu} \int_{\mathbb{R}^N} |\nabla w^{(n)}(\xi)|^2 \mathrm{d}\xi - \sum_{n=1}^{\nu} \int_{\mathbb{R}^N} |\nabla w^{(n)}(\xi)|^2 \mathrm{d}\xi + o(1)$$
$$= \sum_{n=1}^{\nu} \int_{\mathbb{R}^N} |\nabla w^{(n)}(\xi)|^2 \mathrm{d}\xi + o(1).$$

Since ν is arbitrary, we have (7.52).

7. It follows from (7.52) that $\beta_\nu \to 0$ as $\nu \to \infty$. Repeating the argument in [113] with only trivial modifications, one can show, for a suitably renamed sequence, that the series $\mathbf{S}_k$ unconditionally converges in $H^{1,2}(M)$ and that this convergence is uniform in k.

8. It remains to show that defect of compactness is indeed given by the sequence $(\mathbf{S}_k)$. Let (j_k) be a sequence in $\mathbb{N}$, $j_k \to \infty$ and let (y_k) be a sequence in M. Without loss of generality, consider two cases.

Case A. For each $n \in \mathbb{N}$, pairs of sequences $(j_k^{(n)})$, (j_k) and $(y_k^{(n)})$, (y_k) satisfy the condition (7.47). Then $2^{-j_k\frac{N-2}{2}} S_k \circ e_{y_k}(2^{-j_k}\cdot) \to 0$ a. e., because this is true for each term in the series of $\mathbf{S}_k$ by Lemma 7.8.1, and the series $\mathbf{S}_k$ is uniformly convergent. On the other hand, if, on a renamed subsequence, one has $2^{-j_k\frac{N-2}{2}} u_k \circ e_{y_k}(2^{-j_k}\cdot) \to w$ a. e., then, necessarily, $\|\nabla w\|_2 \le \beta_\nu$ for every $\nu \in \mathbb{N}$, that is, $w = 0$. Therefore, $2^{-j_k\frac{N-2}{2}}(u_k - \mathbf{S}_k) \circ e_{y_k}(2^{-j_k}\cdot) \to 0$ a. e. in this case.

Case B. For some $\ell \in \mathbb{N}$, $j_k - j_k^{(\ell)} = m \in \mathbb{Z}$ and $2^{j_k} d(y_k^{(\ell)}, y_k)$ is bounded. Then, repeating the argument of Step 4, we have

$$2^{-j_k\frac{N-2}{2}}(u_k - \mathbf{S}_k^{(\ell)} \circ e_{y_k}(2^{-j_k}\cdot)) \to 0 \quad \text{a. e.},$$

while by (AO), Lemma 7.8.1 and the uniform convergence of the series $\mathbf{S}_k$,

$$2^{-j_k\frac{N-2}{2}}((\mathbf{S}_k - \mathbf{S}_k^{(\ell)}) \circ e_{y_k}(2^{-j_k}\cdot)) \to 0 \quad \text{a. e.},$$

from which follows $2^{-j_k\frac{N-2}{2}}(u_k - \mathbf{S}_k) \circ e_{y_k}(2^{-j_k}\cdot) \to 0$ a. e. in this case as well.

Then, by Theorem 7.7.1 we have $u_k - \mathbf{S}_k \to 0$ in L^{2^*}.

9. Let $q \in (2, 2^*)$ and note that each term in $\mathbf{S}_k$ vanishes in $L^q(M)$. Since the series for $\mathbf{S}_k$ converges uniformly in $H^{1,2}(M)$, it also converges uniformly in $L^q(M)$, and thus it vanishes in $L^q(M)$. Note that $u_k \to 0$ in $L^p(M)$, and since (u_k) is bounded in $H^{1,2}(M)$, $u_k \to 0$ in $L^q(M)$ as well. Then

$$\|u_k - \mathbf{S}_k\|_q \le \|u_k\|_q + \|\mathbf{S}_k\|_q \to 0. \qquad \square$$

We have the following consequences of decoupling of concentration in Theorem 7.8.2 in terms of L^q-norms, $q \in (2, 2^*]$.

Proposition 7.8.3. *Let u_k and $w^{(n)}$, $n \in \mathbb{N}$, be provided by Theorem 7.8.2. Then, with $q = 2^*$,*

$$\int_M |u_k|^q \mathrm{d}v_g \to \sum_{n\in\mathbb{N}} \int_{\mathbb{R}^N} |w^{(n)}|^q \mathrm{d}\xi.$$

Proof. By (7.51), it suffices to show that

$$\int_M \Big|\sum_{n\in\mathbb{N}} W_k^{(n)}\Big|^q \mathrm{d}v_g \to \sum_{n\in\mathbb{N}} \int_{\mathbb{R}^N} |w^{(n)}|^q \mathrm{d}\xi.$$

Since sequences $(W_k^{(n)})_{k\in\mathbb{N}}$ have asymptotically disjoint supports in the sense of (7.47), and since one can by density of $C_0(\mathbb{R}^N)$ in $\dot{H}^{1,2}(\mathbb{R}^N)$ assume that every profile $w^{(n)}$ has compact support, we can easily see that

$$\int_M \Big|\sum_{n\in\mathbb{N}} W_k^{(n)}\Big|^q \mathrm{d}v_g - \sum_{n\in\mathbb{N}} \int_M |W_k^{(n)}|^q \mathrm{d}v_g \to 0,$$

so it suffices to show that for each $n \in \mathbb{N}$,

$$\int_M |W_k^{(n)}|^q \mathrm{d}v_g \to \int_{\mathbb{R}^N} |w^{(n)}|^q \mathrm{d}\xi.$$

Indeed, if we pass to normal coordinates at $y_k^{(n)}$ and then rescale them, we get

$$\begin{aligned}\int_M |W_k^{(n)}|^q \mathrm{d}v_g &= 2^{j_k^{(n)}N} \int_M |\chi \circ e_{y_k^{(n)}}^{-1}(x)\, w^{(n)}(2^{j_k^{(n)}} e_{y_k^{(n)}}^{-1}(x))|^q \mathrm{d}v_g \\ &= 2^{j_k^{(n)}N} \int_{\Omega_p} |\chi(\xi)\, w^{(n)}(2^{j_k^{(n)}} \xi)|^q \sqrt{g(\xi)}\mathrm{d}\xi \\ &= \int_{\mathbb{R}^N} |\chi(2^{-j_k^{(n)}} \eta)\, w^{(n)}(\eta)|^q \sqrt{g(2^{-j_k^{(n)}} \eta)}\mathrm{d}\eta \to \int_{\mathbb{R}^N} |w^{(n)}(\eta)|^q \mathrm{d}\eta.\end{aligned}$$

Taking the limit at the last step is possible by Lebesgue dominated convergence theorem, once we take into account that in normal coordinates $g(0) = 1$. □

7.9 Profile decomposition – the limiting case

We will now assume that the parameter $r \in (0, r(M))$, involved in the statement of Theorem 7.3.5 satisfies the constraint $r < r(M)/8$. In the statement below, $\mathbb{1}_{q=2^*}$ assumes value 1 if $q = 2^*$ and 0 otherwise.

Theorem 7.9.1. *Let M be a connected Riemannian manifold of bounded geometry. If (u_k) is a bounded sequence in $H^{1,2}(M)$, there is a renamed subsequence of (u_k), weakly convergent to some $u \in H^{1,2}(M)$, sequences $(\bar{y}_k^{(m)})_{k\in\mathbb{N}}$, $m \in \mathbb{N}$, and $(y_k^{*(n)})_{k\in\mathbb{N}}$ in M, sequences $(j_k^{(n)})_{k\in\mathbb{N}}$, in $\mathbb{N}$, $j_k^{(n)} \to +\infty$ as $k \to \infty$, with $n \in \mathbb{N}$, satisfying the following relations:*

$$u_k - u - \sum_{m\in\mathbb{N}} \bar{W}_k^{(m)} - \sum_{n\in\mathbb{N}} W_k^{*(n)} \to 0 \quad \text{in } L^q(M),\ q \in (2, 2^*), \tag{7.57}$$

where $\bar{W}_k^{(m)}$ are as in Theorem 7.3.5 (relative to sequences $(\bar{y}_k^{(m)})_{k\in\mathbb{N}}$);

$$W_k^{*(n)}(x) = 2^{j_k^{(n)}\frac{N-2}{2}} \chi \circ e^{-1}_{y_k^{*(n)}}(x)\, w^{*(n)}(2^{j_k^{(n)}} e^{-1}_{y_k^{*(n)}}(x)), \quad x \in M, \tag{7.58}$$

where

$$2^{-j_k^{(n)}\frac{N-2}{2}} u_k \circ e_{y_k^{*(n)}}(2^{-j_k^{(n)}}\cdot) \to w^{*(n)} \quad a.e. \text{ in } \mathbb{R}^N, \tag{7.59}$$

(as in Theorem 7.8.2); $d(\bar{y}_k^{(m)}, \bar{y}_k^{(\ell)}) \to \infty$ when $m \neq \ell$ and sequences $(j_k^{(n)})$, $(y_k^{*(n)})$, $(j_k^{(n')})$, $y_k^{*(n')}$ satisfy the condition (7.47); both series in (7.57) converge unconditionally and uniformly in k.

Moreover, with $M_\infty^{(m)}$, $m \in \mathbb{N}$, as in Theorem 7.3.5, we have

$$\begin{aligned}\sum_{n\in\mathbb{N}} \int_{\mathbb{R}^N} |\nabla w^{*(n)}|^2 d\xi + \sum_{m\in\mathbb{N}} \int_{M_\infty^{(m)}} (g_\infty^{(m)}(d\bar{w}^{(m)}, d\bar{w}^{(m)}) + |\bar{w}^{(m)}|^2) dv_{g_\infty^{(m)}} \\ + \int_M (g_x(du, du) + u^2) dv_g \le \int_M (g_x(du_k, du_k) + u_k^2) dv_g + o(1),\end{aligned} \tag{7.60}$$

and

$$\int_M |u_k|^q dv_g \to \mathbb{1}_{q=2^*} \sum_{n\in\mathbb{N}} \int_{\mathbb{R}^N} |w^{*(n)}|^{2^*} d\xi + \sum_{m\in\mathbb{N}} \int_{M_\infty^{(m)}} |\bar{w}^{(m)}|^q dv_{g_\infty^{(m)}} + \int_M |u|^q dv_g \tag{7.61}$$

for every $q \in (2, 2^*]$.

Proof. Apply Theorem 7.3.5 to u_k and let $v_k = u_k - u - \sum_{m\in\mathbb{N}} \bar{W}_k^{(m)}$ be the left hand side of in (7.14) Note that v_k is a bounded sequence in $H^{1,2}(M)$ because so are both u_k and $\sum_{m\in\mathbb{N}} \bar{W}_k^{(m)}$ (for the latter it can be inferred from (7.15)). Apply Theorem 7.8.2 to v_k. Then (7.57) is immediate from combining (7.14) and (7.51). Relation (7.60) follows from substitution of (7.52) for v_k into (7.15).

If $q < 2^*$, relation (7.61) is immediate from (7.16). Consider now the case $q = 2^*$. By Brezis-Lieb Lemma and (7.57), we have

$$\int_M |u_k|^{2^*} dv_g = \int_M |u|^{2^*} dv_g + \int_M \left| \sum_{m\in\mathbb{N}} \bar{W}_k^{(m)} + \sum_{n\in\mathbb{N}} W_k^{*(n)} \right|^{2^*} dv_g + o(1). \tag{7.62}$$

Leaving details to the reader, we sketch the rest of the argument. Excising small neighborhoods of the concentration points $y_k^{(n}$ of the bubbles $W_k^{*(n)}$ we have

$$\int_M \left|\sum_{m\in\mathbb{N}} \bar{W}_k^{(m)} + \sum_{n\in\mathbb{N}} W_k^{*(n)}\right|^{2^*} dv_g = \int_M \left|\sum_{m\in\mathbb{N}} \bar{W}_k^{(m)}\right|^{2^*} dv_g + \int_M \left|\sum_{n\in\mathbb{N}} W_k^{*(n)}\right|^{2^*} dv_g + o(1).$$

Then the first term in the right-hand side evaluates by the argument for (7.16) in the proof of Theorem 7.3.5, which extends to the case $q = 2^*$ with no further modifications, while the second term evaluates as in Proposition 7.8.3. □

Corollary 7.9.2. *Let M be a manifold of negative curvature with bounded geometry (in particular a hyperbolic space). If (u_k) is a sequence in $H^{1,2}(M)$ satisfying $\int_M g_x(du_k, du_k)dv_g \le C$, then it has a renamed subsequence satisfying the assertions of Theorem 7.9.1.*

Remark 7.9.3. If M is a noncompact homogeneous space (in particular, $\mathbb{R}^N$ or the hyperbolic space $\mathbb{H}^N$) of dimension greater than 2. Then, in face of Theorem 7.6.1, Theorem 7.9.1 holds with $M_\infty^{(m)} = M$ for every $m \in \mathbb{N}$, and with $\bar{W}_k^{(m)} = \bar{w}^{(m)} \circ \eta_k^{(m)}$, where $w^{(m)} = \text{w-lim}\, u_k \circ \eta_k^{(m)^{-1}}$, $\eta_k^{(m)}$ are discrete sequences of isometries on M, and the sequences $\eta_k^{(m)^{-1}} \circ \eta_k^{(m')}$ are discrete whenever $m \ne m'$.

7.10 Compactness in presence of symmetries

Lemma 7.10.1. *Let M be a manifold of bounded geometry and let Y be a (ε, r)-discretization of M, $0 < \varepsilon < r$. Then for any $R > 0$ there exists $n_R \in \mathbb{N}$, such that $\#(Y \cap B(x, R)) \le n_R$ for every $x \in M$.*

Proof. By definition, $\#(Y \cap B(x, R))$ cannot exceed the maximal number of disjoint balls of radius $\varepsilon/2$ contained in $B(x, R + \varepsilon)$, which is finite by (10.39). □

Definition 7.10.2. A (ε, r)- discretization Y of a Riemannian manifold M is called an orbital discretization if there exist nonempty subsets $Y_i \subset Y$, $i \in \mathbb{N}$, such that:
(a) $Y = \bigcup_{i=1}^\infty Y_i$ and $Y_i \cap Y_j = \emptyset$ for $i \ne j$,
(b) $\#Y_i \le \#Y_{i+1} < \infty$, $i \in \mathbb{N}$,
(c) $\lim_{i\to\infty} \#Y_i = \infty$.

We shall write then $Y \in \mathcal{O}_{\varepsilon,r}(M)$. The sets Y_i will be called quasi-orbits.

The term *orbital discretization* will be justified in the next subsection when we discretize group orbits on a manifold.

Lemma 7.10.3. *Let Y be an orbital discretization. For every $R > 0$ and $j \in \mathbb{N}$, there exists $\bar{\imath}(R, j) \in \mathbb{N}$ such that for all $i \ge \bar{\imath}(R, j)$ and for every $x \in Y_i$, there exists a subset $Y_i(x) \subset Y_i$ satisfying:*

(i) $x \in Y_i(x)$,
(ii) $d(y,z) > R$ *whenever* $y, z \in Y_i(x)$, $y \neq z$,
(iii) $\#Y_i(x) \geq j$.

Proof. For $j = 1$ conditions (i)–(iii) hold tautologically when $Y_i(x) = \{x\}$. We assume now that $j \geq 2$. Let n_R be as in Lemma 7.10.1 and let $i_0 \in \mathbb{N}$ be such that $\#Y_i > jn_R$ for any $i \geq i_0$. Such i_0 always exists by property (c) in the definition of the orbital discretization. Let $y_0 = x$ and let us choose recursively $y_{k+1} \in Y_i$, $k = 0, \dots, j-2$, such that $y_{k+1} \notin B(y_\ell, R)$, $\ell = 0, \dots, k$. This is possible since the balls $B(y_\ell, R)$, $\ell = 0, \dots, k$ contain altogether not more than $(k+1)n_R$ points of Y_i, and this number is less than jn_R, and thus less than $\#Y_i$. Obviously, $d(y_k, y_\ell) > R$ whenever $k \neq \ell$. We set $Y_i(x) = \{y_k\}_{k=0,\dots,j-1}$. □

Corollary 7.10.4. *Let Y be an orbital discretization. Then* $\lim_{i\to\infty} \operatorname{diam} Y_i = \infty$.

Definition 7.10.5. Let $Y \in \mathcal{O}_{\varepsilon,r}(M)$, $r < r(M)$. Let $i \in \mathbb{N}$ and $\lambda \geq 1$. A function $f \in L^1_{\text{loc}}(M)$ is called (i,λ)-quasi-symmetric relative to Y if for every $\ell \geq i$,

$$\max_{x \in Y_\ell} \int_{B(x,r)} |f(y)| dv_g \leq \lambda \min_{x \in Y_\ell} \int_{B(x,r)} |f(y)| dv_g. \tag{7.63}$$

We shall write then $f \in \mathcal{S}_{Y,i,\lambda}(M)$.

Remark 7.10.6. 1. For any Y, i and λ, the set $\mathcal{S}_{Y,i,\lambda}(M)$ contains infinitely many linearly independent functions from $H^{1,p}(M)$. In particular, it has the following functions. Let $\varphi_x \in H^{1,p}(M) \setminus \{0\}$ be supported in $B(x, \varepsilon/2)$, $x \in Y_\ell$, and define

$$f = \sum_{x \in Y_\ell} \frac{\varphi_x}{\int_M |\varphi_x| dv_g}, \quad \ell \geq i.$$

2. For any Y, i, and λ, the set $H^{1,p}(M) \cap \mathcal{S}_{Y,i,\lambda}(M)$ is closed with respect to the weak convergence in $H^{1,p}(M)$, since all the quantities in the relation (7.63) are weakly continuous in $H^{1,p}(M)$.

Theorem 7.10.7. *Let M be complete, noncompact, connected, Riemannian manifold of bounded geometry. Let $Y \in \mathcal{O}_{\varepsilon,r}(M)$. Let $1 < p < N = \dim M$, $p < q < p^*$, $i \in \mathbb{N}$, and $\lambda \geq 1$. If a set $K \subset H^{1,p}(M) \cap \mathcal{S}_{Y,i,\lambda}(M)$ is bounded in $H^{1,p}(M)$, then it is relatively compact in $L^q(M)$.*

Proof. By reflexivity, it is sufficient to show that if (u_k) is a sequence in $H^{1,p}(M) \cap \mathcal{S}_{Y,i,\lambda}(M)$ weakly convergent to zero in $H^{1,p}(M)$, then $u_k \to 0$ in $L^q(M)$. Assume that this is not the case. Then by Theorem 7.1.4 and the Banach–Alaoglu theorem there is a sequence $y_k \in Y$ and a function $w \in H^{1,p}(\Omega_r)$, $w \neq 0$, such that $u_k \circ e_{y_k} \rightharpoonup w \neq 0$ in $H^{1,p}(\Omega_r)$. Note that if the sequence (y_k) has a bounded subsequence, it has a constant subsequence and (u_k) has a nonzero weak limit, which contradicts the assumption. So we can assume that $y_k \in Y_{\ell_k}$ with $\ell_k > i$ and $\ell_k \to \infty$. Since the manifold has bounded geometry, for all k large enough we have the following inequality:

$$\int_{B(y_k,r)} |u_k|\,\mathrm{d}v_g \geq C \int_{\Omega_r} |u_k \circ e_{y_k}|\mathrm{d}\xi \geq C \int_{\Omega_r} |w|\mathrm{d}\xi \stackrel{\text{def}}{=} \alpha > 0. \tag{7.64}$$

The functions u_k are of the quasisymmetry class $\mathcal{S}_{Y,i,\lambda}(M)$, so by the Hölder inequality and (7.63), for k large enough we have for every $x \in Y_\ell$, $\ell \geq i$,

$$\begin{aligned}\int_{B(x,r)} |u_k|^q \mathrm{d}v_g &\geq C\Big(\int_{B(x,r)} |u_k|\mathrm{d}v_g\Big)^q \\ &\geq C\lambda^{-q}\Big(\int_{B(y_k,r)} |u_k|\mathrm{d}v_g\Big)^q \geq C\lambda^{-q}\alpha^q \stackrel{\text{def}}{=} \beta > 0.\end{aligned} \tag{7.65}$$

Let us apply Lemma 7.10.3 with $R = 2r$ and for each $j \in \mathbb{N}$ choose k_j such that $\ell_{k_j} \geq \bar{i}(2r, j)$. This gives

$$\int_M |u_k|^q \mathrm{d}v_g \geq \sum_{x \in Y_{\ell_{k_j}}(y_k)} \int_{B(x,r)} |u_k|^q \mathrm{d}v_g \geq j\beta. \tag{7.66}$$

Since j is arbitrarily large, we have a contradiction that proves the theorem. □

To study compactness of embedding of spaces invariant with respect to a group action, it is natural to consider a specific kind of orbital discretizations, namely those associated with group orbits. Let G be a compact connected group of isometries of a complete Riemannian manifold M. Then $H_G^{1,p}(M)$ will denote a subspace of $H^{1,p}(M)$ consisting of all G-invariant functions.

Definition 7.10.8. We say that a continuous action of a group G on a complete Riemannian manifold M is *coercive* if for every $t > 0$, the set

$$O_t = \{x \in M : \ \operatorname{diam} Gx \leq t\}$$

is bounded.

Example 7.10.9. Let $M = \mathbb{R}^N$ and let $G = O(n_1) \times \cdots \times O(N_m)$, $n_1 + \cdots + n_m = N$, $m \in \mathbb{N}$. Then G is coercive if and only if $n_i \geq 2$ for every $i = 1, \dots, m$.

Remark 7.10.10. If the sectional curvature of M is nonpositive and the compact connected group G of isometries fixes some point, then G is coercive if and only if G has no other fixed point; see [109, Proposition 3.1]. An example of a compact connected coercive group without fixed points (see [109, the end of Section 3]) is $M = S^1 \times \mathbb{R}^n$ (a Riemannian product of the unit circle and the Euclidean space), $n \geq 2$, and $G = S^1 \times SO(n)$ acting on M by the formulas $(e^{i\varphi}, h)(e^{i\psi}, x) = (e^{i(\varphi+\psi)}, h(x))$, $e^{i\varphi}, e^{i\psi} \in S^1$, $h \in SO(n)$, and $x \in \mathbb{R}^n$.

Proposition 7.10.11. *Let G be a compact connected group of isometries acting coercively on the manifold M. Then there exists an orbital discretization $Y \in \mathcal{O}_{\varepsilon,2\varepsilon}(M)$ such that any quasi-orbit Y_i is a subset of a distinct orbit of G.*

Proof. Let $\widetilde{M}$ be a union of all principal orbits of the group G. The set $\widetilde{M}$ is a dense open subset of M. On the coset space $\widetilde{M}/G$, one can introduce a Riemannian structure such that the projections $p : \widetilde{M} \to \widetilde{M}/G$ have the following property:

$$d_{\widetilde{M}/G}(p(x), p(y)) = d_M(Gx, Gy)$$

where the distances are taken on respective manifolds (see, e. g., [56]). Let $\widetilde{Y} = \{Gx_\ell\}_{\ell\in\mathbb{N}}$ be an $(\varepsilon, 2\varepsilon)$-discretization of $\widetilde{M}/G$ with $\varepsilon < r(M)/3$. Let $\dot{Y}_\ell$ be an $(\varepsilon, 2\varepsilon)$-discretization of the orbit Gx_ℓ in M. Then $Y = \bigcup_{\ell=1}^{\infty} \dot{Y}_\ell$ is an $(\varepsilon, 2\varepsilon)$-discretization of M. Let $\{Y_i\}$ be the family $\{\dot{Y}_\ell\}$ reordered by the number of elements in $\dot{Y}_\ell$. Then $Y = \bigcup_i Y_i$ is obviously a $(\varepsilon, 2\varepsilon)$-discretization of M. We prove that it is an orbital discretization. Conditions (a) and (b) are satisfied by the construction. The condition (c) is a consequence of the coercivity of the action of G as follows. Let $R > 0$. By the coercivity, all sets Y_i of diameter not exceeding R lie in a bounded set O_R. However, only finitely many elements of Y may lie in O_R. So there exists $i_R \in \mathbb{N}$ such that diameter of Gx_ℓ is greater then R whenever $\ell \geq i_R$. The orbits Gx_ℓ are connected since G is connected, therefore, $\#Y_\ell \to \infty$. □

Taking into account the above proposition, one can apply Theorem 7.10.7 to sets of quasisymmetric functions related to the action of a group G of isometries of M. In particular, it can be applied to the subspaces $H_G^{1,p}(M)$ of $H^{1,p}(M)$ consisting of all G-symmetric functions.

Theorem 7.10.12. *Let G be a compact, connected group of isometries of a N-dimensional noncompact connected Riemannian manifold M of bounded geometry. Let $1 < p < N$ and $p < q < p^*$. Then the subspace $H_G^{1,p}(M)$ is compactly embedded into $L^q(M)$ if and only if G is coercive.*

Proof. Sufficiency in the theorem follows from Theorem 7.10.7 with the orbital discretization given by Proposition 7.10.11 since $H_G^{1,p}(M) \subset H^{1,p}(M) \cap \mathcal{S}_{Y,1,1}$. Note that by isometry

$$\int_{B(x,r)} |f(y)| dv_g = \int_{B(z,r)} |f(y)| dv_g, \quad z \in Gx$$

if $f \in H_G^{1,p}(M)$.

Proof of necessity. If G is not coercive, there exists $R > 0$ and a discrete sequence (x_k) in M such that $Gx_k \subset B(x_k, R)$. Let $r \in (0, r(M))$ and let $\psi \in C_0^\infty(\Omega_r) \setminus \{0\}$ be a nonnegative function. Let us replace x_k with a renumbered subsequence such that distance between any two terms in the sequence will be greater than $2(R+r)$. Let

$$\psi_k = \int_G \psi \circ e_{x_k}^{-1}(\eta\cdot)\, d\mu_G(\eta),$$

where the Haar measure μ_G of G is normalized to the value 1. By the Young inequality, taking into account that G is a group of isometries on M and that M is of bounded geometry, we have

$$\begin{aligned}\|\psi_k\|_{H^{1,p}(M)} &\le \int_G \|\psi \circ e_{x_k}^{-1}(\eta\cdot)\|_{H^{1,p}(M)} d\mu_G(\eta)\\ &= \int_G \|\psi \circ e_{x_k}^{-1}\|_{H^{1,p}(M)} d\mu_G(\eta)\\ &= \|\psi \circ e_{x_k}^{-1}\|_{H^{1,p}(M)} \le C\|\psi\|_{H^{1,p}(\Omega_r)}.\end{aligned}$$

Note that the supports of the functions ψ_k are disjoint and, therefore,

$$\|\psi_m - \psi_n\|_{L^q(M)}^q = \|\psi_m\|_{L^q(M)}^q + \|\psi_n\|_{L^q(M)}^q \ge 2\inf_k \|\psi_k\|_{L^q}^q.$$

Furthermore,

$$\begin{aligned}v_g(B(x_k, R+r))^{1-1/q}\|\psi_k\|_{L^q} &\ge \int_M \psi_k\, dv_g\\ &= \int_G\int_M \psi \circ e_{x_k}^{-1}(\eta\cdot)\, dv_g\, d\mu_G(\eta) = \int_G\int_M \psi \circ e_{x_k}^{-1}\, dv_g\, d\mu_G(\eta)\\ &= \int_M \psi \circ e_{x_k}^{-1}\, dv_g \ge C\int_{\Omega_r} \psi\, d\xi > 0.\end{aligned}$$

Thus, since $\sup_{k\in\mathbb{N}} v_g(B(x_k, R+r)) < \infty$ by the bounded geometry, $\|\psi_k\|_{L^q(M)}$ is bounded away from zero. Therefore, we have a sequence, bounded in $H^{1,p}(M)$ and discrete in $L^q(M)$, and so the embedding $H^{1,p}(M) \hookrightarrow L^q(M)$ is not compact. □

7.11 Bibliographic notes

The earliest profile decomposition for the Sobolev space of a compact manifold, relative to the limit Sobolev embedding, was proved, to our best knowledge, by Struwe [119] (see also [39]), and Theorem 7.9.1 is its natural generalization to general sequences and noncompact manifolds.

The "spotlight lemma" (Theorem 7.1.4) was proved in [110] for $p = 2$ and in [111] for general p, and its counterpart for the limiting Sobolev embedding, Theorem 7.7.1, in [101]. Both are possibly found elsewhere in literature. Theorem 7.3.5 is proved in

[110], and Theorems 7.8.2 and 7.9.1 are proved in [101]. Similar results for $H^{1,p}(M)$ with $p \in (1, \infty)$ can be proved by analogous arguments. Condition of bounded geometry (also called smooth bounded geometry to distinguish from bounds on only some derivatives of the Riemannian curvature) is quite restrictive, and metric spaces at infinity (which are generally no longer Riemannian manifolds) emerge on the grounds of Gromov's compactness theorem using pointed Gromov-Hausdorff convergence. Under the stronger pointed C^m-convergence (see [99, Chapter 10]), however, Gromov-Hausdorff limits of sequences of Riemannian manifolds remain Riemannian manifolds, so it should be possible to extend assertions of Theorem 7.3.5 and Theorem 7.9.1 to a larger class of manifolds, and in modified form to an even larger class.

Compactness of embeddings of radial subspaces of Sobolev spaces into L^p is due to Strauss [115]. For multiradial subspaces of Sobolev spaces see Lions, [81]. Hebey and Vaugon [66] obtained compactness of local Sobolev embeddings in presence of symmetries basing on effective dimension of the quotient manifold, that yields a correspondingly higher critical exponent and thus compactness. In this chapter we give a necessary and sufficient condition for compactness of subcritical embeddings on non-compact manifolds, based on [111].

8 Functions of bounded variation. Sobolev spaces on fractals

This chapter addresses two separate topics. The first three sections address the question if one can extend Solimini's profile decomposition to $H^{1,1}$, dealing with the weak convergence issues arising in this non-reflexive space and arriving at a profile decomposition in the space of functions of bounded variations instead. The remaining three sections deal with energy spaces on fractal blowups, where concentration profiles emerge at functions defined on blowups-at-infinity, resembling the situation described in the previous chapter in the case of manifolds. The difference, however, is that evem if fractal blowups generally do not possess a group of global translations, they admit local translations of balls of any radius, thus sparing a need in a gluing argument used in the previous chapter.

8.1 Cocompactness of the embedding $\dot{\mathrm{BV}} \hookrightarrow L^{1^*}(\mathbb{R}^N)$

The first difficulty in describing defect of compactness of sequences in a non-reflexive Banach space in terms of weak⋆-convergence is that weak⋆-topology is defined only if the space is a conjugate of another Banach space and, furthermore, if the weak⋆-compactness in this space implies sequential weak⋆-compactness (which is true for separable Banach spaces, but not in general).

In particular, in the case of $L^1(\mathbb{R}^N)$, weak⋆-topology is not defined, while a sequence of normalized characteristic functions in $L^1(\mathbb{R}^N)$, $(\frac{\mathbb{1}_{A_n}}{|A_n|})_{n\in\mathbb{N}}$, where $A_n = [-\frac{1}{n}, \frac{1}{n}]^N$, has no weakly convergent subsequence. Instead, it converges to the point mass at the origin in the weak⋆ sense in a larger space of finite signed measures, which it a conjugate of the Banach space $C_0(\mathbb{R}^N)$. Similarly, it is beneficial to regard the space $\dot{H}^{1,1}(\mathbb{R}^N)$ as a subspace of the space of measurable functions whose weak derivative, rather than a L^1- function, is a finite signed measure. This space is known as the space of functions of bounded variation $\dot{\mathrm{BV}}(\mathbb{R}^N)$. It contains, of course, functions that are qualitatively different from those in $H^{1,1}(\mathbb{R}^N)$. In particular, while every element in $\dot{H}^{1,1}(\mathbb{R}^N)$ is represented by a function with a connected range, this is not the case for $\dot{\mathrm{BV}}(\mathbb{R}^N)$, which contains characteristic functions whose range is $\{0, 1\}$. For basic properties of the space $\dot{\mathrm{BV}}(\mathbb{R}^N)$, see the Appendix, Section 10.2. In particular, we note a continuous embedding $\dot{\mathrm{BV}}(\mathbb{R}^N) \hookrightarrow L^{1^*}(\mathbb{R}^N)$, $1^* \overset{\text{def}}{=} \frac{N}{N-1}$, and the notation of the norm $\|D\cdot\|$ (see (10.33)).

Since bounded sequences in $\dot{H}^{1,1}(\mathbb{R}^N)$, as the example above shows, are expected to have concentration profiles in $\dot{\mathrm{BV}}(\mathbb{R}^N)$, the version of the profile decomposition of Solimini for $\dot{H}^{1,1}(\mathbb{R}^N)$ studied in this chapter is, in fact, stated for bounded sequences in $\dot{\mathrm{BV}}(\mathbb{R}^N)$.

https://doi.org/10.1515/9783110532432-008

In what follows, we assume that $N > 1$, and consider the $\dot{\mathrm{BV}}(\mathbb{R}^N)$. equipped with the rescaling group $\mathcal{G}^{N-1}$ (cf. (1.8)), whose elements $g_{j,y} : u \mapsto 2^{(N-1)j}u(2^j(\cdot - y))$ are linear isometries on $\dot{\mathrm{BV}}(\mathbb{R}^N)$ and on $L^{\frac{N}{N-1}}(\mathbb{R}^N)$.

The proof of the theorem below repeats much of the proof of Theorem 3.2.1 but with a different argument in evaluation of sums of BV-seminorms over lattices. We use below notions of $\mathcal{G}$-weak* convergence and $\mathcal{G}$-* cocompactness defined by repeating the definitions of $\mathcal{G}$-weak convergence and of cocompactness verbatim, but with weak convergence replaced by weak*-convergence. Throughout this chapter, following the prevailing convention for measure spaces, we will call weak*-convergence in $\dot{\mathrm{BV}}(\mathbb{R}^N)$ weak convergence, and use the notation $\rightharpoonup$ rather than $\overset{*}{\rightharpoonup}$. Consistently with that, we will write $\overset{\mathcal{G}^{N-1}}{\rightharpoonup}$ rather than $\overset{\mathcal{G}^{N-1}*}{\rightharpoonup}$ for $\mathcal{G}^{N-1}$-weak* convergence, and call $\mathcal{G}^{N-1}$-*-cocompactness $\mathcal{G}^{N-1}$-cocompactness.

Theorem 8.1.1. *The embedding $\dot{\mathrm{BV}}(\mathbb{R}^N) \hookrightarrow L^{1^*}(\mathbb{R}^N)$ is $\mathcal{G}^{N-1}$ cocompact, namely if, for any sequence (j_k, y_k) in $\mathbb{Z} \times \mathbb{R}^N$, $g_{j_k,y_k}u_k \rightharpoonup 0$ then $u_k \to 0$ in $L^{1^*}(\mathbb{R}^N)$.*

Proof. Let (u_k) be a bounded sequence in $\dot{\mathrm{BV}}(\mathbb{R}^N)$ such that for any sequence (j_k, y_k) in $\mathbb{Z} \times \mathbb{R}^N$, $g_{j_k,y_k}u_k \rightharpoonup 0$.

1. Assume first that $\sup_{k\in\mathbb{N}} \|u_k\|_\infty < \infty$ and $\sup_{k\in\mathbb{N}} \|u_k\|_1 < \infty$. Then, using the L^∞-boundedness of (u_k), we have (cf. (10.35))

$$\int_{(0,1)^N} |u_k|^{1^*} \mathrm{d}x \le C\Big(\|Du_k\|_{(0,1)^N} + \int_{(0,1)^N} |u_k|\mathrm{d}x\Big)\Big(\int_{(0,1)^N} |u_k|\mathrm{d}x\Big)^{1-1/1^*}.$$

Repeating this inequality for the domain of integration $(0,1)^N + y$, $y \in \mathbb{Z}^N$, and adding the resulting inequalities over all $y \in \mathbb{Z}^N$, we have

$$\int_{\mathbb{R}^N} |u_k|^{1^*} \mathrm{d}x \le C(\|Du_k\|_{\mathbb{R}^N} + \|u_k\|_{1,\mathbb{R}^N})\Big(\sup_{y\in\mathbb{Z}^N} \int_{(0,1)^N} |u_k(\cdot - y)|\mathrm{d}x\Big)^{1/N}. \tag{8.1}$$

Here, we use the fact that the sum $\sum_{y\in\mathbb{Z}^N} \|Du_k\|_{(0,1)^N+y}$ can be split into 3^N sums of variations over unions of cubes with disjoint closures, each of these sums, as follows from Definition 10.2.3, bound by $\|Du_k\|_{\mathbb{R}^N}$, which implies $\sum_{y\in\mathbb{Z}^N} \|Du_k\|_{(0,1)^N+y} \le 3^N\|Du_k\|_{\mathbb{R}^N}$.

The last term in (8.1) converges to zero, since by the assumption $g_{j_k,y_k}u_k \rightharpoonup 0$ we have $u_k(\cdot - y_k) \to 0$ in $L^1((0,1)^N)$ for any sequence (y_k) in $\mathbb{R}^N$.

2. We now abandon the restrictions imposed in the previous step on the sequence (u_k). Let $\chi \in C_0^\infty((\frac{1}{2^{N-1}}, 4^{N-1}))$ be such that $\chi(t) = t$ whenever $t \in [1, 2^{N-1}]$. Let $\chi_j(t) = 2^{(N-1)j}\chi(2^{-(N-1)j}|t|)$, $j \in \mathbb{Z}$, $t \in \mathbb{R}$, and note that $\|\chi_j'\|_\infty = \|\chi'\|_\infty$. Consider now a general sequence (u_k) in $\dot{\mathrm{BV}}(\mathbb{R}^N)$ satisfying $g_{j_k,y_k}u_k \rightharpoonup 0$ for any $(j_k, y_k) \subset \mathbb{Z} \times \mathbb{R}^N$. By (10.34), we have

$$\int_{\mathbb{R}^{(N)}} \chi_j(u_k)^{1^*} \mathrm{d}x \le C\|D\chi_j(u_k)\|\Big(\int_{\mathbb{R}^{(N)}} \chi_j(u_k)^{1^*} \mathrm{d}x\Big)^{1/N}.$$

Let us sum up the inequalities over $j \in \mathbb{Z}$. Note that by (10.36) $\|D\chi_j(u_k)\| \le \|\chi'\|_\infty \|Du_k\|_{A_{kj}}$ where $A_{kj} = \{x \in \mathbb{R}^N : |u_k| \in (2^{(j-1)(N-1)}, 2^{(j+2)(N-1)})\}$. Furthermore, one can break all the integers j into six disjoint sets $J_1, \dots, J_6$, such that, for any $m \in \{1,2,3,4,5,6\}$, all functions $\chi_j(u_k)$, $j \in J_m$, have pairwise disjoint supports. Consequently, $\sum \|Du_k\|_{A_{kj}} \le 6\|Du_k\|$. We have therefore

$$\int_{\mathbb{R}^N} |u_k|^{1^*} \mathrm{d}x \le C\|Du_k\| \sup_{j\in\mathbb{Z}} \left(\int \chi_j(u_k)^{1^*} \mathrm{d}x \right)^{1/N}.$$

It suffices now to show that for any sequence (j_k) in $\mathbb{Z}$, $\chi_{j_k}(u_k) \to 0$ in L^{1^*}. Taking into account invariance of the L^{1^*}-norm under operators $g_{j,y}$, it suffices to show that $\chi(2^{j_k(N-1)}|u_k(2^{j_k}\cdot)|) \to 0$ in L^{1^*}, but this is immediate from the assumption $g_{j_k,y_k}u_k \rightharpoonup 0$ and the argument of the step 1, once we take into account that for sequences uniformly bounded in L^∞, L^{1^*}-convergence follows from L^1 convergence. □

Corollary 8.1.2. *The embedding $\dot{H}^{1,1}(\mathbb{R}^N) \hookrightarrow L^{1^*}(\mathbb{R}^N)$ is $\mathcal{G}^{N-1}$-cocompact.*

Proof. The statement is immediate once we note that $\mathcal{G}^{N-1}$ acts isometrically on $\dot{H}^{1,1}(\mathbb{R}^N)$ and that $C_0(\mathbb{R}^N) \subset L^1(\mathbb{R}^N)^*$. □

8.2 Profile decomposition in $\dot{\mathrm{BV}}$

Theorem 8.2.1. *Let (u_k) be a bounded sequence in $\dot{\mathrm{BV}}(\mathbb{R}^N)$. For each $n \in \mathbb{N}$, there exist $w^{(n)} \in \dot{\mathrm{BV}}(\mathbb{R}^N)$, and sequences $(j_k^{(n)})_{k\in\mathbb{N}}$ in $\mathbb{Z}$ and $(y_k^{(n)})$ in $\mathbb{R}^N$ with $j_k^{(1)} = 0$, $y_k^{(1)} = 0$, satisfying*

$$|j_k^{(n)} - j_k^{(m)}| + (2^{j_k^{(m)}} + 2^{j_k^{(n)}})|y_k^{(n)} - y_k^{(m)}| \to \infty \quad \text{whenever } m \ne n,$$

such that for a renumbered subsequence, $g^{-1}_{j_k^{(n)},y_k^{(n)}} u_k \rightharpoonup w^{(n)}$, as $k \to \infty$,

$$r_k \stackrel{\text{def}}{=} u_k - \sum_n g_{j_k^{(n)},y_k^{(n)}} w^{(n)} \to 0 \quad \text{in } L^{\frac{N}{N-1}}(\mathbb{R}^N), \tag{8.2}$$

where the series $\sum_n g_{j_k^{(n)},y_k^{(n)}} w^{(n)}$ converges in $\dot{\mathrm{BV}}(\mathbb{R}^N)$ unconditionally and uniformly in k, and

$$\sum_{n\in\mathbb{N}} \|Dw^{(n)}\| + o(1) \le \|Du_k\|. \tag{8.3}$$

Proof. Without loss of generality, we may assume that $u_k \rightharpoonup 0$ (otherwise, one may pass to a weakly convergent subsequence and subtract the weak limit). Observe that if $u_k \overset{\mathcal{G}^{N-1}}{\rightharpoonup} 0$, the theorem is proved with $r_k = u_k$ and $w^{(n)} = 0$, $n \in \mathbb{N}$. Otherwise, consider the expressions of the form $w^{(1)} = \text{w-lim}\, g^{-1}_{j_k^{(1)},y_k^{(1)}} u_k$. The sequence u_k is bounded, $\mathcal{G}^{N-1}$ is a group of isometries, so the sequence $g^{-1}_{j_k^{(1)},y_k^{(1)}} u_k$ has a weakly convergent subsequence. Since we assume that u_k is not $\mathcal{G}^{N-1}$-vanishing, there exists necessarily a sequence $(j_k^{(1)}, y_k^{(1)})$ such that, evaluated on a suitable subsequence, $w^{(1)} \neq 0$. Let $v_k^{(1)} = u_k - g_{j_k^{(1)},y_k^{(1)}} w^{(1)}$, and observe that $g^{-1}_{j_k^{(1)},y_k^{(1)}} v_k^{(1)} = g^{-1}_{j_k^{(1)},y_k^{(1)}} u_k - w^{(1)} \rightharpoonup 0$. If $v_k^{(1)} \overset{\mathcal{G}^{N-1}}{\rightharpoonup} 0$, the assertion of the theorem is verified with $r_k = v_k^{(1)}$. If not – we repeat the argument above – there exist, necessarily, a sequence $(j_k^{(2)}, y_k^{(2)})$ and a $w^{(2)} \neq 0$ such that, on a renumbered subsequence, $w^{(2)} = \text{w-lim}\, g^{-1}_{j_k^{(2)},y_k^{(2)}} v_k^{(1)}$. Let us set $v_k^{(2)} = v_k^{(1)} - g_{j_k^{(2)},y_k^{(2)}} w^{(2)}$. Then we will have

$$g^{-1}_{j_k^{(2)},y_k^{(2)}} v_k^{(2)} = g^{-1}_{j_k^{(2)},y_k^{(2)}} v_k^{(1)} - w^{(2)} \rightharpoonup 0.$$

If we assume that $g^{-1}_{j_k^{(1)},y_k^{(1)}} g_{j_k^{(2)},y_k^{(2)}} w^{(2)} \not\rightharpoonup 0$ or, equivalently, that $|j_k^{(1)} - j_k^{(2)}| + (2^{j_k^{(1)}} + 2^{j_k^{(2)}})|y_k^{(1)} - y_k^{(2)}|$ has a bounded subsequence, then passing to a renamed subsequence we will have $g^{-1}_{j_k^{(1)},y_k^{(1)}} g_{j_k^{(2)},y_k^{(2)}} \to g_{j_0,y_0}$ in the sense of strong operator convergence, for some $j_0 \in \mathbb{Z}$, $y_0 \in \mathbb{R}^N$. Then

$$\begin{aligned} w^{(2)} &= \text{w-lim}\, g^{-1}_{j_k^{(2)},y_k^{(2)}} v_k^{(1)} \\ &= \text{w-lim}(g^{-1}_{j_k^{(2)},y_k^{(2)}} g_{j_k^{(1)},y_k^{(1)}}) g^{-1}_{j_k^{(1)},y_k^{(1)}} v_k^{(1)} \\ &= \text{w-lim}\, g^{-1}_{j_0,y_0} g^{-1}_{j_k^{(1)},y_k^{(1)}} v_k^{(1)} = 0, \end{aligned}$$

a contradiction that proves that $g^{-1}_{j_k^{(1)},y_k^{(1)}} j_k^{(2)}, y_k^{(2)} \rightharpoonup 0$ or, equivalently, $|j_k^{(1)} - j_k^{(2)}| + (2^{j_k^{(1)}} + 2^{j_k^{(2)}})|y_k^{(1)} - y_k^{(2)}| \to \infty$. Then we also have $g^{-1}_{j_k^{(2)},y_k^{(2)}} g_{j_k^{(1)},y_k^{(1)}} \rightharpoonup 0$.

Recursively, we define

$$v_k^{(n)} = v_k^{(n-1)} - g_{j_k^{(n)},y_k^{(n)}} w^{(n)} = u_k - g_{j_k^{(1)},y_k^{(1)}} w^{(1)} - \cdots - g_{j_k^{(n)},y_k^{(n)}} w^{(n)},$$

where $w^{(n)} = \text{w-lim}\, g^{-1}_{j_k^{(n)},y_k^{(n)}} v_k^{(n-1)}$, calculated on a successively renumbered subsequence. We subordinate the choice of $(j_k^{(n)}, y_k^{(n)})$, and thus the extraction of a subsequence for every given n, to the following requirements. For every $n \in \mathbb{N}$, we set

$$W_n = \{w \in \dot{BV}(\mathbb{R}^N) \setminus \{0\} : \exists (j_k, y_k) \subset \mathbb{Z} \times \mathbb{R}^N, (k) \subset \mathbb{N}^{\mathbb{N}} : g^{-1}_{j_m,y_m} v^{(n)}_{k_m} \rightharpoonup w\},$$

with the weak convergence holding up to extraction of a subsequence, and

$$t_n = \sup_{w \in W_n} \|Dw\|.$$

Note that $t_n \leq \sup \|u_k\| < \infty$. If for some n, $t_n = 0$, the theorem is proved with $r_k = v_k^{(n-1)}$. Otherwise, we choose a $w^{(n+1)} \in W_n$ such that $\|Dw^{(n+1)}\| \geq \frac{1}{2} t_n$ and the sequence $(j_k^{(n+1)}, y_k^{(n+1)})$ is chosen so that on a subsequence that we renumber, $g^{-1}_{j_k^{(n+1)}, y_k^{(n+1)}} v_k^{(n)} \rightharpoonup w^{(n+1)}$. An argument analogous to the one brought above for $n = 1$ shows that $g^{-1}_{j_k^{(p)}, y_k^{(p)}} g_{j_k^{(q)}, y_k^{(q)}} \rightharpoonup 0$ or, equivalently,

$$|j_k^{(p)} - j_k^{(q)}| + (2^{j_k^{(p)}} + 2^{j_k^{(q)}})|y_k^{(p)} - y_k^{(q)}| \to \infty \tag{8.4}$$

whenever $p \neq q$, $p, q \leq n$.

Let us show (8.3). Let $n \in \mathbb{N}$ and let $(j_k^{(i)}, y_k^{(i)})_k$, $w^{(i)}$, and $(v_k^{(i)})_k$, $i = 1, \ldots, n$, be defined as above. Let $v^{(i)} \in C_0^\infty(\mathbb{R}^N, \mathbb{R}^N)$, $\|v^{(i)}\|_\infty \leq 1$, $i = 1, \ldots, n$, and set $S_k^{(n)} = \sum_{i=1}^n g_{j_k^{(i)}, y_k^{(i)}} w^{(i)}$, $V_k^{(n)} = \sum_{i=1}^n 2^{(1-N)j_k^{(i)}} g_{j_k^{(i)}, y_k^{(i)}} v^{(i)}$. (To clarify the construction, the operator $2^{(1-N)j/2} g_{j,y}$ is the $L^2(\mathbb{R}^N)$-adjoint of $g_{j,y}^{-1}$.) Then, noting that $\|V_k^{(n)}\|_\infty \leq 1$ and taking into account (8.4), we have

$$\begin{aligned} \|Du_k\| &\geq \int v_k^{(n)} \operatorname{div} V_k^{(n)} dx + \int S_k^{(n)} \operatorname{div} V_k^{(n)} dx \\ &= \sum_{i=1}^n \int g_{-j_k^{(i)}, -y_k^{(i)}} v_k^{(n)} \operatorname{div} v^{(i)} dx + \sum_{i=1}^n \int w^{(i)} \operatorname{div} v^{(i)} dx. \end{aligned}$$

Since the first term converges to zero by construction, while $v^{(i)}$ is arbitrary, we have $\|Du_k\| \geq \sum_{i=1}^n \|Dw^{(i)}\| + o_{k\to\infty}(1)$. Since n is arbitrary, the lower bound in (8.3) follows.

Note now that $\sum_{i=1}^\infty t_i \leq 2\|Du_k\| + o(1)$. Furthermore, $\|DS_k^{(n)}\| \leq \sum_{i=1}^n t_i + o(1)$, and on a suitable subsequence we have $\|DS_k^{(n)}\| \leq 2\sum_{i=1}^n t_i$ and, furthermore, the inequality remains true even if one omits an arbitrary subset of terms in the sum $S_k^{(n)}$. Consequently, by an elementary diagonalization argument, on a suitable subsequence, series S_k^∞ converges in $\dot{BV}(\mathbb{R}^N)$ unconditionally and uniformly in k. This together with (8.4) implies that $u_k - S_k^\infty \overset{\mathcal{G}^{N-1}}{\rightharpoonup} 0$, which by Theorem 8.1.1 implies (8.2). Finally, the second inequality in (8.3) follows from convergence of S_k^∞ and the triangle inequality for norms. □

8.3 Sample minimization problems

Let $\alpha_N > 0$ be such that $w \overset{\text{def}}{=} \alpha_N \mathbb{1}_B$, where B is a unit ball, satisfies $\|Dw\| = 1$. Then it is known that w is a maximizer for the problem

$$c_0 = \sup_{u \in \dot{BV}(\mathbb{R}^N): \|Du\|=1} \int_{\mathbb{R}^N} |u|^{1^*} dx.$$

By scaling invariance, $w_R = R^{1-N} \alpha_N \mathbb{1}_{B_R}$ is also a maximizer.

Theorem 8.3.1. *Let $F : \mathbb{R} \to \mathbb{R}$ be a continuous function such that the following supremum is positive and is attained:*

$$0 < m \stackrel{\text{def}}{=} \sup_{s\in\mathbb{R}} F(s)/|s|^{1^*} = F(t)/|t|^{1^*} \quad \text{for some } t \in \mathbb{R}. \tag{8.5}$$

Then the maximum in the relation

$$c = \sup_{u\in\dot{\mathrm{BV}}(\mathbb{R}^N):\|Du\|=1} \int_{\mathbb{R}^N} F(u)\mathrm{d}x$$

is attained at $u = w_R$ with $R = (\frac{\alpha_N}{t})^{\frac{1}{N-1}}$.

Proof. Since $F(u) \le m|u|^{1^*}$, we have $c \le mc_0$. On the other hand, comparing the supremum with the value of the functional at w_R we have $c \ge \int_{\mathbb{R}^N} F(w_R)\mathrm{d}x = F(t)|B_R| = m|t|^{1^*}|B_R| = m\int_{\mathbb{R}^N} |\alpha_N R^{1-N}\mathbb{1}_{B_R}|^{1^*}\mathrm{d}x = mc_0$. Therefore, $c = mc_0$ and is attained at w_R. □

Theorem 8.3.2. *Let $0 < \lambda < N-1$. Then the minimum in*

$$\kappa = \inf_{u\in\dot{\mathrm{BV}}(\mathbb{R}^N):\int_{\mathbb{R}^N}|u|^{1^*}\mathrm{d}x=1} \|Du\| - \lambda \int_{\mathbb{R}^N} \frac{|u|}{|x|}\mathrm{d}x$$

is attained.

Proof. The proof is based on a standard use of profile decomposition and may be abbreviated. Let (u_k) be a minimizing sequence. Applying Theorem 8.2.1 and noting that there exists a subset of indices $I \subset \mathbb{N}$ such that $\int_{\mathbb{R}^N} \frac{|u_k|}{|x|}\mathrm{d}x \to \sum_{n\in I}\int \frac{|w^{(n)}|}{|x|}\mathrm{d}x$ (provided that the functions $w^{(n)}$ are rescaled, as it is always possible, by application of constant operator $g_{j_n,y_n} \in \mathcal{G}^{N-1}$), we have using the notation,

$$J(u) = \|Du\| - \lambda \int_{\mathbb{R}^N} \frac{|u|}{|x|}\mathrm{d}x,$$

and recalling (8.3),

$$J(u_k) \ge \sum_{n\in I} J(w^{(n)}) + \sum_{n\notin I} \|Dw^{(n)}\| + o(1). \tag{8.6}$$

On the other hand, from the iterated Brezis–Lieb lemma (Theorem 4.7.1) follows:

$$\int_{\mathbb{R}^N} |u_k|^{1^*} = \sum_{n\in\mathbb{N}} \int_{\mathbb{R}^N} |w^{(n)}|^{1^*} + o(1). \tag{8.7}$$

Moreover, each $w^{(n)}$ necessarily minimizes the respective functional, namely J if $n \in I$ and $\|D\cdot\|$ if $n \notin I$, over the functions $u \in \dot{\mathrm{BV}}(\mathbb{R}^N)$ satisfying $\int_{\mathbb{R}^N}|u|^{1^*} = \int_{\mathbb{R}^N}|w^{(n)}|^{1^*}$. In particular, $w^{(n)}$ for $n \notin I$ are multiples of the characteristic function of some ball, which (taken with scalar multiples) are clearly not minimizers for the functional J. From the standard convexity argument, relations (8.6) and (8.7) imply that, necessarily, $w^{(n)} = 0$ for all $n \in \mathbb{N}$ except $n = m$ with some $m \in I$. Thus, $\int_{\mathbb{R}^N} |w^{(m)}|^{1^*} dx = \int_{\mathbb{R}^N} |u_k|^{1^*} dx = 1$ and $J(w^{(m)}) \leq J(u_k) = \kappa + o(1)$. This implies that $w^{(m)}$ is a minimizer. □

8.4 Fractals and fractal blowups

In this and the subsequent sections, we consider loss of compactness in fractal blowups—the noncompact metric structures produced by iteration sequences of inversed constituent maps of a fractal (expansion maps). There is generally an uncountable family of different blowups of the same fractal, parametrized by the infinite words of indices that determine the sequence of expansion maps. Similar to manifolds, nontrivial isometry groups on fractal blowups generally do not exist, and concentration profiles, produced by local isometries, emerge as functions on different blowups. Existence of local isometries with uniform properties, by analogy with manifolds, require a counterpart of the condition of bounded geometry, which in the case of fractals is a condition that all constituent maps scale the fractal measure, as well as the fractal energy, by the same factor. In this setting, a uniform family of local isometries on the blowup can be produced as a compositions of two maps, a "zoom-in" composition of M constituent maps of the fractal and a "zoom-out" composition of the first M members of the blow-up sequence.

Let us define a subset of the class of pcf (post-critically finite) fractals, introduced by Kigami [72], as well as correspondent energy spaces following [117]. An essential restriction below is that the constituent maps of the fractal are to have the same scaling factor. The class includes the Sierpinski gasket.

Let $\psi_i : \mathbb{R}^n \to \mathbb{R}^n$, $i \in \{1, \dots, N\}$, be the contractive similitudes, satisfying

$$|\psi_i(x) - \psi_i(y)| \leq \alpha^{-1}|x - y| \tag{8.8}$$

with some $\alpha > 1$ and assume that there is an open set $U \subset \mathbb{R}^N$ such that

$$U \subset \bigcup_i \psi_i(U). \tag{8.9}$$

There exists a unique compact set $\Omega \subset \mathbb{R}^n$ satisfying

$$\Omega = \bigcup_i \psi_i(\Omega) \tag{8.10}$$

and there is a unique Borel regular measure μ on Ω such that for every integrable $u : \Omega \to \mathbb{R}$,

$$\int_\Omega u \, d\mu = \alpha^{-d} \sum_{i=1}^N \int_\Omega u \circ \psi_i \, d\mu \tag{8.11}$$

where $d = \frac{\log N}{\log \alpha}$. The set Ω is called then a self-similar fractal. An equivalent form of (8.11) is

$$\mu(A) = \alpha^{-d} \sum_{i=1}^{N} \mu(\psi_i^{-1}(A) \cap \Omega). \tag{8.12}$$

Let $\partial\Omega$ be the set of fixed points of ψ_k, $k = 1, \dots, N_0$ with some $N_0 \le N$. We assume that Ω is connected and satisfies the finite ramification condition

$$\psi_i(\Omega) \cap \psi_j(\Omega) \subset \psi_i(\partial\Omega) \cap \psi_j(\partial\Omega) \quad \text{whenever } i \neq j. \tag{8.13}$$

For every function $u : \Omega \to \mathbb{R}$, a positive quadratic form (energy) $E(u) \in [0, +\infty]$, is defined (see [72, 117]), satisfying for every $i \in \{1, \dots, N\}$, $u \in H$, $u \circ \psi_i \in H$, and

$$E(u) = \rho \sum_{i=1}^{N} E(u \circ \psi_i) \tag{8.14}$$

with some $\rho > 0$. Domain $\mathcal{D}$ of $E(u)$ consists of functions for which $E(u) < \infty$. The Sobolev space $H^1(\Omega)$, which we in what follows abbreviate as H, is defined as the linear space $\mathcal{D} \cap L^2(\Omega)$, equipped with the norm

$$\|u\|^2 = E(u) + \|u\|_{2,\mu}^2. \tag{8.15}$$

By definition, H is continuously imbedded into $L^2(\Omega, \mu)$. Moreover, it is compactly imbedded into $L^p(\Omega)$ for all $p \in [1, \infty)$ if $d \le 2$ and for $p \in [1, \frac{2d}{d-2})$ if $d > 2$. In particular, there exists $C > 0$ such that

$$\left(\int_{\Omega} |u|^p \mathrm{d}\mu \right)^{\frac{2}{p}} \le C\left(E(u) + \int_{\Omega} |u|^2 \mathrm{d}\mu \right), \quad u \in H. \tag{8.16}$$

In what follows, we assume that $d < 2$, in which case H is also continuously imbedded into $C(\Omega)$. The space H_0 of functions in H vanishing on $\partial\Omega$ is a proper subspace of H. The functions in H admit continuous restrictions to and continuous extensions from the sets $\psi_i(\Omega)$. The latter are also continuous operators $H_0 \to H_0$. As long as it is not ambiguous, we will not distinguish in notations between the functions and their extensions, respectively, restrictions. In such terms, one has, in particular,

$$E(u \circ \psi_i^{-1} \circ \psi_j) = 0 \quad \text{whenever } i \neq j. \tag{8.17}$$

An infinite blowup Ω^I of Ω, relative to a sequence $I = \{i_1, i_2, \dots\}$, $i_k \in \{1, \dots, N\}$, is the monotone increasing union

$$\bigcup_{M=1}^{\infty} \Omega_M^I, \quad \text{where } \Omega_M^I := \Phi_M^I(\Omega) \text{ and } \Phi_M^I := \psi_{i_1}^{-1} \circ \dots \circ \psi_{i_M}^{-1}, \; M \in \mathbb{N}. \tag{8.18}$$

For the sake of consistency, we set $\Omega_0 = \Omega$ and $\Phi_0 = \mathrm{id}$.

The measure μ and the functional E can be extended to Ω^I and to functions thereupon by self-similarity, as follows. The measure μ induces a measure

$$\mu_M^I = \alpha^{dM}\mu \circ {\Phi_M^I}^{-1} \quad \text{on } \Omega_M^I, \quad M \in \mathbb{N}. \tag{8.19}$$

From (8.13) and (8.11) easily follows that the measures μ_M^I and μ_{M+1}^I coincide on Ω_M^I, $M = 0, 1, \dots$. This defines, by $\mu_{M+1}^I|_{\Omega_M} = \mu_M^I$, a measure on a generator set of a σ-algebra on the whole Ω^I, and thus, a Borel measure on Ω^I.

A similar construction yields an energy functional for the blowup. For a finite blowup Ω_M^I, we set

$$E_M^I(u) = \rho^{-M} E(u \circ \Phi_M^I), \tag{8.20}$$

whenever $u \in H_M^I := \{v \circ {\Phi_M^I}^{-1}, v \in H\}$.

Note that if $u \in H_{0,M}^I := \{v \circ {\Phi_M^I}^{-1}, v \in H_0\}$ then the extension of u by zero to Ω_{M+1}^I is an element of $H_{0,M+1}^I$ (we will extend the adopted convention not to distinguish in notation between u and its extension to this instance). From (8.14) and (8.17), $E_M^I(u) = E_{M+1}^I(u)$. This defines $E^I(u)$ for any $u \in H_0^I := \bigcup_{M\in\mathbb{N}} H_{0,M}^I$. The Hilbert space H^I is defined as the completion of H_0^I with respect to the norm

$$\|u\|_I := \left(E^I(u) + \int_{\Omega^I} |u|^2 d\mu^I\right)^{1/2}, \quad I \in \{1, \dots, N\}^{\mathbb{N}}.$$

8.5 Sobolev inequality and cocompactness on fractals

Let $J = \{j_1, j_2, \dots\}$, $j_k \in \{1, \dots, N\}$, $\Phi_M^J = \psi_{j_1}^{-1} \circ \dots \psi_{j_M}^{-1}$, and let

$$\eta_{I,J,M} \overset{\text{def}}{=} \Phi_M^I \circ {\Phi_M^J}^{-1} : \Omega_M^J \to \Omega_M^I. \tag{8.21}$$

Let $I, J \in \{1, \dots, N\}^{\mathbb{N}}$, $M \in \mathbb{N}$, let

$$\mathcal{J}_M^I \overset{\text{def}}{=} \{\eta_{I,J,M}(\Omega) \mid J \in \{1, \dots, N\}^{\mathbb{N}}\}$$

let

$$\mathcal{J}^I \overset{\text{def}}{=} \bigcup_{M\in\mathbb{N}} \mathcal{J}_M^I.$$

Lemma 8.5.1. *Let $I, J \in \{1, \dots, N\}^{\mathbb{N}}$. The collection of sets $\mathcal{J}^I$ is a covering of Ω^I. Furthermore, for every integrable function w on (Ω^I, μ^I),*

$$\int_{\Omega^I} w d\mu^I = \sum_{\eta_{I,J,M}(\Omega)\in\mathcal{J}^I} \int_{\eta_{I,J,M}(\Omega)} w d\mu^I = \sum_{\eta_{I,J,M}(\Omega)\in\mathcal{J}^I} \int_{\Omega} w \circ \eta_{I,J,M} d\mu, \tag{8.22}$$

and for every $u \in H^I$

$$E^I(u) = \sum_{\eta_{I,J,M}(\Omega)\in \mathcal{J}^I} E(u \circ \eta_{I,J,M}), \tag{8.23}$$

where the terms in the last two sums, corresponding to J, M resp. J', M' such that $\eta_{I,J,M}|_\Omega = \eta_{I,J',M'}|_\Omega$, are repeated only once.

Proof. Let $x \in \Omega^I$. By definition of Ω^I, there exist $M \in \mathbb{N}$ and $y \in \Omega$, such that $x \in \Phi^I_M y$. Furthermore, by (8.10) there is a $i \in \{1,\dots,N\}$ such that $y \in \Phi^{J^{-1}}_M \Omega$ for some J. This proves that $\mathcal{J}^I$ is a covering.

By density, it suffices to prove (8.22) for functions from $H^I_{0,M}$, $M \in \mathbb{N}$, that is, to show that for every μ^I-measurable function w on Ω^I_M,

$$\int_{\Omega^I_M} w\,d\mu^I_M = \sum_{J\in\{1,\dots,N\}^M} \int_{\eta_{I,J,M}(\Omega)} w\,d\mu^I_M. \tag{8.24}$$

Let $v = w \circ \Phi^I_M$. Then (8.24) is equivalent to

$$\int_\Omega v\,d\mu = \sum_{J\in\{1,\dots,N\}^M} \int_{\Phi^{J^{-1}}\Omega} v\,d\mu = \sum_{J\in\{1,\dots,N\}^M} \int_{\psi_{j_M}\circ\dots\circ\psi_{j_1}\Omega} v\,d\mu.$$

The last relation easily follows from (8.10) and (8.13).

It suffices to prove (8.23) for functions in $H^I_{0,M}$, $M \in \mathbb{N}$, that is, to show

$$E^I_M(u) = \sum_{J\in\{1,\dots,N\}^M} E(u \circ \eta_{I,J,M}), \quad \text{for } u \in H^I_{0,M}. \tag{8.25}$$

If we set $u = w \circ \Phi^I_M$, then (8.25) is equivalent to

$$E(v) = \rho^{-M} \sum_{J\in\{1,\dots,N\}^M} E(v \circ \Phi^{J^{-1}}_M), \quad \text{for } v \in H_0,$$

which in turn is the Mth iteration of (8.14). □

Lemma 8.5.2. *Let $I, J \in \{1,\dots,N\}^N$, and let $\Omega' \subset \Phi^J_M(\Omega)$ be a μ^J-measurable set. For any μ^I-measurable function $u : \Omega^I \to \mathbb{R}$,*

$$\int_{\eta_{I,J,M}(\Omega')} u\,d\mu^I = \int_{\Omega'} u \circ \eta_{I,J,M}\,d\mu^J. \tag{8.26}$$

Proof. Using (8.19),

$$\int_{\eta_{I,J,M}(\Omega')} u\,d\mu^I = \int_{\Phi^I_M \Phi^{J^{-1}}_M(\Omega')} u\,d\mu^I_M$$

$$= \alpha^{-Md} \int_{\Phi_M^{J^{-1}}(\Omega')} u \circ \Phi_M^I \,d\mu = \int_{\Omega} u \circ \Phi_M^I \Phi_M^{J^{-1}} \,d\mu_M^J.$$

with understanding that the composition $u \circ \eta_{I,J,M}$, although not defined on the whole Ω^I, is defined on the domain of the integration. □

Corollary 8.5.3. *For any μ^I-measurable function $u : \Omega^I \to \mathbb{R}$,*

$$\int_{\eta_{I,J,M}\Omega} u \,d\mu^I = \int_{\Omega} u \circ \eta_{I,J,M} \,d\mu. \tag{8.27}$$

Lemma 8.5.4. *Let $I, J \in \{1, \dots, N\}^{\mathbb{N}}$, $M \in \mathbb{N}$. For every $u \in H^I_{0,M}$,*

$$E_M^J(u \circ \eta_{I,J,M}) = E_M^I(u).$$

Proof. By (8.20) and the definition of $\eta_{I,J,M} = \Phi_M^I \circ \Phi_M^{J^{-1}}$,

$$E_M^J(u \circ \eta_{I,J,M}) = \rho^{-M} E(u \circ \Phi_M^I) = E_M^I(u).$$ □

Proposition 8.5.5. *Let $p > 2$. The following Sobolev inequality holds true:*

$$\left(\int_{\Omega^I} |u|^p d\mu^I\right)^{\frac{2}{p}} \le C\left(E^I(u) + \int_{\Omega^I} |u|^2 d\mu^I\right), \quad u \in H^I. \tag{8.28}$$

Proof. It suffices to consider $u \in H_0$. From (8.16) for $u \circ \eta_{I,J,M}$ and Corollary 8.5.3 for $J \in \{1, \dots, N\}^{\mathbb{N}}$, $M \in \mathbb{N}$, follows

$$\left(\int_{\eta_{I,J,M}(\Omega)} |u|^p \,d\mu^I\right)^{\frac{2}{p}} \le C\left(E(u \circ \eta_{I,J,M}|_{\Omega}) + \int_{\eta_{I,J,M}(\Omega)} |u|^2 d\mu^I\right), \quad u \in H_0^I. \tag{8.29}$$

Add the inequalities above over $J \in \{1, \dots, N\}^M$, use Lemma 8.5.1, and then subadditivity of the left-hand side. □

Lemma 8.5.6. *Let $u_k \in H^I$ be a bounded sequence and assume that for every sequence (Ω_k) in $\mathcal{J}^I$, $\Omega_k = \eta_{I,J_k,M_k}(\Omega)$, $u_k \circ \eta_{I,J_k,M_k}|_{\Omega} \to 0$ in $L^p(\Omega, \mu)$. Then $u_k \to 0$ in $L^p(\Omega^I, \mu^I)$.*

Proof. From (8.29), it is immediate for all $u \in H^I$ that

$$\int_{\eta_{I,J,M}(\Omega)} |u|^p \,d\mu^I$$

$$\le C(E(u \circ \eta_{I,J,M})|_{\Omega}) + \int_{\eta_{I,J,M}(\Omega)} |u|^2 d\mu^I)\left(\int_{\eta_{I,J,M}(\Omega)} |u|^p \,d\mu^I\right)^{1-\frac{2}{p}}.$$

Adding the inequalities above for $\eta_{I,J,M}\Omega \in \mathcal{J}^I$ and using Lemma 8.5.1 together with subadditivity of the left-hand side we obtain, setting $u = u_k$,

$$\int_{\Omega^I} |u_k|^p \, d\mu^I$$
$$\leq C\Big(E(u_k) + \int_{\Omega^I} |u_k|^2 d\mu^I\Big) \sup_{\eta^{-1}_{I,J,M}(\Omega)\in\mathcal{J}^I} \Big(\int_{\Omega} |u_k \circ \eta_{I,J,M}|^p \, d\mu\Big)^{1-\frac{2}{p}}.$$

Let $\Omega_k = \eta^{-1}_{I,J_k,M_k}(\Omega) \in \mathcal{J}^I$, be such that

$$\int_{\Omega} |u_k \circ \eta_{I,J_k,M_k}|^p \, d\mu \geq \frac{1}{2} \sup_{\eta^{-1}_{I,J,M}(\Omega)\in\mathcal{J}^I} \int_{\Omega} |u_k \circ \eta_{I,J,M}|^p \, d\mu.$$

Then, by the assumption of the lemma,

$$\int_{\Omega^I} |u_k|^p \, d\mu^I$$
$$\leq C\Big(E(u_k) + \int_{\Omega^I} |u_k|^2 d\mu^I\Big)\Big(\int_{\Omega} |u_k \circ \eta_{I,J_k,M_k}|^p \, d\mu\Big)^{1-\frac{2}{p}} \to 0.$$ □

8.6 Minimizers on fractal blowups

Proposition 8.6.1. *Let $p > 2$ and let*

$$c^I = \inf\Big\{E^I(u) + \int_{\Omega^I} |u|^2 d\mu^I : \quad u \in H^I, \int_{\Omega^I} |u|^p = 1\Big\}. \tag{8.30}$$

Then for every $I, J \in \{1, \dots, N\}^{\mathbb{N}}$, $c^I = c^J$.

Proof. It suffices to show that $c^I \geq c^J$. Let $\epsilon > 0$ and let $u_\epsilon \in H^I_0$ be such that $\int_{\Omega^I} |u_\epsilon|^p d\mu^I = 1$ and $E^I(u) + \int_{\Omega^I} |u_\epsilon|^2 d\mu^I \leq c^I + \epsilon$. By definition of H^I_0, there exists $M_\epsilon \in \mathbb{N}$ such that $u_\epsilon \in H^I_{0,M_\epsilon}$. Let $v_\epsilon = u_\epsilon \circ \eta_{I,J,M_\epsilon}$. Then by Lemma 8.5.4 and Lemma 8.5.2, we have $E^J(v_\epsilon) = E^I(u_\epsilon)$, $\int_{\Omega^J} |v_\epsilon|^2 d\mu^J = \int_{\Omega^I} |u_\epsilon|^2 d\mu^I$ and $\int_{\Omega^J} |v_\epsilon|^p d\mu^J = \int_{\Omega^I} |u_\epsilon|^p d\mu^I = 1$. Consequently, $c^J \leq c^I + \epsilon$. Since ϵ, I, and J are arbitrary, the lemma follows. □

Due to the proposition above, we may denote the common value of constants c^I, $I \in \{1, \dots, N\}^{\mathbb{N}}$, as c^Ω. Note that $c^\Omega > 0$ due to (8.29).

Theorem 8.6.2. *Let Ω be a self-similar fractal equipped with the energy E as defined in Section 8.4. Let Ω^I, $I \in \{1, \dots, N\}^{\mathbb{N}}$, be its blowup with correspondent energy E^I, as defined in Section 8.4, and let $p > 2$. Then there exists $J \in \{1, \dots, N\}^{\mathbb{N}}$ such that the minimum in (8.30) with $I = J$ is attained.*

Proof. The proof consists of three steps. On the first step one moves an Ω-sized "spotlight" $\eta_{I,J,M}(\Omega)$ to find a weak limit of the minimizing sequence in restriction to the spotlight domain. At this step, we also obtain the multiindex $J \in \{1,\dots,N\}^N$ from the sequence of spotlight shifts $\eta_{I,J,M}$.

On the second step, we expand the size of the spotlight to the blowup Ω^J, which is generally different from Ω^I, and which becomes a domain of the weak limit for a shifted sequence of u_k.

The third step is a standard concentration compactness argument based on the Brezis–Lieb lemma for functions on Ω^J.

Step 1. Let $u_k \in H_0^I$ be a minimizing sequence for (8.30), that is, $\int_{\Omega^I} |u_k|^p \mathrm{d}\mu^I = 1$ and $E^I(u_k) + \int_{\Omega^I} |u_k|^2 \mathrm{d}\mu^I \to c^\Omega$. Since u_k does not converge to zero in $L^p(\Omega^I, \mu^I)$, by Lemma 8.5.6, there is a sequence of $J_k \in \{1,\dots,N\}^N$, $M_k \in \mathbb{N}$, such that $u_k \circ \eta_{I,J_k,M_k}$ does not converge in $L^p(\Omega,\mu)$ to zero, and since the local Sobolev imbedding (8.16) is compact, the sequence does not converge to zero weakly in H. It is bounded, however, in H due to Lemma 8.5.1 and Lemma 8.5.4. Thus, there exists a $w_1 \in H$, such that on a renumbered subsequence, $u_k \circ \eta_{I,J_k,M_k}|_\Omega \rightharpoonup w_1 \neq 0$ in H.

Step 2. Consider the sequence of maps

$$\eta_{I,J_k,M_k} = \Phi^I_{M_k} \circ \psi_{j^k_{M_k}} \circ \dots \circ \psi_{j^k_1} : \Omega^{J_k}_{M_k} \to \Omega^{I_k}_{M_k}. \tag{8.31}$$

We recall that we consider all functions of the class H_0^I extended by zero to all of Ω^I. Without loss of generality, as both composition chains $\Phi^I_{M_k}$ and $\Phi^J_{M_k}$ may be lengthened with mutually cancelling terms, we may assume that the values of renamed M_k are so large that $u_k \in H^I_{0,M_k}$. In more detail, assume first that $u_k \in H^I_{0,M_k+m_k}$, with some $m_k \in \mathbb{N}$, set $j_{M_k+m} \overset{\text{def}}{=} i_{M_k+m}$, $m = 1,\dots,m_k$ and let $\Phi_{M_k+m_k} := \psi_{j_{M_k+m_k}} \circ \dots \circ \psi_1$. Then $\eta_{I,J_k,M_k} = \Phi^I_{M_k+m_k} \Phi^J_{M_k+m_k}$. The map

$$\eta_{I,J_k,M_k+m_k} : \Omega^{J_k}_{M_k+m_k} \to \Omega^{I_k}_{M_k+m_k}$$

is an extension of the map $\eta_{I,J_k,M_k} : \Omega^{J_k}_{M_k} \to \Omega^{I_k}_{M_k}$. As we rename $M_k + m_k$ as M_k, the map η_{I,J_k,M_k+m_k}, acquires the notation η_{I,J_k,M_k} of the map it extended.

There is a renamed subsequence J_k^1 where $j_{1,k} \in \{1,\dots,N\}$ is constant, to be denoted as j_1. Moreover, if for a given $m \in \mathbb{N}$ there is a subsequence J_k^m where $j_{1,k},\dots,j_{m,k}$ are constant, then it has an extraction where $j_{m+1,k}$ is constant as well. Let $J := (j_1, j_2, \dots)$. Finally, rename $J_k^{M_k}$ as J_k so that $j_{i,k} = j_i$ for $i = 1,2,\dots,M_k$ (so that the componentwise limit of J_k is J).

The map η_{I,J_k,M_k} is defined then as a map $\Omega^J_{M_k} \to \Omega^I_{M_k}$ (since the components of J_k with $k > M_k$ are not involved in the definition of η_{I,J_k,M_k}) and $u_k \circ \eta_{I,J_k,M_k} : \Omega^J_{M_k} \to \mathbb{R}$ is a bounded sequence in H^J. Then, on a renamed subsequence, $u_k \circ \eta_{I,J_k,M_k} \rightharpoonup w$ in H^J. Due to the step 1, $w|_\Omega = w_1 \neq 0$.

Step 3. Let $v_k := u_k \circ \eta_{I,J_k,M_k} - w$. By Lemma 8.5.2 and the Brezis–Lieb lemma (Corollary 1.3.3),

$$\begin{aligned} 1 = \lim \int_{\Omega^I} |u_k|^p d\mu^I &= \lim \int_{\Omega^J} |u_k \circ \eta_{I,J_k,M_k}|^p d\mu^J \\ &= \lim \int_{\Omega^J} |v_k|^p d\mu^J + \int_{\Omega^J} |w|^p d\mu^J. \end{aligned} \tag{8.32}$$

We also have, since $v_k \rightharpoonup 0$ in H^J, using the scalar products of, respectively H^I and H^J,

$$c^\Omega = \lim \|u_k\|_I^2 = \lim \|u_k \circ \eta_{I,J_k,M_k}\|_J^2 = \lim \|v_k\|_J^2 + \|w\|_J^2. \tag{8.33}$$

Let $t := \lim \|v_k\|_{p,\mu^J}^p$. Then $\|w\|_{p,\mu^J}^p = 1 - t$, and by (8.30), from (8.33) follows:

$$c^\Omega \geq c^\Omega t^{p/2} + c^\Omega (1-t)^{p/2},$$

which is true only if $t = 1$ (which is impossible since $w_1 \neq 0$, and thus $w \neq 0$) or $t = 0$. Therefore, $\|w\|_{p,\mu^J}^p = 1$, which easily yields that w is a minimizer for (8.30) with $I = J$. □

8.7 Bibliographic notes

The first three sections are based on the paper [9], with corrections to both the statement and the proof of Theorem 8.2.1. A similar profile decomposition, with a weaker remainder and limited to sequences of characteristic functions of sets (but on a general Riemannian manifold) is given in [95, Lemma 2.2] (cf. also references to the prior work of Nardulli therein).

The remaining three sections deal with the scalar field equation on fractals (see [45], Falconer and Hu [46], and Matzeu [89]) associated with a Dirichlet form ([72, 93, 117]) and extended by self-similarity to fractal blowups (see [116]). The presentation is based on [106].

9 Sample applications to variational problems

This chapter contains a very small selection of variational problems where profile decompositions in Sobolev spaces are used to prove existence of critical points, or to specify the structure of blowups for critical sequences. The main technical steps in finding critical points of variational problems is to identify a minimax statement that yields a critical sequence and to prove convergence of the critical sequence (the Palais–Smale condition). The most elementary minimax statements are upper/lower bound of the functional or mountain pass geometry. A common, although far from optimal, sufficient condition to have a norm bound for a critical sequence for a semilinear elliptic functional is the Ambrosetti–Rabinowitz condition [10]. Convergence of the critical sequences is assured by compactness argument, and in absence of compactness, by means of profile decomposition.

9.1 Nonlinear Schrödinger equation with positive mass

Let $N \geq 2$. Let $p \in (1, 2^* - 1)$, and let $f : \mathbb{R} \to \mathbb{R}$ be a continuous function satisfying:
(f1) $|f(s)| \leq C|s|^p$,
(f2) $\lim_{s\to 0} \frac{f(s)}{s} = 0$,
(f3) $\lim_{|s|\to\infty} \frac{|f(s)|}{|s|} = \infty$.

Let

$$F(s) = \int_0^s f(t)\mathrm{d}t \tag{9.1}$$

and let

$$\varphi(u) = \int_{\mathbb{R}^N} F(u)\mathrm{d}x. \tag{9.2}$$

Functional φ has a Fréchet derivative in $H^{1,2}(\mathbb{R}^N)$ given by $v \mapsto \int_{\mathbb{R}^N} f(u)v\,\mathrm{d}x$, which by (f1), (f2) is a continuous map $H^{1,2}(\mathbb{R}^N) \to H^{1,2}(\mathbb{R}^N)$. In other words, we write $\varphi \in C^1(H^{1,2}(\mathbb{R}^N))$. Moreover, φ' remains continuous when the domain and the target space are equipped with the weak topology. We will consider the following C^1-functional on $H^{1,2}(\mathbb{R}^N)$:

$$J_{V,F}(u) = \frac{1}{2}\int_{\mathbb{R}^N} (|\nabla u|^2 + V(x)u^2)\mathrm{d}x - \int_{\mathbb{R}^N} F(u)\mathrm{d}x, \quad u \in H^{1,2}(\mathbb{R}^N). \tag{9.3}$$

Let $V \in L^\infty(\mathbb{R}^N)$, $\lim_{|x|\to\infty} V(x) = V_\infty > 0$, and let us use the equivalent Sobolev norm:

$$\|u\|_{1,2,V} = \left(\int_{\mathbb{R}^N} (|\nabla u|^2 + V(x)u^2)\mathrm{d}x\right)^{\frac{1}{2}}. \tag{9.4}$$

https://doi.org/10.1515/9783110532432-009

The functional $J_{V,F}$ has a continuous Fréchet derivative on $H^{1,2}(\mathbb{R}^N)$, with $J'_{F,V}(u) = u - \varphi'(u)$, so that critical points of $J_{V,F}$ are defined by

$$-\Delta u + V(x)u = f(u) \quad \text{a.e. in } \mathbb{R}^N. \tag{9.5}$$

Lemma 9.1.1. *Assume* (f1)–(f3). *Then the function $t \mapsto \gamma(t)$, $t \in [0,\infty)$, given by*

$$\gamma(t) = \sup_{\frac{1}{2}\|u\|^2_{1,2,V}=t} \int_{\mathbb{R}^N} F(u)dx, \tag{9.6}$$

is locally Lipschitz continuous, nondecreasing and satisfies

$$\gamma(t)/t \to 0 \quad \text{as } t \to 0 \tag{9.7}$$

and

$$\gamma(t)/t \to \infty \quad \text{as } t \to \infty. \tag{9.8}$$

Proof. 1. It is easy to see that γ is locally Lipschitz continuous since φ' is bounded on every ball.

2. To prove the monotonicity, let $\Omega = (0,1)^N$ and let $v_k \in H_0^{1,2}(\Omega) \setminus \{0\}$, $k \in \mathbb{N}$, considered as extended by zero to functions in $H^{1,2}(\mathbb{R}^N)$, satisfy $v_k \rightharpoonup 0$ and $\|v_k\|_{1,2,V} = 1$. For example, one can choose functions $v_k(x) = \prod_{i=1}^N \sin(k\pi x_i)$ normalized in the norm (9.4). Let $t, \tau \geq 0$, and let u satisfy $\frac{1}{2}\|u\|^2_{1,2,V} = t$. Then, using expansion by bilinearity and $v_k \rightharpoonup 0$, we have $\frac{1}{2}\|u + \tau v_k\|^2_{1,2,V} \to t + \frac{1}{2}\tau^2$, while $\varphi(u + \tau v_k) \to \varphi(u)$. Therefore, $\gamma(t + \frac{1}{2}\tau^2) \geq \varphi(u)$. Taking the supremum over all u with $\frac{1}{2}\|u\|^2_{1,2,V} = t$, we have $\gamma(t + \frac{1}{2}\tau^2) \geq \gamma(t)$ with an arbitrary τ.

3. Relations (9.7) and (9.8) follow from (f2) and (f3), respectively. □

Definition 9.1.2. A sequence (u_k) in a Banach space E is called a critical sequence for a C^1-functional J on E if $J'(u_k) \to 0$ in E^* and $J(u_k) \to c$ for some $c \in \mathbb{R}$.

Lemma 9.1.3. *Assume* (f1)–(f3). *There exists a number $C > 0$ such that any profile decomposition with respect to group of shifts $\mathcal{G}_{\mathbb{Z}^N}$, for every critical sequence (u_k) of the functional* (9.3), *satisfying $\|u_k\|_{1,2,V_\infty} \leq L$, $L > 0$, has at most $M = CL^2 + 1$ terms. Furthermore, with a convention that a sum over an empty set equals zero,*

$$J_{V,F}(u_k) \to J_{V,F}(w^{(1)}) + \sum_{n=2}^{M} J_{V_\infty,F}(w^{(n)}). \tag{9.9}$$

Proof. Note that if $u_k(\cdot - y_k) \rightharpoonup w$ and $|y_k| \to \infty$, then by (f1) function w satisfies $-\Delta w + V_\infty w = f(w)$ and, in particular, by the Pohozhaev identity ([133, Corollary B4]), $\|w\|^2_{1,2,2^*V_\infty} = 2^* \int_{\mathbb{R}^N} F(w)dx$, which gives $2\tau \leq 2^*\gamma_\infty(\tau)$ where $\tau = \frac{1}{2}\|w\|^2_{1,2,V_\infty}$, and

$$\gamma_\infty(t) \stackrel{\text{def}}{=} \sup_{\frac{1}{2}\|u\|^2_{1,2,V_\infty}=t} \int_{\mathbb{R}^N} F(u)dx, \quad t \in [0,\infty).$$

Note that y_∞ is a function of the form (9.6), so by Lemma 9.1.1 the infimum in

$$t_\infty \stackrel{\text{def}}{=} \inf\{t > 0 : 2^* y_\infty(t) \geq 2t\} \tag{9.10}$$

is taken on a nonempty set and is positive. Thus $\|w\|^2_{1,2,V_\infty} \geq 2t_\infty$. Then by (1.15), if the profile decomposition has at least M terms, and at least $M-1$ terms correspond to $|y_k| \to \infty$. Thus $2(M-1)t_\infty \leq \limsup \|u_k\|^2_{1,2,V_\infty}$, which gives $M \leq 1 + \frac{\limsup \|u_k\|^2_{1,2,V_\infty}}{2t_\infty}$.

Relation (9.9) follows from (1.15) and Theorem 4.7.3. □

Case $V \leq V_\infty$

Theorem 9.1.4. *Let $f : \mathbb{R} \to \mathbb{R}$ be a continuous odd function satisfying* (f1), (f2), *as well as*
(f4) *$\frac{f(t)}{t}$ is increasing on $(0, \infty)$, and*
(f5) *$f(s)s - \mu F(s) \geq 0$ with some $\mu > 2$.*

Let $J_{V,F}$ be the functional (9.3) *with $V \in L^\infty(\mathbb{R}^N)$ and $V_\infty = \lim_{|x|\to\infty} V(x) > 0$. Let Γ be a set of all continuous paths $\gamma : [0,\infty) \to H^{1,2}(\mathbb{R}^N)$ such that $\gamma(0) = 0$ and $J_{V,F}(\gamma(s)) \to -\infty$ as $|s| \to \infty$. Let*

$$c_{V,F} = \inf_{\eta\in\Gamma} \sup_{s\geq 0} J_{V,F}(\eta(s)). \tag{9.11}$$

If $V \leq V_\infty$ a.e. on $\mathbb{R}^N$, then the functional $J_{V,F}$ has a critical point u with $J_{V,F}(u) = c_{V,F} > 0$.

Proof. By (f3), the set Γ contains every path of the form $t \mapsto tv$, $v \neq 0$. The functional $J_{V,F}$ has mountain pass geometry, namely $J_{V,F}$ is zero at the origin, by (9.7) it is positive on every sphere of sufficiently small radius, while the path η is necessarily unbounded and thus crosses these spheres. By the standard mountain pass argument (e. g., [133, Theorem 1.15]), functional $J_{V,F}$ has a critical sequence (u_k) in $H^{1,2}(\mathbb{R}^N)$ satisfying $J'_{V,F}(u_k) \to 0$ and $J_{V,F}(u_k) \to c_{V,F} > 0$. By (f5) and the standard argument of Ambrosetti–Rabinowitz, [10], it follows that this sequence is bounded. Consider a renamed subsequence of (u_k) that admits a profile decomposition relative to the group of shifts $\mathcal{G}_{\mathbb{Z}^N}$.

Let us show that $c_{V,F} \leq J_{V,F}(w^{(1)})$ and $c_{V,F} \leq J_{V_\infty,F}(w^{(n)})$ for $n \geq 2$ whenever the respective profile is not zero. This follows from the inequality $c_{V,F} \leq \max_{t>0} J_{V,F}(\eta(t))$, where $\eta(t) = t^{1/2}w^{(n)}(\cdot - y)$, $y \in \mathbb{Z}^N$. Indeed, when $w^{(1)} \neq 0$ set $y = 0$. By (f4) $t \mapsto J_{V_\infty,F}(\eta(t))$ is a strictly concave function, and by the chain rule $t = 1$ is a critical point of $J_{V,F}(\eta(t))$ since $J'_{V,F}(w^{(1)}) = 0$. Therefore, $t = 1$ is a point of maximum and $c_{V,F} \leq J_{V,F}(w^{(1)})$. For $n \geq 0$, when $w^{(n)} \neq 0$, take $|y| \to \infty$. Then

$$|J_{V,F}(\eta(t)) - J_{V_\infty,F}(\eta(t))| \leq t \int_{\mathbb{R}^N} |V(x+y) - V_\infty||w^{(n)}(x)|^2 dx \to 0 \quad \text{as } |y| \to \infty$$

uniformly for t near the maximal value for all y. Thus $c_{V,F} \le \max_{t>0} J_{V_\infty,F}(\eta(t))$, where $\eta(t) = t^{1/2} w^{(n)}$ and the argument above for $w^{(1)}$ applies in this case as well, giving $c_{V,F} \le J_{V_\infty,F}(w^{(n)})$.

Comparing this with (9.9), we conclude that the profile decomposition for (u_k) consists of a single nonzero term, say $w^{(m)}$. Consider two cases.

Case 1: $V = V_\infty$ a. e. In this case, $(u_k(\cdot + y_k^{(m)}))$ is also a critical sequence, which we rename as u_k, which gives us $m = 1$.

Case 2: $V < V_\infty$ on a set of positive measure. Assume that $m \ge 2$. Then we have, taking into account that $w^{(m)} \ne 0$ by the maximum principle,

$$J_{V_\infty,F}(w^{(m)}) \le c_{V,F} \le \max_{t \in [0,\infty)} J_{V,F}(t w^{(m)}) < J_{V_\infty,F}(w^{(m)}),$$

a contradiction, which shows that $m = 1$, and thus $u_k \to w^{(1)}$ in $L^p(\mathbb{R}^N)$, and thus $\varphi'(u_k) \to \varphi'(w^{(1)})$. Since (u_k) is a critical sequence, we have $u_k = (u_k - \varphi'(u_k)) + \varphi'(u_k)$, that is, a sum of two sequences convergent in $H^{1,2}(\mathbb{R}^N)$. Therefore, u_k converges in $H^{1,2}(\mathbb{R}^N)$ to its weak limit $w^{(1)}$. Thus $J'_{V,F}(w^{(1)}) = 0$ and $J(w^{(1)}) = c_{V,F}$. □

Mountain pass solution as a ground state

Let V be as in the previous section and let $F(u) = \frac{1}{p}|u|^p$, $p \in (2, 2^*)$, which obviously satisfies conditions (f1)–(f5).

Let w be a critical point given by Theorem 9.1.4. Consider the minimax value in (9.11), noting that $t \mapsto J_{V,F}(\sqrt{t} w)$ is a concave function and its unique maximum is attained at $t = 1$ because w is a critical point:

$$c_{V,F} \le \max_{t>0} J_{V,F}(\sqrt{t} w) = J_{V,F}(w) = c_{V,F}. \tag{9.12}$$

This implies that the infimum in (9.11) is attained on the path $t \mapsto \sqrt{t} w$ and that $c_{V,F} = \max_{t>0} J_{V,F}(\sqrt{t} w)$. Comparing the maximal value of the functional on the optimal path $t \mapsto \sqrt{t} w$ with other straight-line paths $t \mapsto \sqrt{t} u$, we have

$$\begin{aligned} c_{V,F} &= \inf_{u \in H^{1,2} : \|u\|_p = 1} \max_{t>0} \frac{1}{2} t \int_{\mathbb{R}^N} (|\nabla u|^2 + V u^2) - \frac{1}{p} t^{\frac{p}{2}} \\ &= \frac{p-2}{2p} \Big(\inf_{u \in H^{1,2} : \|u\|_p = 1} \int_{\mathbb{R}^N} (|\nabla u|^2 + V u^2) \Big)^{\frac{p}{p-2}}. \end{aligned} \tag{9.13}$$

In other words, a function $w \in H^{1,2}(\mathbb{R}^N)$ is a critical point given by Theorem 9.1.4 if and only if it is the minimizer for the right-hand side of (9.13) multiplied by $[\frac{2p}{p-2} c_{V,F}]^{\frac{1}{p}}$. For this reason, a positive-valued critical point from Theorem 9.1.4 is called a ground state. From (9.13), we find that if w is a ground state, then so is $|w|$, and by maximum principle $|w|$ has no internal zeros, so w is strictly positive.

When $V(x) = V_\infty > 0$, it is known (see [60]) that w is radial with respect to some point and decreasing, and it is unique up to the sign, once its point of maximum is fixed (see [75]). We will reserve the notation w_∞ for the positive minimizer for (9.13) with $V = V_\infty$ (so that $\|w_\infty\|_p = 1$) centered at the origin. Existence of a minimizer for (9.13) can be also proved directly by an argument similar to one in Example 1.3.7.

Proposition 9.1.5. *Every minimizing sequence (u_k) for the right-hand side of (9.13) has a renamed subsequence of the form $u_k = w_\infty(\cdot + y_k) + r_k$ with $|y_k| \to \infty$ and $r_k \to 0$ in $H^{1,2}(\mathbb{R}^N)$. In particular, w_∞ is a minimizer.*

Proof. Let $\kappa = \inf_{u\in H^{1,2}:\|u\|_p=1} \int_{\mathbb{R}^N}(|\nabla u|^2 + Vu^2)\mathrm{d}x$. Without loss of generality, assume that $u_k \rightharpoonup w \neq 0$ in $H^{1,2}(\mathbb{R}^N)$. Indeed, if $u_k(\cdot + y_k) \rightharpoonup 0$ for any sequence (y_k), then by cocompactness of the embedding $u_k \to 0$ in $L^p(\mathbb{R}^N)$, which contradicts the constraint $\|u_k\|_p = 1$. Then, on a renamed subsequence $u_k(\cdot + y_k) \rightharpoonup w \neq 0$ and w necessarily satisfies $J'_{V_\infty,F}(w) = 0$. By the Brezis–Lieb lemma, $\|w\|_p^p + \|u_k(\cdot - y_k) - w\|_p^p \to 1$. Let $\|w\|_p^p = s$. Since $u_k(\cdot + y_k) \rightharpoonup w$ in $H^{1,2}(\mathbb{R}^N)$, we have

$$\begin{aligned}
\kappa &= \int_{\mathbb{R}^N} (|\nabla u_k|^2 + V_\infty u_k^2)\mathrm{d}x + o(1) \\
&= \int_{\mathbb{R}^N} (|\nabla w|^2 + V_\infty w^2)\mathrm{d}x + \int_{\mathbb{R}^N} (|\nabla(u_k - w)|^2 + V_\infty(u_k - w)^2)\mathrm{d}x + o(1) \\
&\geq \kappa s^{\frac{2}{p}} + \kappa(1-s)^{\frac{2}{p}} + o(1),
\end{aligned}$$

which can be true only if $s = 0$ or $s = 1$. However, the case $s = 0$ is vacuous since $w \neq 0$. Therefore, $u_k \to w$ in $L^p(\mathbb{R}^N)$. Then $u_k = J'_{V_\infty,F}(u_k) + \varphi(u_k) \to w$ in $H^{1,2}(\mathbb{R}^N)$ and w is the ground state. □

Case $V \geq V_\infty$

We will now give an example of an existence result for the critical points of the functional (9.3) with $F(u) = \frac{1}{p}|u|^p$, $p \in (2, 2^*)$, where $V > V_\infty$ on a set of positive measure.

Let w_∞ be the radial positive minimizer in (9.13) with $V = V_\infty$, and let $\alpha \overset{\text{def}}{=} \frac{2p}{p-2}c_{V_\infty,F}$. By calculations in Subsection 9.1, $w^{(\infty)} \overset{\text{def}}{=} \alpha^{\frac{1}{p}} w_\infty$ is the critical point of $J_{V_\infty,F}$ of the ground state type. Let T be such that $J_{V,F}(Tw^{(\infty)}) < 0$, let $\gamma_0(y,t) = tw^{(\infty)}(\cdot - y)$, $y \in \mathbb{R}^N$, $t > 0$, and let, for $R \in (0,\infty)$,

$$\Gamma_R \overset{\text{def}}{=}: \{\gamma \in C(B_R \times [0,T], \mathbb{R}^{N+1}) : \gamma(y,t) = \gamma_0(y,t), \text{ whenever } (y,t) \in \partial(B_R \times (0,T))\}. \tag{9.14}$$

Let us introduce the following map $\eta : H^{1,2}(\mathbb{R}^N) \to \mathbb{R}^{N+1}$:

$$\eta_i(u) = \begin{cases} \int_{\mathbb{R}^N} \frac{x_i}{1+|x|}|u|^p dx, & i = 1, \ldots, N, \\ \int_{\mathbb{R}^N} |u|^p dx, & i = N+1. \end{cases} \tag{9.15}$$

Note that $\eta(\gamma(y,t)) \neq \eta^* \stackrel{\text{def}}{=} (0_N, \alpha)$ when $(y,t) \in \partial(B_R \times (0,T))$. Indeed, if $t = 0$, we have $\eta_{N+1}(\gamma_0(y,0)) = 0 < \eta^*_{N+1} = \alpha$. If $t = T$ we have $\eta_{N+1}(\gamma_0(y,T)) = T^p > \eta^*_{N+1} = \alpha$. If $|y| = R$, since w_∞ is radially symmetric, positive, and decreasing, $(\eta_1, \ldots, \eta_N)(\gamma_0(y,t)) \neq 0_N$. Therefore, for any $\gamma \in \Gamma_R$, degree $\deg(\eta \circ \gamma, B_R \times (0,T), \eta^*)$ is well-defined, and

$$\deg(\eta \circ \gamma, B_R \times (0,t), \eta^*)) = \deg(\eta \circ \gamma_0, B_R \times (0,t), \eta^*) \neq 0. \tag{9.16}$$

We set

$$C_V(R,T) \stackrel{\text{def}}{=} \inf_{\gamma \in \Gamma_R} \max_{y \in B_R, t \in (0,T)} J_{V,F}(\gamma(y,t)). \tag{9.17}$$

We now compare the constant (9.17) with the mountain pass level (9.11).

Lemma 9.1.6. *Under assumptions above, there exists $\delta > 0$, independent of R, such that*

$$C_V(R,T) \geq c_{V_\infty,F} + \delta. \tag{9.18}$$

Proof. Since $\deg(\eta \circ \gamma, B_R \times (0,t), (0,\alpha)) \neq 0$ for all $\gamma \in \Gamma_R$, we have

$$C_V(R,T) \geq \inf_{\|u\|_p^p = \alpha, \int \frac{x}{1+|x|}|u_k|^p dx = 0} J_{V,F}(u) \stackrel{\text{def}}{=} \hat{C}.$$

If $\hat{C}$ coincides with $\inf_{\|u\|_p^p=\alpha} J_{V_\infty,F}(u)$, which equals $c_{V_\infty,F}$ by (9.13), then there exists a minimizing sequence for $c_{V_\infty,F}$ that satisfies the constraint $\int \frac{x}{1+|x|}|u_k|^p dx = 0$. However, by Proposition 9.1.5 every minimizing sequence for $c_{V,F}$ has a renamed subsequence of the form $u_k = w^{(\infty)}(\cdot - y_k) + r_k$, $\|r_k\|_{H^{1,2}(\mathbb{R}^N)} \to 0$, which cannot satisfy the constraint $\int \frac{x}{1+|x|}|u_k|^p dx = 0$ unless $y_k = y_k^{(1)} = 0$. Then $w^{(\infty)} = \alpha^{1/p} w_\infty$ is necessarily a minimizer for $\hat{C}$. However, since $V > V_\infty$ on a set of positive measure, $J_{V,F}(w^{(\infty)}) > J_{V_\infty,F}(w^{(\infty)}) = c_{V_\infty,F}$, which contradicts the assumption.

Thus $\hat{C} > c_{V_\infty,F}$. Since neither value is dependent on R, the lemma is proved. □

Lemma 9.1.7. *Under the assumptions above, one has*

$$C_V(R,T) \leq 2c_{V_\infty,F} + o_{R\to\infty}(1). \tag{9.19}$$

Proof. Consider the path

$$\gamma(y,t)(x) = tw^{(\infty)}\left(x - R\frac{y}{|y|}\right)\cos\left(\frac{\pi|y|}{2R}\right) + tw^{(\infty)}\left(x + R\frac{y}{|y|}\right)\sin\left(\frac{\pi|y|}{2R}\right),$$
$$t \in [0,T], |y| \leq R. \tag{9.20}$$

Then

$$
\begin{aligned}
C_V &\le \max_{|y|\le R, t\in[0,T]} J_{V,F}(\gamma(y,t))\\
&\le \max_{|y|\le R, t\in[0,T]}\left(\frac{1}{2}t^2 J_{V,0}(\gamma(y,1)) - \frac{1}{p}t^p\|\gamma(y,1)\|_p^p\right)\\
&= 2\frac{2p}{p-2}c_{V_\infty,F}\max_{|y|\le R, t\in[0,T]}\left(\frac{1}{2}t^2 - \frac{1}{p}t^p\left[\cos\left(\frac{\pi|y|}{2R}\right)^p + \sin\left(\frac{\pi|y|}{2R}\right)^p\right]\right) + o_{R\to\infty}(1)\\
&= 2\frac{2p}{p-2}c_{V_\infty,F}\max_{t\in[0,T]}\left(\frac{1}{2}t^2 - \frac{1}{p}t^p\frac{1}{2^{\frac{p-2}{2}}}\right) + o_{R\to\infty}(1)\\
&\le 2c_{V_\infty,F} + o_{R\to\infty}(1).
\end{aligned}
$$

□

In the course of the proof of the theorem below we will need, however, a strict inequality in (9.19), which will follow from an upper bound imposed on V.

Theorem 9.1.8. *Let $F(s) = |s|^p$, $p \in (2, 2^*)$, and assume that $V_\infty \le V \le 2^{\frac{p-2}{p}} V_\infty$, where the first inequality is strict on a set of positive measure. There exists a solution $u \in H^{1,2}(\mathbb{R}^N)$ to the equation*

$$
-\Delta u + V(x)u = |u|^{p-2}u, \quad x \in \mathbb{R}^N, \tag{9.21}
$$

satisfying $J_{V,F}(u) = C_V(R,T)$, where $C_V(R,T) > c_{V_\infty,F} > 0$ is given by the relation (9.17) with $R > R_0$ sufficiently large.

Proof. Let (u_k) be a critical sequence for $J_{V,F}$ at the level $C_V(R,T)$. By the standard Ambrosetti–Rabinowitz argument [10], we have the sequence (u_k) bounded in $H^{1,2}(\mathbb{R}^N)$. By Lemma 9.1.3, with some $m \in \mathbb{N}\cup\{0\}$ we have

$$
C_V(R,T) = mc_{V_\infty,F} + J_{V,F}(w^{(1)}), \tag{9.22}
$$

where $w^{(1)}$ is the weak limit of the renamed subsequence of u_k, which necessarily is a critical point of $J_{V,F}$. Assume that we have proved that $w^{(1)} \ne 0$. Then

$$
\begin{aligned}
J_{V,F}(w^{(1)}) &= \frac{p-2}{p}J_{V,0}(w^{(1)})\\
&\ge \frac{p-2}{p}J_{V_\infty,0}(w^{(\infty)})\frac{\|w^{(1)}\|_p^2}{\|w^{(\infty)}\|_p^2}\\
&= J_{V_\infty,F}(w^{(\infty)})\frac{J_{V,0}(w^{(1)})^{\frac{1}{p}}}{J_{V_\infty,0}(w^{(\infty)})^{\frac{1}{p}}},
\end{aligned}
$$

which implies $J_{V,F}(w^{(1)}) \ge J_{V_\infty,F}(w_{(\infty)}) = c_{V\infty,F}$. Combining this with (9.22) and (9.19), we get $(m+1)c_{V_\infty,F} + \delta \le C_V(R,T) \le 2c_{V_\infty,F} + o_{R\to\infty}(1)$, which implies $m < 1$, that is, $m = 0$, provided that the value of R is fixed sufficiently large. Consequently, the profile decomposition of the critical sequence (u_k) consists only of the term $w^{(1)}$, that is, $u_k \to w^{(1)}$ in L^p. Since $J'_{V,F}(u_k) \to 0$ in $H^{1,2}(\mathbb{R}^N)$, we also have $u_k \to w^{(1)}$ in $H^{1,2}(\mathbb{R}^N)$, which implies $J'_{V,F}(w^{(1)}) = 0$ and $J_{V,F}(w^{(1)}) = C_V(R,T)$.

It remains therefore to prove that $w^{(1)} \neq 0$. Assume that $w^{(1)} = 0$. Combining (9.22), (9.18), and (9.19) we have necessarily $C_V(R,T) = 2c_{V_\infty,F}$ with a critical sequence of the form $w_\infty(\cdot - y_k) + w_\infty(\cdot - z_k)$, where $|y_k| \to \infty$, $|z_k| \to \infty$, and $|y_k - z_k| \to \infty$ modulo a remainder vanishing in $H^{1,2}(\mathbb{R}^N)$. On the other hand, evaluating the functional on the path $\gamma_0(y,t) = tw_\infty(\cdot - y)$ and using $V \le 2^{\frac{p-2}{2}} V_\infty$, we have

$$
\begin{aligned}
C_V(R,T) &\le \max_{|y|\le R, t\in[0,T]} J_{V,F}(\gamma_0(y,t)) \\
&\le \max_{|y|\le R, t\in[0,T]} J_{2^{\frac{p-2}{p}} V_\infty,F}(\gamma_0(y,t)) \\
&< \max_{t\in[0,T]} \int_{\mathbb{R}^N} \left[2^{\frac{p-2}{p}} \frac{t^2}{2} (|\nabla w^{(\infty)}|^2 + V_\infty |w^{(\infty)}|^2) - \frac{t^p}{p} |w^{(\infty)}|^p \right] dx,
\end{aligned}
$$

and from an elementary evaluation of the maximum in the last line, analogous to (9.13), one has $C_V(R,T) < 2c_{V_\infty,F}$ uniformly in R, which contradicts (9.19), and thus proves the theorem. □

9.2 Nonlinear Schrödinger equation – zero mass case

Let $N \ge 3$ and let $f : \mathbb{R} \to \mathbb{R}$ be a continuous function, positive on $(0,\infty)$ and negative on $(-\infty, 0)$, satisfying

(f*) $f(2^{\frac{N-2}{2}} s) = 2^{\frac{N+2}{2}} f(s)$, $s \in \mathbb{R}$.

An example of such function is $f(s) = |s|^{2^*-2}s$, and in general (f*) implies that there exist $C_1, C_2 > 0$ such that $C_1|s|^{2^*-1} \le |f(s)| \le C_2|s|^{2^*-1}$. Let F be the primitive of f as in (9.1) and let φ be the functional (9.2). Fréchet derivative φ' of φ in $\dot{H}^{1,2}(\mathbb{R}^N)$ equals $v \mapsto \int_{\mathbb{R}^N} f(u)v dx$, which is by (f*) is a continuous as a map $\dot{H}^{1,2}(\mathbb{R}^N) \to \dot{H}^{1,2}(\mathbb{R}^N)$, as well as a map from $\dot{H}^{1,2}(\mathbb{R}^N)$ equipped with weak topology to $\dot{H}^{1,2}(\mathbb{R}^N)$ equipped with weak topology. Note that φ is invariant with respect to the rescaling group $\mathcal{G}^{\frac{N-2}{2}}$. We will consider now the functional of the form (9.3) with $V \in L^{N/2}(\mathbb{R}^N)$, defined on $\dot{H}^{1,2}(\mathbb{R}^N)$. It has a continuous Fréchet derivative on $\dot{H}^{1,2}(\mathbb{R}^N)$ and $J'_{F,V}(u) = u + [Vu] - \varphi'(u)$, where $[Vu](v) \overset{\text{def}}{=} \int_{\mathbb{R}^N} V(x)uv\, dx$.

Lemma 9.2.1. *Assume* (f*) *and let* $V \in L^{N/2}(\mathbb{R}^N)$. *There exists a number* $C > 0$ *such that every profile decomposition in* $\dot{H}^{1,2}(\mathbb{R}^N)$, *relative to the rescaling group* $\mathcal{G}^{\frac{N-2}{2}}$, *for any critical sequence* (u_k), $\|\nabla u_k\|_2 \le L$ *of the functional* (9.3) *has at most* $M = CL^2 + 1$

terms. Furthermore,

$$J_{V,F}(w^{(1)}) + \sum_{n=2}^{M} J_{0,F}(w^{(n)}) \leq \limsup J_{V,F}(u_k). \tag{9.23}$$

Proof. Note that if $g_k \in \mathcal{G}$, $g_k \rightharpoonup 0$, and $g_k u_k \rightharpoonup w$ then function w satisfies $-\Delta w = f(w)$, and, in particular, by the Pohozhaev identity, $\|\nabla w\|_2^2 = 2^* \int_{\mathbb{R}^N} F(w)\,dx$. Let

$$\gamma(t) = \sup_{\frac{1}{2}\|\nabla u\|_2^2 = t} \int_{\mathbb{R}^N} F(u)\,dx, \quad t \in [0,\infty).$$

By setting $u(x) = v(t^{-\frac{1}{N-2}}x)$, we easily get that

$$\gamma(t) = \gamma(1)t^{\frac{N}{N-2}}. \tag{9.24}$$

From this, it is easy to see that $\|\nabla w\|_2^2 \geq 2^{\frac{N}{2}}(2^*\gamma(1))^{\frac{2-N}{2}}$. Then by (1.15), if the profile decomposition has at least M terms, then at least $M-1$ terms correspond to $g_k \rightharpoonup 0$, and then $(M-1)2^{\frac{N}{2}}(2^*\gamma(1))^{\frac{2-N}{2}} \leq \limsup \|\nabla u_k\|_2^2$, which gives $M \leq 1 + \frac{\limsup \|\nabla u_k\|_2^2}{2^{\frac{N}{2}}(2^*\gamma(1))^{\frac{2-N}{2}}}$.

Relation (9.23) is analogous to (9.9). □

Theorem 9.2.2. *Let $f : \mathbb{R} \to \mathbb{R}$ be a continuous function satisfying conditions* (f*) *and* (f5) *of Theorem 9.1.4. Let Γ be a set of all continuous paths $\gamma : [0,\infty) \to \dot{H}^{1,2}(\mathbb{R}^N)$ such that $\gamma(0) = 0$ and $J_{V,F}(\gamma(s)) \to -\infty$ if $|s| \to \infty$. Let*

$$c = \inf_{\eta\in\Gamma} \sup_{s\geq 0} J_{V,F}(\eta(s)). \tag{9.25}$$

If $V \leq 0$, then the functional $J_{V,F}$ has a critical point u with $J_{V,F}(u) = c > 0$.

Proof. Note that the set Γ contains all paths of the form $t \mapsto v(t^{-1}\cdot)$ with $v \neq 0$ and thus it is nonempty. The functional $J_{V,F}$ has the mountain pass geometry since it is positive on any sphere of sufficiently small radius by (9.24), and since the path η is necessarily unbounded, and thus crosses the sphere. This yields $c > 0$. By the standard mountain pass argument (e. g., [133, Theorem 1.15]), functional $J_{V,F}$ has a critical sequence (u_k) satisfying $J'_{V,F}(u_k) \to 0$ and $J_{V,F}(u_k) \to c$. From (f4) and the argument of Ambrosetti–Rabinowitz, it follows that this sequence is bounded. Consider a renamed subsequence of (u_k) that admits a profile decomposition relative to the rescaling group $\mathcal{G}^{\frac{N-2}{2}}$. Note that for any $n \geq 2$, $w^{(n)}$ is a critical point of $J_{0,F}$. This follows from the weak-to-weak continuity of φ' and of $[Vu]$.

Assume now that $w^{(n)} \neq 0$ for some $n \geq 2$. By definition of the minimax value c we have $c \leq \max_{t>0} J_{V,F}(\eta(t))$, where $\eta(t) = w^{(n)}(t^{-1}(\cdot - y))$, $y \in \mathbb{R}^N$. As we take $|y| \to \infty$, we have

$$|J_{V,F}(\eta(t)) - J_{0,F}(\eta(t))| \leq \int_{\mathbb{R}^N} |V(x+y)||w^{(n)}(t^{-1}x)|^2 dx \to 0$$

uniformly for t near the maximal value for all y, since $V \in L^{N/2}(\mathbb{R}^N)$ (details are left to the reader). We have then

$$c \le \max_{t>0} J_{0,F}(w^{(n)}(t^{-1}\cdot)) = \max_{t>0} \frac{1}{2} t^{N-1} \|\nabla w^{(n)}\|_2 - t^N \varphi(w^{(n)}).$$

By the chain rule, $t = 1$ is a critical point of $J_{0,F}(\eta(t))$ since $J'_{0,F}(w^{(n)}) = 0$. Therefore, $t = 1$ is a point of maximum and $c \le J_{0,F}(w^{(n)})$. Comparing this with (9.23), we conclude that the profile decomposition for (u_k) consists of a single non-zero term $g_k^{(n)} w^{(n)}$. In the exceptional case $V = 0$, the problem is shift-invariant, so $g_k^{(n)^{-1}} u_k$ is also a critical sequence, renaming which as u_k, so that its weak limit is $w^{(n)}$, to be renamed as $w^{(1)}$, we get $u_k \to w^{(1)}$ in $L^{2^*}(\mathbb{R}^N)$. If, however, our assumption $w^{(n)} \ne 0$ were false for all $n \ge 2$, this would imply $u_k \to w^{(1)}$ in $L^{2^*}(\mathbb{R}^N)$.

Consider the case when $V < 0$ on a set of positive measure. Taking into account that $w^{(n)} \ne 0$ by the maximum principle, we have

$$J_{0,F}(w^{(n)}) \le c \le \max_{t\in[0,\infty)} J_{V,F}(w^{(n)}(t^{-1}\cdot)) < J_{0,F}(w^{(n)}),$$

a contradiction, which shows that $u_k \to w^{(1)}$ in $L^{2^*}(\mathbb{R}^N)$ also in the case $V \ne 0$.

Then $\varphi'(u_k) \to \varphi'(w^{(1)})$ and $[Vu_k] \to [Vw^{(1)}]$. Since (u_k) is a critical sequence, we can write (u_k) a sum of two sequences convergent in $\dot H^{1,2}(\mathbb{R}^N)$: $(J'_{V,F}(u_k)) = (u_k + [Vu_k] - \varphi'(u_k))$ and $(\varphi'(u_k) - [Vu_k])$. This implies that u_k converges in $\dot H^{1,2}(\mathbb{R}^N)$ to its weak limit $w^{(1)}$. Therefore, $J'_{V,F}(w^{(1)}) = 0$ and $J_{V,F}(w^{(1)}) = c$. □

9.3 Equations with finite symmetry

Let $N \ge 2$, let $p \in (2, 2^*)$, and let $J_{V,F}$ be the functional (9.3) with $F = \int_{\mathbb{R}^N} |u|^p dx$.

Theorem 9.3.1. *Let G be a subgroup of $O(N)$ such that there is $m \in \mathbb{N}$ such that for every $y \in \mathbb{R}^N$, $y \ne 0$, the orbit Gy contains at last m distinct vectors. Let $J_{V,F}$ be the functional (9.3) with $V \in L^\infty(\mathbb{R}^N)$, $\lim_{x\to\infty} V(x) = V_\infty$ and $V \circ \eta = V$ for every $\eta \in G$, and consider $J_{V,F}$ defined on $H_G^{1,2}(\mathbb{R}^N) = \{u \in H^{1,2}(\mathbb{R}^N) : \forall \eta \in G, u \circ \eta = u\}$. Let Γ be a set of all continuous paths $\gamma : [0, \infty \to H_G^{1,2}(\mathbb{R}^N)$ such that $\gamma(0) = 0$ and $J_{V,F}(\gamma(s)) \to -\infty$ if $|s| \to \infty$. Let*

$$c_G = \inf_{\eta\in\Gamma} \sup_{s\ge0} J_{V,F}(\eta(s)). \tag{9.26}$$

If

$$V(x) \le m^{\frac{p}{p-2}} V_\infty \text{ with a strict inequality on a set of positive measure,} \tag{9.27}$$

then $J_{V,F}$ has a critical point $u \in H_G^{1,2}(\mathbb{R}^N)$ satisfying $J_{V,F}(u) = c_G$.

Proof. Similar to the proof of Theorem 9.1.4, the functional $J_{V,F}$ has a bounded critical sequence (u_k) in $H_G^{1,2}(\mathbb{R}^N)$ satisfying $J'_{V,F}(u_k) \to 0$ and $J_{V,F}(u_k) \to c_G > 0$. Consider a renamed subsequence of (u_k) that admits a profile decomposition relative to the group of shifts $\mathcal{G}_{\mathbb{Z}^N}$. Note that any nonzero profile $w^{(n)}$ with $n \geq 2$ equals (as discussed in the previous section), a radially symmetric ground state $w^{(\infty)} \stackrel{\text{def}}{=} \lambda w_\infty$, up to a sign and an Euclidean shift, with $\lambda = (\frac{p-2}{p} c_{V_\infty,F})^{\frac{1}{p}}$. As in the proof of Theorem 9.1.4, we have that if $w^{(1)} \neq 0$, then $c_G \leq J_{V,F}(w^{(1)})$. Comparing this with (9.9), we conclude that the profile decomposition for (u_k) has no terms $w^{(n)}$, $n \geq 2$, that is, $u_k \to w^{(1)}$ and, as in Theorem 9.1.4, $w^{(1)}$ is the required critical point and the theorem is proved. It remains to prove that $w^{(1)} \neq 0$.

Assume that $w^{(1)} = 0$ and that, without loss of generality, $w^{(2)} = w^{(\infty)}$. Since $u \in H_G^{1,2}(\mathbb{R}^N)$ and $u_k(\cdot + y_k^{(2)}) \rightharpoonup w^{(\infty)}$, we also have $u_k(\eta(\cdot + y_k^{(2)})) \rightharpoonup w^{(\infty)}$, $\eta \in G$, and thus $u_k(\cdot + \eta y_k^{(2)}) \rightharpoonup w^{(\infty)} \circ \eta^{-1} = w^{(\infty)}$. Passing to a renamed subsequence, assume that $\frac{y_k^{(2)}}{|y_k^{(2)}|} \to \omega$. By the assumption of the theorem, there exist $\eta_1, \dots, \eta_m \in G$ such that vectors $\eta_1\omega, \dots, \eta_m\omega$ are distinct. Then $|\eta_n y_k^{(2)} - \eta_\ell y_k^{(2)}| \to \infty$ as $k \to \infty$ whenever $n \neq \ell$, and thus a renamed subsequence of (u_k) has a profile decomposition with at least m nonzero terms $w^{(\infty)}(\cdot - \eta_\ell y_k^{(2)})$, $\ell = 1, \dots, m$. Then by (9.9), we have $C_G \geq m c_{V_\infty,F}$. On the other hand,

$$C_G \leq \max_{t>0} J_{V,F}(t w^{(\infty)}) < \max_{t>0} J_{m^{\frac{p}{p-2}} V_\infty,F}(t w^{(\infty)}) = (m^{\frac{p}{p-2}})^{\frac{1}{p-2}} c_{V_\infty,F} = m c_{V_\infty,F},$$

which yields a contradiction. Thus $w^{(1)} \neq 0$ and, consequently, $w^{(1)}$ is the required critical point. □

9.4 Blowups for the Moser–Trudinger functional

We begin with a partial refinement of Corollary 1.4.2 concerning weak continuity properties of the Moser–Trudinger functional

$$J(u) = \int_\Omega e^{\alpha_N |u|^{\frac{N}{N-1}}} dx, \quad \alpha_N = N\omega_N^{\frac{1}{N-1}}. \tag{9.28}$$

Let m_t, $t \in (0,1)$, be the family (3.68) of Moser functions on $H_{0,\text{rad}}^{1,N}(B)$, $B = B_1(0)$:

$$m_t(r) \stackrel{\text{def}}{=} (\omega_N)^{-\frac{1}{N}} \log(1/t)^{\frac{N-1}{N}} \min\left\{\frac{\log(1/r)}{\log(1/t)}, 1\right\}, \quad r = |x|, r, t \in (0,1). \tag{9.29}$$

and consider the following functional on $H_{0,\text{rad}}^{1,N}(B)$:

$$\langle m_t^*, u\rangle \stackrel{\text{def}}{=} \int_B |\nabla m_t|^{N-2} \nabla m_t \cdot \nabla u dx, \quad t \in (0,1). \tag{9.30}$$

An elementary computation shows that the functional m_t^* is continuous. By Lemma 3.11.1, we have

$$\langle m_t^*, u\rangle = \omega_N^{1/N} \log(1/t)^{-\frac{N-1}{N}} u(t), \quad t \in (0,1). \tag{9.31}$$

Proposition 9.4.1. *Let $u_k \in H^{1,N}_{0,\mathrm{rad}}(B)$, $\|\nabla u_k\|_N \le 1$, $u_k \rightharpoonup u$, and let J be the functional (9.28) Then $J(u_k) \to J(u)$, unless the sequence (u_k) has a renamed subsequence such that $u_k - m_{t_k} \to 0$ in $H^{1,N}_{0,\mathrm{rad}}(B)$, with $t_k \to 0$.*

Proof. Let us substitute (9.31) into (9.28). After elementary simplifications one arrives at the following representation:

$$J(u) = \omega_{N-1}\left(\int_0^1 r^{N(1-\langle m_r^*, u\rangle^{\frac{N}{N-1}})} \frac{dr}{r}\right), \tag{9.32}$$

where $u \in H^{1,N}_{0,\mathrm{rad}}(B)$ and $\|\nabla u\|_N = 1$. Assume first that there exists $\varepsilon > 0$ such that $\langle m_t^*, u_k\rangle \le 1 - \varepsilon$. Then $J(u_k) \to J(u)$ by the Lebesgue dominated convergence theorem. The remaining case is when for some $t_k \in (0,1)$, $u_k - m_{t_k} \to 0$ in $H^{1,N}_{0,\mathrm{rad}}(B)$. Assume first that the weak limit u is not zero. Then, necessarily, $u_k \to m_t$ in $H^{1,N}_{0,\mathrm{rad}}(B)$ for some $t \in (0,1)$. This implies the uniform convergence of u_k on $[t,1]$ as well as $\int_B |\nabla u_k|^N dx \to 0$, from which easily follows $J(u_k) \to J(m_t)$. If $u_k = m_{t_k} + o(1) \rightharpoonup 0$ with $t_k \to 1$, an argument repetitive of that for the case $u_k \to m_t$ above will give $J(u_k) \to J(0)$. We have, therefore, with necessity, a renamed subsequence $u_k = m_{t_k} + o(1)$ with $t_k \to 0$. ☐

The rest of this section is dedicated to the structure of Palais–Smale sequences for a semilinear elliptic problem of critical growth in two dimensions. We should note that unlike the critical nonlinearity $\int |u|^{2^*} dx$ when $N > 2$, in the case $N = 2$ the Moser-Trudinger functional lacks scale invariance and, as we seen above, has weak continuity behavior that is absent in $\int |u|^{2^*} dx$ (which lacks weak continuity at any point). In this context, it is not a surprise that critical sequences of the semilinear elliptic functional with a critical nonlinearity have more complex structure in the case $N = 2$ than in the case $N > 2$.

Let B denote an open unit disk in $\mathbb{R}^2$ centered at the origin.

Definition 9.4.2 (Moser–Carleson–Chang tower functions). Let C_+, C_- be closed subsets of $(0,1)$, such that $C = C_+ \cup C_-$ is nonempty, let $A = (0,1) \setminus C$, and let $\mathcal{A} = \{(a_n, b_n)\}_{n\in\mathbb{N}}$ be an enumeration of all connected components of A starting with $a_1 = 0$. A continuous radial function $\mu_{C_+,C_-} \in H^{1,2}_{0,\mathrm{rad}}(B)$ is called a *Moser–Carleson–Chang tower* if

$$\mu_{C_+,C_-}(r) = \begin{cases} \sqrt{\frac{1}{2\pi}\log\frac{1}{r}}, & r \in C_+, \\ -\sqrt{\frac{1}{2\pi}\log\frac{1}{r}}, & r \in C_-, \\ A_n + B_n \log\frac{1}{r}, & r \in (a_n, b_n), A_n, B_n \in \mathbb{R}. \end{cases} \tag{9.33}$$

If $C_- = \emptyset$, we will use the notation μ_C instead.

When the set C_+ consists of a single point $t \in (0,1)$ and $C_- = \emptyset$, the function μ_C is an original Moser function (3.68). When $C \subset (0,1)$ is a closed interval, a function of the form μ_C is found in the proof of existence of extremals for the Moser–Trudinger functional by Carleson and Chang ([28], p. 121) written in the variable $t = \log \frac{1}{r}$.

Let us prove some elementary properties of Moser–Carleson–Chang towers.

Proposition 9.4.3.

(i) *Coefficients A_n, B_n are defined uniquely by continuity at the points $a_n, b_n \in C$. In particular, if $C = C_+$,*

$$A_n = \frac{1}{\sqrt{2\pi}} \frac{\sqrt{\log \frac{1}{a_n}}\sqrt{\log \frac{1}{b_n}}}{\sqrt{\log \frac{1}{a_n}} + \sqrt{\log \frac{1}{b_n}}},$$

$$B_n = \frac{1}{\sqrt{2\pi}} \frac{1}{\sqrt{\log \frac{1}{a_n}} + \sqrt{\log \frac{1}{b_n}}} \tag{9.34}$$

(when $n = 1$ the values in (9.34) are understood in the sense of the limits as $a_1 \to 0$, that is, $A_1 = \sqrt{\frac{1}{2\pi}\log \frac{1}{b_1}}$ and $B_1 = 0$).

(ii) *The function $\mu_{C_+,C_-}(r)$ has continuous derivative at every point of $(0,1)$ except $\{a_n, b_n\}_{(a_n,b_n)\in\mathcal{A}}$.*

(iii) *Let $\mathcal{A}'$ be the set of all intervals $(a,b) \in \mathcal{A}$ where μ_{C_+,C_-} does not change sign, and let $\mathcal{A}'' = \mathcal{A} \setminus \mathcal{A}'$. Then*

$$\|\nabla \mu_{C_+,C_-}\|_2^2 = \frac{1}{4}\int_C \frac{dr}{r\log \frac{1}{r}} + \sum_{(a,b)\in\mathcal{A}'} \frac{\sqrt{\log \frac{1}{a}} - \sqrt{\log \frac{1}{b}}}{\sqrt{\log \frac{1}{a}} + \sqrt{\log \frac{1}{b}}} + \sum_{(a,b)\in\mathcal{A}''} \frac{\sqrt{\log \frac{1}{a}} + \sqrt{\log \frac{1}{b}}}{\sqrt{\log \frac{1}{a}} - \sqrt{\log \frac{1}{b}}}. \tag{9.35}$$

(iv) *The number of zeros of μ_{C_+,C_-} on $(0,1)$ is less or equal than the value of $\|\nabla \mu_{C_+,C_-}\|_2^2$-1.*

(v) *For any choice of C_-, C_+, one has $\|\nabla \mu_{C_+,C_-}\|_2^2 \geq 1$ and the equality holds only if C consists of one point.*

Proof. (i): Values (9.34) for $n \geq 2$ are the unique solutions of continuity conditions at a_n and b_n, $A_n + B_n \log \frac{1}{a_n} = \sqrt{\frac{1}{2\pi}\log \frac{1}{a_n}}$ and $A_n + B_n \log \frac{1}{b_n} = \sqrt{\frac{1}{2\pi}\log \frac{1}{b_n}}$. Since μ_{C_+,C_-} has a finite Sobolev norm, we have, necessarily, $B_1 = 0$, which yields $A_1 = \sqrt{\frac{1}{2\pi}\log \frac{1}{b_1}}$.

(ii): For the sake of simplicity, we consider the case $C = C_+$, the general case is similar. If $(a_{n_k}, b_{n_k}) \subset \mathcal{A}$ and $a_{n_k} \to c$ for some c, then necessarily $b_{n_k} \to c$, from which, by elementary computation, follows $\mu_C'(c) = \lim \mu_C'(a_{n_k}) = \lim \mu_C'(b_{n_k}) = (\sqrt{\frac{1}{2\pi}\log \frac{1}{r}})'|_{r=c}$. Consequently, since μ_C is (by definition) smooth at all internal points of A and of C, the only points in $(0,1)$ where μ_C' is discontinuous are the points a_n and b_n.

(iii) It follows from the direct computation of the right-hand side in

$$\|\nabla\mu_{C_+,C_-}\|_2^2 = 2\pi \int\limits_{C_+,C_-} |\mu'_{C_+,C_-}(r)|^2 r\mathrm{d}r + 2\pi \sum_n \int\limits_{a_n}^{b_n} |\mu'_{C_+,C_-}(r)|^2 r\mathrm{d}r.$$

(iv): The terms in (9.35) corresponding to the set $\mathcal{A}''$ are greater than 1. Furthermore, on the interval $(a,1) \in \mathcal{A}$, one has necessarily $\mu_{C_+,C_-}(r) = \pm\frac{\log\frac{1}{r}}{\sqrt{2\pi\log\frac{1}{a}}}$ and the contribution of this interval to (9.35) is

$$\int\limits_{r\in(a,1)} |\nabla\mu_{C_+,C_-}|^2 = 1.$$

(v): By the last observation, the sum in (9.35) is greater than or equal to 1, and the equality occurs only if the sum consists of the one term $(a,1)$, a situation corresponding to $\mathcal{A} = \{(0,a),(a,1)\}$, that is, to the Moser function m_a. □

We consider critical nonlinearities as defined in [3]. Without loss of generality, we restrict the consideration to the factor b in the exponent equal to 4π, since the general case can be always recovered by replacing the variable u with a suitable scalar multiple. Let $f \in C(\mathbb{R})$ and $F(s) = \int_0^s f(t)\mathrm{d}t$.

Definition 9.4.4. We say that a continuous function $f : \mathbb{R} \to \mathbb{R}$ is of the 4π-critical growth, if $f(s) = 8\pi g(s)e^{4\pi s^2}$ and for any $\delta > 0$,

$$\lim_{|s|\to\infty} g(s)e^{-\delta s^2} = 0.$$

We will study the functional

$$J(u) = \frac{1}{2}\int\limits_\Omega |\nabla u|^2\mathrm{d}x - \frac{1}{8\pi}\int\limits_\Omega F(u)\mathrm{d}x, \quad u \in H^{1,2}_{0,\mathrm{rad}}(B). \tag{9.36}$$

We write $\frac{1}{8\pi}f(t) = g(t)e^{4\pi t^2}$, and we will use the following assumptions:

(g0) $\lim_{|t|\to\infty}\frac{g'(t)}{g(t)t} = 0$;

(g1) There is a $T > 0$ such that $\inf_{t\geq T} g(t) > 0$ and $\sup_{t\leq -T} g(t) < 0$;

(g2) $\lim_{|t|\to\infty}\frac{F(t)}{f(t)t} = 0$.

Remark 9.4.5. Examples of $g(t)$ can be found in [3]. They include $g(t) = t$.

We use the fact that u_k is a critical sequence for the functional (9.36) in order to make the expansion (4.44) more specific, namely, to verify that every asymptotic profile (4.41) is a Moser–Carleson–Chang tower and that the expansion (4.44) has finitely many terms. This is stated at the end of this section as Theorem 9.4.7.

Theorem 9.4.6. *Assume that function f be of 4π-critical growth and satisfies* (g0) *and* (g1). *Let u_k be a critical sequence of* (9.36). *Then every concentration profile $w^{(n)}$, $n \geq 2$, of (u_k), given by* (4.41), *equals a function $\mu_{C_+^{(n)},C_-^{(n)}}$ with some disjoint closed sets $C_+^{(n)}, C_-^{(n)} \subset (0,1)$, as given by Definition* 9.4.2.

Proof. Let us derive first the equation satisfied by the limit (4.41). Index n in the argument below is fixed and will be omitted. Since u_k is a critical sequence, and operators (3.66) are isometries on $H^{1,2}_{0,\mathrm{rad}}(B)$,

$$(g_{s_k}J'(u_k), v)_{H_0^{1,2}(B)} \to 0. \tag{9.37}$$

Since $gs_k u_k \rightharpoonup w$ by (4.41), we have $g_{s_k}(u_k - J'(u_k)) \rightharpoonup w$, which in turn implies that $gs_k(-\Delta)^{-1}\frac{f(u_k)}{8\pi} \rightharpoonup w$, where $(-\Delta)^{-1}$ is the inverse of the Dirichlet Laplacian on B (responsible for representation of the scalar product of L^2 as a bilinear form in $H_0^{1,2}$). An elementary computation shows the following identity:

$$\Delta g_s\psi = s^2 r^{2j-2} g_s\Delta\psi.$$

Taking now $\psi = (-\Delta)^{-1}\frac{f}{8\pi}$, we have

$$\Delta g_{s_k}\Delta^{-1}\frac{f}{8\pi} = s_k^2 r^{2j_k-2} g_{s_k}\frac{f}{8\pi},$$

from which it immediately follows that

$$s_k^{3/2} r^{2j_k-2}\frac{1}{8\pi}f(u_k(r^{s_k}) \rightharpoonup -\Delta w \tag{9.38}$$

in the sense of $H^{-1,2}(B)$.

Recall that $\frac{1}{8\pi}f(s) = g(s)e^{4\pi s^2}$. It easily follows from (g0) and (g1) that

$$\lim_{|s|\to\infty}\frac{\log|g(s)|}{s^2} = 0. \tag{9.39}$$

Note that $|w(r)| \leq \sqrt{\frac{1}{2\pi}\log\frac{1}{r}}$ for all $r \in (0,1]$. Indeed, if for some $a \in (0,1]$ a converse inequality is true, then for all k sufficiently large, $4\pi u_k(\rho^{s_k})^2 - 2s_k\log\frac{1}{\rho}$ will be bounded away from zero when r is in some neighborhood of a and $\theta \in S^1$, and thus, taking into account (9.39), we have the left-hand side in (9.38) uniformly convergent to ∞ on an interval. Taking a positive test function supported on an interval, we arrive at a contradiction, since $-\Delta w$ is a distribution.

Let $C_1 = \{r \in (0,1] : |w(r)| = \sqrt{\frac{1}{2\pi}\log\frac{1}{r}}\}$. Since w is continuous on $(0,1]$, the set C_1 is relatively closed in $(0,1]$. Since $w \in H^{1,2}(B)$ and $\sqrt{\frac{1}{2\pi}\log\frac{1}{r}} \notin H^{1,2}(B)$, $C_1 \neq (0,1]$. Thus the complement of C_1 in $(0,1]$ is an at most countable union of open intervals of $(0,1]$. Let $\mathcal{A}$ be an enumeration of all such intervals. If $(a,b) \in \mathcal{A}$, then $w(a) = \pm\sqrt{\frac{1}{2\pi}\log\frac{1}{a}}$,

$w(b) = \pm\sqrt{\frac{1}{2\pi}\log\frac{1}{b}}$ and $|w(r)| < \sqrt{\frac{1}{2\pi}\log\frac{1}{r}}$ for $r \in (a,b)$. From (9.38) it follows that w is harmonic on (a,b), and, as a radial function, it has the form $A + B\log\frac{1}{r}$, $A, B \in \mathbb{R}$, and the values of A and B are uniquely defined by the values $w(a)$ and $w(b)$. Let $C = C_1 \setminus \{1\}$. It remains to show that w is constant in a neighborhood of zero and is harmonic in a neighborhood of 1. Assume first that $\mathcal{A}$ is infinite, and thus countable in this case. Then, by elementary calculations already mentioned in the proof of Proposition 9.4.3, we have that w satisfies (9.35), which in turn shows that the set $\mathcal{A}''$ of intervals in $\mathcal{A}$, where the function w changes sign, is finite. Setting $\sigma_n = 1$ when $a_n = 0$, $\sigma_n = +\infty$ when $b_n = 1$ and

$$\sigma_n \overset{\text{def}}{=} \frac{\sqrt{\log\frac{1}{a_n}}}{\sqrt{\log\frac{1}{b_n}}}$$

otherwise, we have

$$\|\nabla w\|_2^2 \geq \sum_{(a_n,b_n)\in\mathcal{A}'} \frac{\sigma_n - 1}{\sigma_n + 1}. \tag{9.40}$$

Note that the sequence σ_n is bounded, otherwise the sum above would have infinitely many terms greater than 1/2, say $\sigma_n \leq M - 1$, $M > 0$. Then from the relation above it is immediate that $\sigma_n \to 1$, and

$$\prod_n \sigma_n \leq C \prod_{(a_n,b_n)\in\mathcal{A}'} \sigma_n \leq Ce^{\sum_n(\sigma_n - 1)} \leq Ce^{M\sum_n \frac{\sigma_n - 1}{\sigma_n + 1}} \leq Ce^{M\|\nabla w\|_2^2} = \hat{C} < \infty.$$

Let ν be any finite subset of $\mathbb{Z}$ such that $(a_n, b_n)_{n\in\nu}$ are ordered by n and none of a_n is zero. Then

$$\frac{\max_{n\in\nu}\sqrt{\log\frac{1}{a_n}}}{\min_{n\in\nu}\sqrt{\log\frac{1}{b_n}}} \leq \prod_{n\in\nu}\sigma_n \leq \prod_{n\in\mathbb{Z}}\sigma_n \leq \hat{C},$$

from which one immediately concludes that there are no sequences $(a_{n_k}, b_{n_k}) \in \mathcal{A}$ such that $a_{n_k} > 0$ and $a_{n_k} \to 0$ or $b_{n_k} \to 1$. No such sequences exist when the set $\mathcal{A}$ is finite. Thus there exists an $\varepsilon > 0$ such that on the whole interval $(0,\varepsilon)$, respectively, $(\varepsilon, 1)$, the function w is either harmonic or equals $\pm\sqrt{\frac{1}{2\pi}\log\frac{1}{r}}$. The latter, however, cannot occur since this contradicts $w \in H_0^{1,2}(B)$. □

We derive the specific form of profile decomposition given by Theorem 4.8.1 to critical sequences of the functional (9.36).

Theorem 9.4.7. *Let J be the functional* (9.36) *with f of critical growth satisfying* (g0), (g1), *and* (g2). *Let $u_k \in H_{0,\mathrm{rad}}^{1,2}(B)$ be a bounded sequence such that $J'(u_k) \to 0$ and $J(u_k) \to c$. Then the sequence u_k has a renumbered subsequence of the following form:*

There exists an $m \in \mathbb{N}$, $m \le 2c$, *sequences* $(s_k^{(1)}), \dots, (s_k^{(m)})$ *of positive numbers, convergent to zero except* $s_k^{(1)} = 1$, *and closed sets* $C_\pm^{(1)}, \dots, C_\pm^{(m)} \in (0,1)$, *such that*

$$\left|\log \frac{1}{s_k^{(p)}} - \log \frac{1}{s_k^{(q)}}\right| \to \infty \quad \textit{whenever } p \ne q, \tag{9.41}$$

$$u_k - \sum_{j=1}^{m} g_{s_k^{(j)}} \mu_{C_+^{(j)}, C_-^{(j)}} \to 0 \quad \textit{in } \exp L^2, \tag{9.42}$$

and

$$\|\nabla u_k\|_2^2 \to \sum_j \|\nabla \mu_{C_+^{(j)}, C_-^{(j)}}\|_2^2.$$

If Z_j *is the number of zeros of* $w^{(j)} = \mu_{C_+^{(j)}, C_-^{(j)}}$, *then* $\sum_{j=1}^m Z_j < 2c - 2m$. *In particular, if* $c \le m$, *all functions* $w^{(j)}$ *are sign definite. Furthermore, if* $\frac{m}{2} = c$, *then for every* $j = 1, \dots, m$, $C^{(j)} = \{t_j\}$ *for some* $t_j \in (0,1)$, *and* $\mu_{C^{(j)}}$ *is a Moser function* m_{t_j}, $j = 1, \dots, m$.

Proof. The statement follows immediately from application of Theorem 9.4.6 to Theorem 4.8.1 and properties of the profiles μ_{C_+, C_-} from Proposition 9.4.3. Note that from (g2) it follows that $\int F(u_k) \to 0$, so $c \ge \frac{1}{2} \sum_j \|\nabla \mu_{C_+^{(j)}, C_-^{(j)}}\|_2^2$. Then relation $\sum_{j=1}^m Z_j < 2c - 2m$ is immediate from Proposition 9.4.3. If $c = m/2$ then, necessarily, each of the norms in the right-hand side equals 1, each $\mu_{C_+^{(j)}, C_-^{(j)}}$ is a Moser function, the inequality becomes an equality, and the resulting convergence of $H_0^{1,2}$-norms in (9.42), $\|\nabla u_k\|_2^2 \to \sum \|\nabla \mu_{C_+^{(j)}, C_-^{(j)}}\|_2^2$, together with convergence in $\exp L^2$ implies $H_0^{1,2}$-convergence. □

Remark 9.4.8. It is natural to ask if any sum of the form as in (9.42) is a critical sequence for the functional (9.36). The answer is negative in the direct sense: the simplest expression of the kind, $\lambda_k g_{s_k} m_t$ with $\lambda_k \to 1$ may or may not be a critical sequence, dependent on the choice of the sequence (λ_k) approximating 1 (see [5]). Rephrasing the question, however, whether every sum of the form (9.42) becomes a critical sequence if appended with a suitable sequence vanishing in $H^{1,2}_{0,\mathrm{rad}}(B)$, the answer is answered positively in [33] in the cases when the sets $C_\pm^{(j)}$ are singletons of when the family $C_\pm^{(j)}{}_j$ consists of a single set which is a closed interval.

9.5 Bibliographic notes

A general introduction to direct methods of calculus of variations one may find in many books, such as Chabrowski [30], Struwe [120], or Willem [133]. Profile decompositions for Palais–Smale sequences of semilinear elliptic functionals containing finitely many terms (Lemmas 9.1.3 and 9.2.1 above), are abundant in literature, for example, Struwe [119] (bounded domain, critical nonlinearity), Lions [85] (unbounded domain, subcritical nonlinearity), Benci and Cerami [17] (unbounded domain, critical nonlinearity). Theorems 9.1.4 and 9.2.2 stem from Proposition 9.1.5, proved by Lions [83], whose different generalizations are too numerous to be cited here. Theorem 9.1.8 uses a baricentric minimax statement inspired by [24] (see also [18] and [29, Proposition 3.3]).

Existence of solution with finite symmetry under relaxed conditions, Theorem 9.3.1, is due to [137].

Proposition 9.4.1 is following [7]. It uses some technical elements from [94] and is related to a step in the proof in [28] for existence of extremals for the functional (9.28), which then shows that the functional attains, on a function of the type (9.33), a higher level than the value where it fails to be weakly continuous. Section 9.4 is the radial case of the corresponding results in [33].

10 Appendix

10.1 Topics in functional analysis

Weak compactness

A Banach space is locally compact if and only if it is finite-dimensional. Infinite dimensional Banach spaces possess weaker compactness properties.

Theorem 10.1.1 (Banach–Alaoglu theorem). *A closed ball in a Banach space E, which is a conjugate of another normed vector space, is compact in the weak*-topology of E.*

We recall that if F is a Banach space, then weak*-topology on $E = F^*$ is generated by sets $\{x \in E : \langle x, \varphi\rangle_F < a\}$, $a \in \mathbb{R}$, $\varphi \in F$, and a sequence $(x_n)_{n\in\mathbb{N}}$ in E is weak*-convergent to a point $x \in E$ if and only if $\langle \varphi, x_n\rangle \to \langle \varphi, x\rangle$ for every $\varphi \in F$.

Theorem 10.1.1 does not imply that every bounded sequence in E has a weak*-convergent subsequence. This is true, however, if weak*-topology is metrizable on the bounded subsets of E, which is the case, in particular, if E is reflexive or separable:

Theorem 10.1.2. *Let $(x_n)_{n\in\mathbb{N}}$ be a bounded sequence in a Banach space E.*
(i) *If E is reflexive, then $(x_n)_{n\in\mathbb{N}}$ has a weakly convergent subsequence.*
(ii) *If E is a conjugate of another normed vector space and is separable, then $(x_n)_{n\in\mathbb{N}}$ has a weak*-convergent subsequence.*

For more details the reader may refer, for example, to [23, Chapter 3].

Uniformly convex and uniformly smooth Banach spaces

Definition 10.1.3. A normed vector space X is called uniformly convex if the following function, called the *modulus of convexity of X*, is strictly positive for all $\varepsilon > 0$:

$$\delta(\varepsilon) = \inf_{x,y\in X,\, \|x\|=\|y\|=1,\, \|x-y\|=\varepsilon} 1 - \left\|\frac{x+y}{2}\right\|, \quad \varepsilon \in [0,2].$$

The function $\varepsilon \mapsto \delta(\varepsilon)/\varepsilon$ is nondecreasing on $(0,2]$, ([48] Proposition 3, p. 122), and thus $\varepsilon \mapsto \delta(\varepsilon)$ is strictly increasing if $\delta(\varepsilon) > 0$.

Uniform convexity is equivalent to the property

$$x, y \in X,\ \|x\| \le 1,\ \|y\| \le 1 \Longrightarrow \left\|\frac{x+y}{2}\right\| \le 1 - \delta(\|x-y\|), \tag{10.1}$$

[48, Lemma 4, p. 124].

https://doi.org/10.1515/9783110532432-010

From (10.1), it follows that for any two elements $u, v \in X$ which satisfy $\|u\| \leq \|v\|$ and $v \neq 0$,

$$\left\| \frac{u+v}{2} \right\| \leq \|v\| \left(1 - \delta\left(\frac{\|u-v\|}{\|v\|} \right) \right). \tag{10.2}$$

This in turn implies that for every two elements $u, v \in X$ that are not both zero one has

$$\left\| \frac{u+v}{2} \right\| \leq C_1 - C_2 \delta\left(\frac{\|u-v\|}{C_2} \right) \tag{10.3}$$

whenever $C_1 \geq \max\{\|u\|, \|v\|\}$ and $C_2 \geq \max\{\|u\|, \|v\|\}$. When $C_1 = C_2 = \max\{\|u\|, \|v\|\}$ relation (10.3) is exactly (10.2), up to an interchange of u and v.

One calls a Banach space X uniformly smooth if for every $\varepsilon > 0$ there exists $\delta > 0$ such that $\|x + y\| + \|x - y\| \leq 2 + \varepsilon\|y\|$ whenever $\|x\| = 1$ and $\|y\| \leq \delta$, $x, y \in X$. The conjugate space X^* is uniformly convex if and only if X is uniformly smooth, see [80, Proposition 1.e.2]. If X is uniformly convex, then the norm of X, as a function $\phi : x \mapsto \|x\|$, considered on the unit sphere $S_1 = \{x \in X, \|x\| = 1\}$, is uniformly Gateaux differentiable, which immediately implies that ϕ' is a uniformly continuous function $S_1 \to S^*{}_1$; see [80, p. 61].

If one considers ϕ as a function on the whole X, by homogeneity one has $\phi'(x) = \phi'(x/\|x\|) \in S^*{}_1$ for all $x \neq 0$, and it is easy to see that $\phi'(x)$ coincides with the uniquely defined duality conjugate x^* of x relative to the modulus $\|x\|$ (i. e., with normalization $\langle x^*, x\rangle = \|x\|$). We summarize this in the following statement.

Lemma 10.1.4. *Let X be a uniformly convex and uniformly smooth Banach space. Then the map $x \mapsto x^*$, where x^* is a duality conjugate element of x relative to the modulus $\|x\|$, is a continuous map $X \setminus \{0\} \to X^*$ with respect to the norm topologies on X and X^* and is in fact uniformly continuous on all closed subsets of $X \setminus \{0\}$.*

In uniformly convex spaces, one also has an important connection between weak convergence and convergence in the norm.

Proposition 10.1.5 ([23, Proposition 3.32]). *Let E be a uniformly convex space. If $(u_k)_{k\in\mathbb{N}}$ is a sequence in E, such that $u_k \rightharpoonup u$ and $\|u_k\| \to \|u\|$, then $u_k \to u$ in E.*

Schauder basis

For more details on Schauder bases, we refer to [79]. A Schauder basis $\{b_n\}_{n\in\mathbb{N}}$ of a Banach space E is a sequence of elements of E such that for every element $x \in E$ there exists a unique sequence α_n of scalars such that

$$x = \sum_{n=0}^{\infty} \alpha_n b_n,$$

where convergence is in the norm of E.

Every expansion in Schauder basis is unconditionally convergent. It follows from the Banach–Steinhaus theorem that the linear mappings P_n defined by

$$x = \sum_{k=0}^{\infty} \alpha_k b_k \longrightarrow P_n(x) = \sum_{k=0}^{n} \alpha_k b_k$$

are uniformly bounded in norm by some constant C.

Let b_n^* denote the coordinate functionals, where b_n^* assigns to every vector $x \in E$ the coordinate α_n of x in the above expansion. Each b_n^* is a bounded linear functional on E. Indeed, for every vector $x \in E$,

$$\begin{aligned}|b_n^*(x)|\,\|b_n\| &= |\alpha_n|\,\|b_n\| \\ &= \|\alpha_n b_n\| = \|P_n(x) - P_{n-1}(x)\| \le 2C\|x\|.\end{aligned}$$

Functionals b_n^* are called bi-orthogonal functionals associated with the basis b_n. When the basis b_n is normalized, one has a uniform bound $\|b_n^*\|_{E^*} \le 2C$. The famous question of Banach if separable Banach spaces always admit a Schauder basis was answered negatively by Paul Enflo [43].

The real and the convex interpolation methods

Let (X_0, X_1) be two Banach spaces continuously embedded into some Hausdorff topological vector space (such spaces are called a Banach couple or a compatible couple). Their interpolation by the real method can be defined with help of the Peetre K-functional

$$K(x, t; X_0, X_1) = \inf\{\|x_0\|_{X_0} + t\|x_1\|_{X_1} \ : \ x = x_0 + x_1,\ x_0 \in X_0,\ x_1 \in X_1\}. \tag{10.4}$$

The interpolated space $X_{\theta,q}$ is the space of all elements in $X_0 + X_1$ for which the following norm is finite:

$$\|x\|_{\theta,q;K} = \left(\int_0^{\infty} (t^{-\theta} K(x, t; X_0, X_1))^q \frac{dt}{t}\right)^{\frac{1}{q}}, \quad 0 < \theta < 1, 1 \le q < \infty, \tag{10.5}$$

$$\|x\|_{\theta,\infty;K} = \sup_{t>0}\ t^{-\theta} K(x, t; X_0, X_1), \quad 0 \le \theta \le 1. \tag{10.6}$$

One has ([2, p. 216, Corollary 7.17], [19, p. 46] or [128, pp. 25–26]) the following inclusion:

$$(X_0, X_1)_{\theta,q_0} \subset (X_0, X_1)_{\theta,q_1} \quad \text{for all } \theta \in (0,1) \text{ and } 1 \le q_0 \le q_1 \le \infty. \tag{10.7}$$

Let now (A_0, A_1) be a compatible couple of Banach spaces (over the field $\mathbb{C}$) and define $\Phi = \Phi(A_0, A_1)$ as the space of all functions f of the complex variable $z = x + iy$ with values in $A_0 + A_1$ such that:

(a) f is bounded and continuous on the strip $0 \leq x \leq 1$;
(b) f is analytic from the strip $0 < x < 1$ into $A_0 + A_1$;
(c) f is continuous on the line $x = 0$ into A_0 and $\|f(iy)\|_{A_0} \to 0$ as $|y| \to \infty$;
(d) f is continuous on the line $x = 1$ into A_1 and $\|f(1+iy)\|_{A_1} \to 0$ as $|y| \to \infty$.

The space Φ endowed with a norm

$$\|f\|_\Phi = \max\Big\{\sup_{y\in\mathbb{R}}\|f(iy)\|_{A_0}, \sup_{y\in\mathbb{R}}\|f(1+iy)\|_{A_1}\Big\}$$

is a Banach space. The complex interpolation spaces A_θ, $\theta \in (0,1)$, are defined as

$$A_\theta = [A_0, A_1]_\theta = \{u \in A_0 + A_1 : \ u = f(\theta) \text{ for some } f \in \Phi\}$$

(see Calderón [26]), and they are Banach spaces with norms

$$\|u\|_{A_\theta} = \inf\{\|f\|_\Phi : f(\theta) = u\}.$$

Interpolation estimates

Theorem 10.1.6 (see, e. g., [2, pp. 220–221]). *Let (A_0, A_1) and (B_0, B_1) be two Banach couples. Let $T : A_0 + A_1 \to B_0 + B_1$ be a linear operator continuous as a map $A_0 \to B_0$ and as a map $A_1 \to B_1$. Then*

$$\|T\|_{(A_0,A_1)_{\theta,p}\to(B_0,B_1)_{\theta,p}} \leq \|T\|^{\theta}_{A_0\to B_0}\|T\|^{1-\theta}_{A_1\to B_1}, \tag{10.8}$$

and

$$\|T\|_{[A_0,A_1]_\theta\to[B_0,B_1]_\theta} \leq \|T\|^{\theta}_{A_0\to B_0}\|T\|^{1-\theta}_{A_1\to B_1} \tag{10.9}$$

for every $p \in [1,\infty]$ and every $\theta \in (0,1)$.

For elements $a \in A_0 \cap A_1$ the following estimates hold:

$$\|a\|_{[A_0,A_1]_\theta} \leq \|a\|^{1-\theta}_{A_0}\|a\|^{\theta}_{A_1} \tag{10.10}$$

and

$$\|a\|_{(A_0,A_1)_{\theta,p}} \leq c_{\theta,p}\|a\|^{1-\theta}_{A_0}\|a\|^{\theta}_{A_1}. \tag{10.11}$$

Examples of interpolated spaces

1. Interpolation of Lebesgue spaces (see, e. g., [2, Chapter 7]):

$$(L^{p_0}, L^{p_1})_{\theta,p} = [L^{p_0}, L^{p_1}]_\theta = L^p \quad \text{for all } 1 \le p_0 < p_1 \le \infty \text{ and } \theta \in (0,1), \tag{10.12}$$

with $\frac{1}{p} = \frac{1-\theta}{p_0} + \frac{\theta}{p_1}$.

2. Sobolev spaces by interpolation (see [2, p. 250]):

$$H^{s,p}(\mathbb{R}^N) = [H^{m,p}(\mathbb{R}^N), L^p(\mathbb{R}^N)]_{s/m}, \quad m \in \mathbb{N}, p \in (1,\infty), 0 < s < m. \tag{10.13}$$

Note that all choices of m as above give the same space.

3. Interpolation of Besov spaces (see [19] for the homogeneous case, and [128, p. 186] or [2, p. 230]) for the inhomogeneous case): for each $s_0, s_1 \in (0,\infty)$, $p \in (1,\infty)$ and $q \in [1,\infty]$, $s_0 \ne s_1$, $s_\theta = \theta s_1 + (1-\theta)s_0$, one has

$$\dot{B}^{s_\theta,p,q}(\mathbb{R}^N) = (\dot{H}^{s_0,p}(\mathbb{R}^N), \dot{H}^{s_1,p}(\mathbb{R}^N))_{\theta,q}, \tag{10.14}$$

and

$$B^{s_\theta,p,q}(\mathbb{R}^N) = (H^{s_0,p}(\mathbb{R}^N), H^{s_1,p}(\mathbb{R}^N))_{\theta,q}. \tag{10.15}$$

4. Lorentz spaces are obtained by the real-method interpolation of Lebesgue spaces:

$$(L^{p_0}(\mathbb{R}^N), L^{p_1}(\mathbb{R}^N))_{\theta,q} = L^{p_\theta,q}(\mathbb{R}^N), \quad \frac{1}{p_\theta} = \frac{\theta}{p_1} + \frac{1-\theta}{p_0}. \tag{10.16}$$

10.2 Function spaces with scale invariance

In this section, we summarize properties of Banach spaces of functions (or classes of equivalence of functions modulo polynomials) on $\mathbb{R}^N$ whose norms are invariant with respect to the rescaling group (3.1). We follow here [51, 128], and [2].

Sobolev spaces

Sobolev spaces $\dot{H}^{m,p}(\mathbb{R}^N)$, $m \in \mathbb{N}$, $p \in [1,\infty]$ are defined as completions of $C_0^\infty(\mathbb{R}^N)$ with respect to the norm $\|\nabla^m u\|_p$, where

$$\nabla^m u = \{\nabla^\alpha\}_{|\alpha|=m} = \left\{\frac{\partial^m u}{\partial x_1^{\alpha_1}, \dots, \partial x_N^{\alpha_N}}\right\}_{\alpha_1+\dots+\alpha_N=m}$$

is the collection of all partial derivatives of u of order m. This definition extends to the case of $\dot{H}^{0,p}(\mathbb{R}^N) = L^p(\mathbb{R}^N)$. The space $\dot{H}^{m,p}(\mathbb{R}^N)$ is continuously embedded into the space of distributions only if $N > pm$, in which case one has the limiting Sobolev embedding $\dot{H}^{m,p}(\mathbb{R}^N) \hookrightarrow L^q(\mathbb{R}^N)$ with $1/p - 1/q = m/N$.

A common extension of the Sobolev spaces to fractional values of m, which retains the name Sobolev spaces (otherwise called potential Sobolev spaces or spaces of Bessel potentials), denoted $\dot{H}^{s,p}(\mathbb{R}^N)$, $s \in \mathbb{R}$, $p \in [1, \infty]$, is characterized by the norm $\|(-\Delta)^{\frac{s}{2}} u\|_p$.

A different extension, Sobolev–Slobodecki spaces, denoted $\dot{W}^{s,p}(\mathbb{R}^N)$, $s > 0$, $s \notin \mathbb{N}$, $p \in [1, \infty)$, are defined with the help of the Slobodecki–Gagliardo seminorm

$$[f]_{\theta,p,\Omega} = \left(\int_\Omega \int_\Omega \frac{|f(x) - f(y)|^p}{|x - y|^{\theta p + n}} \, dx \, dy \right)^{\frac{1}{p}}, \quad 1 \le p < \infty, \theta \in (0,1),$$

and are characterized by the norm

$$\|f\|_{W^{s,p}(\Omega)} \overset{\text{def}}{=} \|f\|_{H^{\lfloor s \rfloor,p}(\Omega)} + \sup_{|\alpha| = \lfloor s \rfloor} [\nabla^\alpha f]_{s - \lfloor s \rfloor, p, \Omega}, \quad s > 0.$$

Sobolev–Slobodecki spaces coincide with the real interpolation spaces of Sobolev spaces, that is, in the sense of equivalent norms the following holds:

$$\dot{W}^{s,p}(\Omega) = (\dot{H}^{k,p}(\Omega), \dot{H}^{k+1,p}(\Omega))_{s - \lfloor s \rfloor, p}, \quad k \in \mathbb{N}, s \in (k, k+1). \tag{10.17}$$

When $p = 2$ and $s > 0$ Sobolev and Sobolev–Slobodecki spaces coincide.

Corresponding inhomogeneous spaces are defined, respectively, as $H^{s,p} = \dot{H}^{s,p} \cap L^p$ and $W^{s,p} = \dot{W}^{s,p} \cap L^p$.

Besov and Triebel–Lizorkin spaces via Littlewood–Paley decomposition

The Littlewood–Paley family of operators $\{P_j\}_{j \in \mathbb{Z}}$ is based on existence of a family of functions $\{\varphi_j\}_{j \in \mathbb{Z}}$ with the following properties:

$$\operatorname{supp} \varphi_j \subset \{\xi \in \mathbb{R}^N : 2^{j-1} \le |\xi| \le 2^{j+1}\}, \tag{10.18}$$

$$\sum_{j \in \mathbb{Z}} \varphi_n(\xi) = 1 \quad \text{for all } \xi \in \mathbb{R}^N \setminus \{0\}, \tag{10.19}$$

$$\varphi_j(\xi) = \varphi_0(2^{-j}\xi) \quad \text{for all } \xi \in \mathbb{R}^N \text{ and } j \in \mathbb{Z}, \tag{10.20}$$

$$\varphi_{j-1}(\xi) + \varphi_j(\xi) + \varphi_{j+1}(\xi) = 1 \quad \text{for all } \xi \in \operatorname{supp} \varphi_j. \tag{10.21}$$

Then $P_j : \dot{H}^{1,p}(\mathbb{R}^N) \to \mathbb{R}$, $j \in \mathbb{Z}$, are given by

$$P_j u = \mathcal{F}^{-1} \varphi_0(2^{-j} \cdot) \mathcal{F} u, \tag{10.22}$$

where $\mathcal{F}$ is the Fourier transform. Homogeneous Besov and Triebel–Lizorkin spaces are characterized by the following norms (cf [128, (4) in Chapter 5] or [51, (5.2)]), respectively,

$$\|u\|_{\dot{B}^{s,p,q}} = \|(\|2^{js}P_j u\|_{L^p})_{j\in\mathbb{Z}}\|_{\ell^q}, \quad s \in \mathbb{R},\ 1 \le p \le \infty,\ 1 \le q \le \infty,$$
$$\|u\|_{\dot{F}^{s,p,q}} = \|\|(2^{js}P_j u)_{j\in\mathbb{Z}}\|_{\ell^q}\|_{L^p}, \quad s \in \mathbb{R},\ 1 \le p < \infty,\ 1 \le q \le \infty.$$

Strictly speaking, the expressions above, evaluated on tempered distributions (space $\mathcal{S}'$), are seminorms vanishing on polynomials, so the homogenous Besov and Triebel-Lizorkin spaces are initially defined as quotient spaces. However, they can be realized as functional spaces, in particular by means of embeddings into known function spaces or via wavelet decompositions.

Inhomogeneous Besov spaces with $s > 0$, $p, q \in [1, \infty]$, can be identified as intersection of homogeneous Besov spaces with Lebesge spaces: $B^{s,p,q}(\mathbb{R}^N) = \dot{B}^{s,p,q}(\mathbb{R}^N) \cap L^p(\mathbb{R}^N)$ ([19, Theorem 6.3.2]).

Sobolev spaces $H^{s,p}(\mathbb{R}^N)$ and Sobolev–Slobodecki spaces $W^{s,p}(\mathbb{R}^N)$ are identified as subfamilies of Triebel–Lizorkin spaces:

$$\dot{H}^{s,p}(\mathbb{R}^N) = \dot{F}^{s,p,2}(\mathbb{R}^N), \quad 1 < p < \infty, s > 0, \tag{10.23}$$
$$\dot{W}^{s,p}(\mathbb{R}^N) = \dot{F}^{s,p,p}(\mathbb{R}^N)[= \dot{B}^{s,p,p}(\mathbb{R}^N)], \quad 1 < p < \infty, s > 0, s \notin \mathbb{N}. \tag{10.24}$$

Space $\dot{F}^{0,\infty,2}$ is identified as the space of bounded mean oscillations BMO. The following refined Sobolev inequality is due to Gerard, Meyer, and Oru [59]:

$$\|u\|_q \le \|(-\Delta)^{\frac{s}{2}}u\|_p^{p/q}\|u\|_{\dot{B}^{s-N/p,\infty,\infty}}^{1-p/q}, \quad 1 < q < p < \infty, \text{ and } s = N/q - N/p. \tag{10.25}$$

Embeddings of Besov and Triebel–Lizorkin spaces

The first of the following embeddings is immediate from the definition, and the other is an elementary consequence of order between ℓ^p-norms:

$$\dot{B}^{s,p,p}(\mathbb{R}^N) = \dot{F}^{s,p,p}(\mathbb{R}^N), \quad s \in \mathbb{R}, p \in [1, \infty), \tag{10.26}$$
$$\dot{B}^{s,p,a}(\mathbb{R}^N) \hookrightarrow \dot{B}^{s,p,b}(\mathbb{R}^N), \quad 1 \le a < b \le \infty, p \in [1, \infty], s \in \mathbb{R}. \tag{10.27}$$

Other embeddings are of the same character as Sobolev embeddings: in heuristic terms, they trade smoothness for integrability. For $p, q, a, b \in [1, \infty]$, $s, t \in \mathbb{R}$,

$$\dot{B}^{s,p,a}(\mathbb{R}^N) \hookrightarrow \dot{B}^{t,q,a}(\mathbb{R}^N), \quad s - N/p = t - N/q, s > t, \tag{10.28}$$
$$\dot{F}^{s,p,a}(\mathbb{R}^N) \hookrightarrow \dot{F}^{t,q,b}(\mathbb{R}^N), \quad s - N/p = t - N/q, s > t, q < \infty. \tag{10.29}$$

There is also a continuous embedding, derived from embeddings into Lebesgue spaces and interpolation (10.16),

$$\dot{B}^{s,p,a}(\mathbb{R}^N) \hookrightarrow L^{p_s^*,a}, \quad p \in (1, N/s), a \in [1, \infty]. \tag{10.30}$$

Wavelet characterization of Besov spaces

We follow the presentation of the topic from [92, Section 6.10] and [13].

Definition 10.2.1. Let $\mathcal{G}_0$ be a set of linear bijective isometries on a Banach space E and let $\psi \in E$. A set

$$\mathcal{W}(\psi, \mathcal{G}_0) = \{g\psi\}_{g \in \mathcal{G}_0}$$

is called a wavelet basis of E with a mother wavelet ψ if it forms a Schauder basis for E.

For $s \in \mathbb{R}$ and $1 \le p, q \le \infty$, Besov spaces $\dot{B}^{s,p,q}(\mathbb{R}^N)$ have normalized wavelet bases with the same mother wavelet and with a set of rescalings

$$\mathcal{G}_0 = \{u \mapsto 2^{rj} u(2^j \cdot -k)\ j \in \mathbb{Z}, k \in \mathbb{Z}^N\} \subset \mathcal{G}^r, \quad r = \frac{N - sp}{p}.$$

Let $c_{j,k}(u)$, $j \in \mathbb{Z}$, $k \in \mathbb{Z}^N$, denote coefficients of the wavelet expansion of an element u from one of the spaces above, that is,

$$u = \sum_{j \in \mathbb{Z}, k \in \mathbb{Z}^N} c_{j,k}(u) 2^{rj} \psi(2^j \cdot -k). \tag{10.31}$$

Remark 10.2.2. Expansion (10.31) uses the same wavelets for any values of N, p, q, or s, but with different normalization, which dependent only on the value of $r = \frac{N}{p} - s$, so the coefficients $c_{j,k}(u)$ in (10.31) with different values of s, p, q, and N depend only on the value of $r = \frac{N}{p} - s$.

Then equivalent norms in Besov spaces can be expressed in terms of the wavelet coefficients as:

$$\|u\|_{\dot{B}^{s,p,q}} = \|(\|(c_{j,k}(u))_{k \in \mathbb{Z}^N}\|_{\ell^p})_{j \in \mathbb{Z}}\|_{\ell^q}. \tag{10.32}$$

Space $\dot{\mathrm{BV}}(\mathbb{R}^N)$ – functions of bounded variation

For a comprehensive exposition on functions of bounded variations, we refer the reader to the book [11]. We consider the case $N \ge 2$.

Definition 10.2.3. The space of functions of bounded variation $\dot{\mathrm{BV}}(\mathbb{R}^N)$ is the space of all measurable functions $u : \mathbb{R}^N \to \mathbb{R}$ vanishing at infinity (i. e., $\forall \varepsilon > 0\ |\{x \in \mathbb{R}^N : |u(x)| > \varepsilon\}| < \infty$) such that

$$\|Du\| \overset{\text{def}}{=} \sup_{v \in C_0^\infty(\mathbb{R}^N;\mathbb{R}^N):\|v\|_\infty = 1} \int_{\mathbb{R}^N} u \operatorname{div} v\, dx < \infty. \tag{10.33}$$

The $\dot{BV}(\mathbb{R}^N)$-norm can be interpreted as the total variation $\|Du\|$ of the measure associated with the derivative ∇u (in the sense of distributions on $\mathbb{R}^N$). If $u \in C_0^1(\mathbb{R}^N)$, then the right-hand side in (10.33) by integration by parts equals $\int_{\mathbb{R}^N} |\nabla u| dx$. The value of the total variation of Du on a measurable set $A \subset \mathbb{R}^N$ we denote as $\|Du\|_A$.

The space $\dot{BV}(\mathbb{R}^N)$ is a conjugate space and therefore is complete. We follow the convention that calls the weak* convergence in the space of bounded variation *weak convergence*. The space $\dot{BV}(\mathbb{R}^N)$ is separable and, therefore, each bounded sequence in $\dot{BV}(\mathbb{R}^N)$ has a weakly (i. e., weak*-) convergent subsequence. We use the following properties of $\dot{BV}(\mathbb{R}^N)$:

1. A sequence (u_k) in $\dot{BV}(\mathbb{R}^N)$ is weakly convergent to u if and only if $u_k \to u$ in $L^1_{\mathrm{loc}}(\mathbb{R}^N)$ and the weak derivatives $\partial_i u_k$, $i = 1, \dots, N$ converge to $\partial_i u$ weakly as finite measures on $\mathbb{R}^N$.
2. Density of $C_0^\infty(\mathbb{R}^N)$ in *strict* topology (note that the closure of $C_0^\infty(\mathbb{R}^N)$ in the norm topology is $\dot{H}^{1,1}(\mathbb{R}^N)$). One says that u_k converges strictly to u if $\|u_k - u\|_{1^*} \to 0$ and $\|Du_k\| \to \|Du\|$. Consequently, the rescaling group $\mathcal{G}^{N-1}$ (3.1) extends by this density property to a group of isometries on $\dot{BV}$.
3. V. Maz'ya's inequality (often referred to as Sobolev, Aubin–Talenti or Gagliardo–Nirenberg inequality) [90]:

$$NV_N^{1/N}\|u\|_{1^*} \le \|Du\|, \tag{10.34}$$

   where $1^* = \frac{N}{N-1}$ and V_N is the volume of the unit ball in $\mathbb{R}^N$. A local version of this inequality is

$$\|u\|_{1^*,\Omega} \le C(\|Du\|_\Omega + \|u\|_{1,\Omega}), \tag{10.35}$$

   where $\Omega \subset \mathbb{R}^N$ is a bounded domain with sufficiently regular, say locally C^1-boundary.
4. Hardy inequality:

$$\|Du\| \ge (N-1)\int_{\mathbb{R}^N} \frac{|u|}{|x|} dx .$$

   (It follows from the Hardy inequality in $\dot{H}^{1,1}(\mathbb{R}^N)$ and the density of $C_0^\infty(\mathbb{R}^N)$ in $\dot{BV}(\mathbb{R}^N)$ with respect to the strict convergence, if one first replaces $1/|x|$ with its $L^N(\mathbb{R}^N)$-approximations from below.)
5. Local compactness: for any set $\Omega \subset \mathbb{R}^N$ of finite Lebesgue measure, $\dot{BV}(\mathbb{R}^N)$ is compactly embedded into $L^1(\Omega)$ and any sequence weakly convergent to zero in $\dot{BV}(\mathbb{R}^N)$ converges to zero in $L^1(\Omega)$.
6. Chain rule (a cruder version of a more refined statement due to Vol'pert, see [11, Remark 3.98]): let $\varphi \in C^1(\mathbb{R})$. Then for every $u \in \dot{BV}(\mathbb{R}^N)$,

$$\|D\varphi(u)\| \le \|\varphi'\|_\infty \|Du\|. \tag{10.36}$$

10.3 Manifolds of bounded geometry

We list some general properties of Riemannian manifolds of bounded geometry. For basic definitions and notation, see Chapter 7.

Lemma 10.3.1 ([41]). *Let M be a Riemannian manifold of bounded geometry and let $0 < r < r(M)$. If $k \in \mathbb{N}$, then there exists a constant C_k dependent on the curvature bounds and r but independent of $x \in M$, which bounds the C^k-norm of components g_{ij} of the metric tensor g and its inverse g^{ij} in any normal coordinate system of radius not exceeding r at any point $x \in M$.*

Manifolds of bounded geometry have the following properties (see [107] for the first assertion and [65] for the second one):

Lemma 10.3.2. *If the manifold M has bounded geometry and $0 < r < r(M)$, then for any $\alpha \in \mathbb{N}_0^N$ there exists a constant $C_\alpha > 0$, such that*

$$\left|d^\alpha(e_y^{-1} \circ e_x)(\xi)\right| \le C_\alpha \quad \text{whenever } x, y \in M, \text{ and } B(x,r) \cap B(y,r) \ne \emptyset.$$

Moreover, there exists $\lambda > 0$ such that for any $y \in M$, $x \in B(y,r)$,

$$\lambda^{-1}\delta_{ij} \le g_{ij}(e_y(x)) \le \lambda\delta_{ij}. \tag{10.37}$$

The following corollary is the immediate consequence of Lemma 10.3.1 above.

Corollary 10.3.3. *Let $p \in (0,\infty)$ and $r \in (0, r(M))$. There exists a constant $C > 1$ such that for any $x \in M$,*

$$C^{-1} \int\limits_{B(x,r)} |u|^p \mathrm{d}v_g \le \int\limits_{\Omega_r} |u \circ e_x|^p \mathrm{d}\xi \le C \int\limits_{B(x,r)} |u|^p \mathrm{d}v_g, \tag{10.38}$$

and

$$C^{-1} \int\limits_{B(x,r)} g_x(du, du)\mathrm{d}v_g \le \int\limits_{\Omega_r} \sum_{i=1}^N \left|\frac{\partial}{\partial x_i}(u \circ e_x)\right|^2 \mathrm{d}\xi \le C \int\limits_{B(x,r)} g_x(du, du)\mathrm{d}v_g.$$

A related global estimate that follows from Bishop–Gromov theorem (see [65, Theorem 1.1]) says that whenever $0 < r < R$, there is a $C(r,R) > 0$ such that

$$v_g(B(x,R)) \le C(r,R) v_g(B(y,r)) \quad \text{for any } x \in M, y \in B(x,R). \tag{10.39}$$

Let us also recall a technical but useful equivalent norm of $H^{1,2}(M)$; cf. [64] or [129, Chapter 7].

Lemma 10.3.4. *Let Y be a (ε, r)-discretization of an N-dimensional manifold M with bounded geometry, $r \in (0, r(M))$ and let $\{\chi_i\}$ be a partition of unity subordinated to the covering $\{B(y_i, r)\}_{y_i \in Y}$ and satisfying* (7.1). *Then*

$$|||f|||_{H^{1,2}(M)} \stackrel{\text{def}}{=} \left(\sum_i \|\chi_i f \circ \exp_{y_i} \|^2_{H^{1,2}(\mathbb{R}^N)} \right)^{1/2} \tag{10.40}$$

is an equivalent norm in $H^{1,2}(M)$. Moreover,

$$\|f\|_{H^{1,2}(M)} \sim |||f|||_{H^{1,2}(M)} \sim \left(\sum_i \|\chi_i f\|^2_{H^{1,2}(M)} \right)^{1/2} .$$

Below is a particular case of the gluing theorem from Gallier et al. [55, Theorem 3.1].

Definition 10.3.5 ([55, Definition 3.1], [54, Definition 8.1]). A *set of gluing data* is a triple $(\{\Omega_i\}_{i \in \mathbb{N}_0}, \{\Omega_{ij}\}_{i,j \in \mathbb{N}_0}, \{\psi_{ji}\}_{(i,j) \in \mathbb{K}})$ satisfying the following properties:

(1) For every $i \in \mathbb{N}_0$, the set Ω_i is a nonempty open subset of $\mathbb{R}^N$ and the sets $\{\Omega_i\}_{i \in \mathbb{N}_0}$ are pairwise disjoint;
(2) For every pair $i, j \in \mathbb{N}_0$, the set Ω_{ij} is an open subset of Ω_i. Furthermore, $\Omega_{ii} = \Omega_i$ and $\Omega_{ji} \neq \emptyset$ if and only if $\Omega_{ij} \neq \emptyset$;
(3) $\mathbb{K} = \{(i,j) \in \mathbb{N}_0 \times \mathbb{N}_0 : \Omega_{ij} \neq \emptyset\}$, $\psi_{ji} : \Omega_{ij} \to \Omega_{ji}$ is a diffeomorphism for every $(i,j) \in \mathbb{K}$, and the following conditions hold:
   (a) $\psi_{ii} = \mathrm{id}|_{\Omega_i}$, for all $i \in \mathbb{N}_0$,
   (b) $\psi_{ij} = \psi_{ji}^{-1}$, for all $(i,j) \in \mathbb{K}$,
   (c) For all $i, j, k \in \mathbb{N}_0$, if $\Omega_{ji} \cap \Omega_{jk} \neq \emptyset$, then $\psi_{ij}(\Omega_{ji} \cap \Omega_{jk}) = \Omega_{ij} \cap \Omega_{ik}$, and $\psi_{ki}(x) = \psi_{kj} \circ \psi_{ji}(x)$, for all $x \in \Omega_{ij} \cap \Omega_{ik}$;
(4) For every pair $(i,j) \in \mathbb{K}$, with $i \neq j$, for every $x \in \partial\Omega_{ij} \cap \Omega_i$ and every $y \in \partial\Omega_{ji} \cap \Omega_j$, there are open balls V_x and V_y centered at x and y so that no point of $V_y \cap \Omega_{ji}$ is the image of any point of $V_x \cap \Omega_{ij}$ by ψ_{ji}.

Each set Ω_i is called *parametrization domain* or *p-domain*, each nonempty set Ω_{ij} is called a *gluing domain*, and each map ψ_{ij} is called *transition map* or *gluing map*.

Theorem 10.3.6 ([55, Theorem 3.1]). *For every set of gluing data,*

$$(\{\Omega_i\}_{i \in \mathbb{N}_0}, \{\Omega_{ij}\}_{i,j \in \mathbb{N}_0}, \{\psi_{ji}\}_{(i,j) \in \mathbb{K}}),$$

there exists a N-dimensional smooth manifold M an atlas $(U_i, \tau_i)_i$ of M such that $\tau_i(U_i) = \Omega_i$, whose transition maps are $\tau_j \circ \tau_i^{-1} = \psi_{ji} : \Omega_{ij} \to \Omega_{ji}$. $i, j \in \mathbb{N}_0$.

Remark 10.3.7. Note that the theorem does not provide any specifics about the maps τ_i which are obviously not uniquely defined.

Corollary 10.3.8. *Let $0 < \rho < r < a$ and let $\Omega_\rho \subset \Omega_r \subset \Omega_a$ be balls in $\mathbb{R}^N$ centered at the origin with radius ρ, r, and a, respectively. Let $\{\widetilde{\psi}_{ij}\}_{i,j\in\mathbb{N}_0}$ be a family of smooth open maps $\widetilde{\psi}_{ij} : \Omega_r \to \Omega_a$. Assume that a family $\{\psi_{ji} = \widetilde{\psi}_{ji}|_{\Omega_\rho}\}_{i,j\in\mathbb{N}_0}$ satisfies the following conditions:*
(i) $\psi_{ii} = \mathrm{id}$, $i \in \mathbb{N}_0$;
(ii) ψ_{ji} *is a diffeomorphism between* $\Omega_{ij} \stackrel{\text{def}}{=} \psi_{ij}(\Omega_\rho) \cap \Omega_\rho$ *and* Ω_{ji}, $i,j \in \mathbb{N}_0$, *whenever* $\Omega_{ji} \neq \emptyset$;
(iii) $\psi_{ij} = \psi_{ji}^{-1}$ *on* Ω_{ji}, *whenever* $\Omega_{ji} \neq \emptyset$, $i,j \in \mathbb{N}_0$;
(iv) $\psi_{ij}(\Omega_{ji} \cap \Omega_{jk}) = \Omega_{ij} \cap \Omega_{ik}$, *and* $\psi_{ki}(x) = \psi_{kj} \circ \psi_{ji}(x)$ *for all* $x \in \Omega_{ij} \cap \Omega_{ik}$, $i,j,k \in \mathbb{N}_0$;
(v) *for all* $(i,j) \in \mathbb{K} \stackrel{\text{def}}{=} \{(i,j) \in \mathbb{N}_0\times\mathbb{N}_0 : \Omega_{ij} \neq \emptyset\}$ *and all* $x \in \partial\Omega_{ij}\cap\Omega_\rho$, $\psi_{ji}(x) \in \partial\Omega_{ji}\cap\partial\Omega_\rho$.

Then there exists a smooth differential manifold M with an atlas $\{(U_i, \tau_i)\}_{i\in\mathbb{N}_0}$, such that $\tau_i(U_i) = \Omega_\rho$ for any $i \in \mathbb{N}_0$ and whose transition maps $\tau_j \circ \tau_i^{-1}$ are $\psi_{ji} : \Omega_{ij} \to \Omega_{ji}$. $i,j \in \mathbb{N}_0$.

Proof. Fix an enumeration $(z_i)_{i\in\mathbb{N}_0}$ of the lattice $3a\mathbb{Z}^N \subset \mathbb{R}^N$. Set $\Omega_i' \stackrel{\text{def}}{=} z_i + \Omega_\rho$, $i \in \mathbb{N}_0$, and $\Omega_{ij}' \stackrel{\text{def}}{=} \Omega_{ij} + z_i$, $\psi_{ij}' \stackrel{\text{def}}{=} \psi_{ij}(\cdot - z_j) + z_i$, for $(i,j) \in \mathbb{K}$. The corollary is immediate from Theorem 10.3.6 once we show that $(\{\Omega_i'\}_{i\in\mathbb{N}_0}, \{\Omega_{ij}'\}_{i,j\in\mathbb{N}_0}, \{\psi_{ij}'\}_{(i,j)\in\mathbb{K}})$ is a set of gluing data according to Definition 10.3.5. Conditions of the definition verify as follows.

Condition (1) is immediate since $3a > 2\rho$.

Condition (2). The sets Ω_{ij} (and thus Ω_{ij}') are open since the maps ψ_{ji} are open. The relation $\Omega_{ij}' \subset \Omega_i'$ follows from $\Omega_{ij} \subset \Omega_\rho$ in (ii). By (i), we have $\Omega_{ii} = \Omega_\rho$, and thus $\Omega_{ii}' = \Omega_i'$. If $\Omega_{ij}' \neq \emptyset$, then $\Omega_{ij} \neq \emptyset$, and since ψ_{ij} is the inverse of ψ_{ji}, $\Omega_{ji} \stackrel{\text{def}}{=} \psi_{ji}(\Omega_\rho\cap\psi_{ij}\Omega_\rho) = \psi_{ji}\Omega_{ij} \neq \emptyset$. Thus $\Omega_{ji}' \neq \emptyset$.

Conditions (3)*: properties* (a), (b), and (c) *are immediate, respectively, from* (i), (iii), and (iv).

Condition (4). Let $x \in \partial\Omega_{ij}' \cap \Omega_\rho(z_i)$ and $y \in \partial\Omega_{ji}' \cap \Omega_\rho(z_j)$. Then $\tilde{x} = x - z_i \in \partial\Omega_{ij} \cap \Omega_\rho$ and $\tilde{x} = x - z_j \in \partial\Omega_{ji} \cap \Omega_\rho(z_j)$. By assumption (v), we have $\tilde{y} \neq \psi_{ji}(\tilde{x})$. In consequence, there exist Euclidean balls $\Omega(\tilde{x},\varepsilon)$ and $\Omega(\tilde{y},\varepsilon)$ such that no point of $\Omega(\tilde{y},\varepsilon) \cap \Omega_\rho$ is an image of $\Omega(\tilde{x},\varepsilon) \cap \Omega_\rho$. □

10.4 Concentration compactness – traditional approach

Defect of compactness of sequences in functional spaces can be described in terms of sequences of measures, rather than as profile decompositions. In the pioneering work of Lions, defect of compactness was identified in two types: type I, studied in [83], related to the group of shifts $\mathcal{G}_{\mathbb{R}^N}$, and type II, studied in [84], related to the rescaling group (actions of translations and dilations) $\mathcal{G}^r$.

Concentration compactness I is used in applications involving subcritical Sobolev embeddings $H^{1,p}(\mathbb{R}^N) \hookrightarrow L^q(\mathbb{R}^N)$, $1 < p < N$, $p < q < p^*$, which admit profile decomposition with the scaling group of shifts $\mathcal{G}_{\mathbb{R}^N}$. We quote it in a slightly refined version, [30, Theorem 8.7.1].

Theorem 10.4.1 (Concentration compactness I). *Let (u_k) be a bounded sequence in $H^{1,p}(\mathbb{R}^N)$, $1 < p < N$, and let $\rho_k = |u_k|^p$ with $\int_{\mathbb{R}^N} \rho_k \mathrm{d}x = \lambda > 0$ for all k. Then there exists a renamed subsequence satisfying one of the three following possibilities:*

(i – Tightness) *There exists a sequence (y_k) in $\mathbb{R}^N$, such that (ρ_k) is tight, that is, for every $\varepsilon > 0$ there exists $R \in (0, \infty)$ such that*

$$\int_{B_R(y_k)} \rho_k \mathrm{d}x \geq \lambda + \varepsilon,$$

(ii – Vanishing)

$$\sup_{y \in \mathbb{R}^N} \int_{B_R(y)} \rho_k \mathrm{d}x \longrightarrow 0 \quad \text{as } k \to \infty,$$

(iii – Dichotomy) *There exists $\alpha \in (0, \lambda)$, such that for all $\varepsilon > 0$ there exists $k_0 \in \mathbb{N}$ and bounded sequences $(u_k^{(1)})$ and $(u_k^{(2)})$ in $H^{1,p}(\mathbb{R}^N)$ satisfying for $k \geq k_0$,*

$$\|u_k - (u_k^{(1)} + u_k^{(2)})\|_q \leq \delta_q(\varepsilon), \quad q \in [p, p^*),$$

with $\delta_q(\varepsilon) \to 0$ as $\varepsilon \to 0$,

$$\left| \int_{\mathbb{R}^N} |u_k^{(1)}|^p \mathrm{d}x - \alpha \right| < \varepsilon,$$

$$\left| \int_{\mathbb{R}^N} |u_k^{(2)}|^p \mathrm{d}x - (\lambda - \alpha) \right| < \varepsilon,$$

$$\operatorname{dist}(\operatorname{supp} u_k^{(1)}, \operatorname{supp} u_k^{(2)}) \to \infty \quad \text{as } k \to \infty,$$

and

$$\liminf_{k \to \infty} \int_{\mathbb{R}^N} (|\nabla u_k|^p - |\nabla u_k^{(1)}|^p - |\nabla u_k^{(2)}|^p) \mathrm{d}x \geq 0.$$

Concentration compactness II, [84, formula (1.15)], describes defect of compactness and is used in applications related to the embedding $\dot{H}^{1,p}(\mathbb{R}^N) \hookrightarrow L^{p^*}$, $1 < p < N$, relative to the scaling group $\mathcal{G}^r$, $r = \frac{N-p}{p}$. We cite a significantly improved version of Chabrowski, [30, Sections 9.2–9.3], that, in particular, addresses concentration at infinity, which, in terms of profile decomposition, accounts for concentrations $t_k^r w(t_k(\cdot - y_k))$ with $|y_k| \to \infty$. We will denote as $S_{N,p}$ the best constant in the limiting Sobolev inequality on $\mathbb{R}^N$, $1 \leq p < N$, $\|\nabla u\|_p \geq S_{N,p} \|u\|_{p^*}$.

Theorem 10.4.2 (Concentration compactness II). *Let (u_k) be a sequence in $\dot{H}^{1,p}(\mathbb{R}^N)$, $1 < p < N$, weakly convergent to u, and such that:*

(i) $\mu_k \stackrel{\text{def}}{=} |\nabla u_k|^p \mathrm{d}x$ *is weak*-convergent to a measure* μ;
(ii) $\nu_k \stackrel{\text{def}}{=} |u_k|^{p^*} \mathrm{d}x$ *is weak*-convergent to a measure* ν;
(iii) $\mu_\infty \stackrel{\text{def}}{=} \lim_{R\to\infty} \limsup_{k\to\infty} \int_{|x|>R} |\nabla u_k|^p \mathrm{d}x$;

and
(iv) $\nu_\infty \stackrel{\text{def}}{=} \lim_{R\to\infty} \limsup_{k\to\infty} \int_{|x|>R} |u_k|^{p^*} \mathrm{d}x$.

Then there exist an at most countable index set J, *sequences* $(x_j)_{j\in J}$, *in* $\mathbb{R}^N$; $(\mu_j)_{j\in J}$, $0 < \mu_j < \infty$; *and* $(\nu_j)_{j\in J}$, $0 < \nu_j < \infty$, *such that*

$$\nu = |u|^{p^*} \mathrm{d}x + \sum_{j\in J} \nu_j \delta_{x_j}, \tag{10.41}$$

$$\mu \geq |\nabla u|^p \mathrm{d}x + \sum_{j\in J} \mu_j \delta_{x_j}, \tag{10.42}$$

where δ_{x_j} *are atomic measures supported at* x_j. *Furthermore,*

$$\limsup_{k\to\infty} \int_{\mathbb{R}^N} |u_k|^{p^*} \mathrm{d}x = \int_{\mathbb{R}^N} |u|^{p^*} \mathrm{d}x + \sum_{j\in J} \nu_j + \nu_\infty, \tag{10.43}$$

$$\limsup_{k\to\infty} \int_{\mathbb{R}^N} |\nabla u_k|^p \mathrm{d}x = \int_{\mathbb{R}^N} |\nabla u|^p \mathrm{d}x + \sum_{j\in J} \mu_j + \mu_\infty, \tag{10.44}$$

$$S_{N,p} \nu_j^{\frac{p}{p^*}} \leq \mu_j \quad \text{and} \quad S_{N,p} \nu_\infty^{\frac{p}{p^*}} \leq \mu_\infty. \tag{10.45}$$

Moreover, if $u = 0$ *and* $\mu(\mathbb{R}^N) S_{N,p} \nu(\mathbb{R}^N)^{\frac{p}{p^*}}$, *then* J *is a singleton and, for some* $\gamma \geq 0$, $\nu = \gamma \delta_{x_0} = S_{N,p}^{-1} \gamma^{\frac{p}{N}} \mu$.

A similar statement by Palatucci and Pisante deals with embeddings of fractional Sobolev spaces $\dot{H}^{s,2}(\Omega)$, where $0 < s < N/2$ and $\Omega \subset \mathbb{R}^N$ is a bounded domain, defined as the completion of $C_0^\infty(\Omega)$, in the $\dot{H}^{s,2}(\mathbb{R}^N)$-norm. For $0 < s < N/2$, this space is continuously embedded into $L^{2_s^*}_{\mathrm{loc}}(\mathbb{R}^N)$, where $2_s^* \stackrel{\text{def}}{=} \frac{2N}{N-2s}$.

Theorem 10.4.3 (Palatucci and Pisante, [98]). *Let* $\Omega \subset \mathbb{R}^N$ *be a bounded domain, and let* $(u_k)_{k\in\mathbb{N}}$ *be a sequence in* $\dot{H}^{s,2}(\Omega)$, $0 < s < N/2$, *weakly convergent to* u, *and such that*

$$|(-\Delta)^{\frac{s}{2}} u_k|^2 dx \stackrel{*}{\rightharpoonup} \mu \quad \text{and} \quad |u_k|^{2_s^*} dx \stackrel{*}{\rightharpoonup} \nu \quad \text{in } \mathcal{M}(\mathbb{R}^N).$$

Then, either $u_k \to u$ *in* $L^{2_s^*}_{\mathrm{loc}}(\mathbb{R}^N)$ *or there exists a (at most countable) set of distinct points* $\{x_n\}_{n\in J} \subset \bar{\Omega}$ *and positive numbers* $\{\nu_n\}_{n\in J}$ *such that we have*

$$\nu = |u|^{2_s^*} dx + \sum_n \nu_n \delta_{x_n}. \tag{10.46}$$

Moreover, there exist a positive measure $\tilde{\mu} \in \mathcal{M}(\mathbb{R}^N)$ supported in $\bar{\Omega}$ and positive numbers $\{\mu_n\}_{n\in J}$ such that

$$\mu = |(-\Delta)^{\frac{s}{2}}u|^2 dx + \tilde{\mu} + \sum_n \mu_n \delta_{x_n}, \quad \nu_n \le S(\mu_n)^{\frac{2_s^*}{2}}, \tag{10.47}$$

where $S = \inf_{\|u\|_{2_s^*}=1} \|(-\Delta)^{\frac{s}{2}}u\|_2$ is the best Sobolev constant in $\mathbb{R}^N$.

Sketch of proof. We sketch how to derive this statement from the profile decomposition (1.16). Without loss of generality, we may assume that $u_k \in C_0^\infty(\Omega)$. Consider a subsequence of (u_k), extended by zero to a sequence in $\dot{H}^{s,2}(\mathbb{R}^N)$, that has the profile decomposition (1.16). By the definition of profile decomposition, $w^{(1)} = u$. The fact that sequence (u_k) is supported in a bounded domain Ω restricts the possible values of $s_k^{(n)}$ and $y_k^{(n)}$ in (1.16). In particular, if for some $n \in \mathbb{N}$, $s_k^{(n)}$ is not bounded from below, then corresponding profile $w^{(n)}$ is zero. This and analogous arguments concerning possible support of $w^{(n)}$, together with passing to subsequence, allow to conclude that, without loss of generality, $s_k^{(n)} \to \infty$ for $n \ge 2$, and $y_k^{(n)} \to x_n$ with some $x_n \in \bar{\Omega}$.

Evaluating $\int_{\mathbb{R}^N} \varphi |u_k|^{2_s^*} dx$ with $\varphi \in C_c(\mathbb{R}^N)$, and taking into account uniform convergence of the series (1.16) and the asymptotic orthogonality (1.12), we arrive at a decoupled sum

$$\begin{aligned}
&\int_\Omega \varphi(x)|u_k(x)|^{2_s^*} dx \\
&\quad = \int_\Omega \varphi(x)|w^{(1)}(x)|^{2_s^*} dx + \sum_{n\ge2} \int_{\mathbb{R}^N} \varphi(x) 2^{Ns_k^{(n)}} |w^{(n)}(2^{s_k^{(n)}}(x - y_k^{(n)}))|^{2_s^*} dx + o(1) \\
&\quad = \int_\Omega \varphi(x)|u(x)|^{2_s^*} dx + \sum_{n\ge2} \varphi(x_n) \int_{\mathbb{R}^N} |w^{(n)}(x)|^{2_s^*} dx + o(1).
\end{aligned}$$

We have arrived at (10.46) with

$$\nu_n = \|w^{(n)}(x)\|_{2_s^*}^{2_s^*}. \tag{10.48}$$

Similar calculations based on (1.15) yield the first relation in (10.47) with

$$\mu_n = \|\nabla w^{(n)}(x)\|_2^2,$$

which, compared with (10.48), provides the second relation in (10.47). □

Classical concentration estimates like those in Theorem 10.4.3 have been extended to the case of unbounded domains with help of the notion of concentration at infinity, that adds further positive measures to the counterparts of (10.46) and (10.47). We refer the reader to [133, Lemma 1.40], that originates in the work of Bianchi, Chabrowski, and Szulkin [21, inequality (1.16)] and Ben-Naoum, Troestler, and Willem [16]. Concentration at infinity, similar to Theorem 10.4.3, can be also interpreted in terms of concentration profiles: when the domain is unbounded, profile decompositions contain blowup terms whose supports may escape to infinity ($|y_k^{(n)}| \to \infty$), or spread over the space ($s_k^{(n)} \to -\infty$).

It is natural to expect that the case of dichotomy in Theorem 10.4.1 allows a further splitting of cluster sequences ($u_k^{(1)}$) and ($u_k^{(2)}$) until they become tight. This idea can be seen as a precursor of profile decomposition, but it was also developed for abstract sequences of measures on metric spaces by Mihai Mariş [87].

Theorem 10.4.4 (Mihai Mariş). *Let* (Ω, d) *be a metric space and* $(\mu_n)_{n\geq 1}$ *a sequence of positive Borel measures on* Ω *such that*

$$M \stackrel{\text{def}}{=} \limsup_{n\to\infty} \mu_n(\Omega) < \infty.$$

Let $\varphi : [0,\infty) \to [0,\infty)$ *be an increasing function such that* $\varphi(s) \leq \frac{s}{2}$ *for all* s *and* $\lim_{s\to\infty} \varphi(s) = \infty$.

Then either $(\mu_n)_{n\in\mathbb{N}}$ *is a vanishing sequence, or there exists an increasing mapping* $j : \mathbb{N} \to \mathbb{N}$ *such that the subsequence* $(\mu_{j(n)})_{n\in\mathbb{N}}$ *satisfies one of the following properties:*

(i) *There are* $k \in \mathbb{N}$, *positive numbers* $m_1, \dots m_k$, *sequences of points* $(x_n^{(i)})_{n\in\mathbb{N}}$ *in* Ω *and increasing sequences of positive numbers* $(r_n^{(i)})_{n\in\mathbb{N}}$ *such that* $r_n^{(i)} \to \infty$ *as* $n \to \infty$, $i \in \{1, \dots, k\}$, *satisfying the following properties:*
   (a) *For each* n *the balls* $B_{r_n^{(i)}}(x_n^{(i)})$, $i \in \{1, \dots, k\}$ *are disjoint.*
   (b) *For each* $i \in \{1, \dots, k\}$, *we have*

$$\mu_{j(n)}(B_{\varphi(r_n^{(i)})}(x_n^{(i)})) \to m_i \quad \text{as } n \to \infty,$$
$$\mu_{j(n)}(B_{r_n^{(i)}}(x_n^{(i)}) \setminus B_{\varphi(r_n^{(i)})}(x_n^{(i)})) \leq \frac{1}{2^{n+i}},$$

   and the sequence of measures $(\mu_{j(n)}|_{B_{r_n^{(i)}}(x_n^{(i)})})_{n\in\mathbb{N}}$ *concentrates around* $(x_n^{(i)})_{n\in\mathbb{N}}$.
   (c) *The sequence of measures* $(\mu_{j(n)}|_{\Omega\setminus\bigcup_{i=1}^k B_{r_n^{(i)}}(x_n^{(i)})})_{n\in\mathbb{N}}$ *is a vanishing sequence.*

(ii) *There are positive numbers* $m_1, \dots m_k, \dots$ *such that* $m_{k+1} \leq 2m_k$, *sequences of points* $(x_n^{(i)})_{n\geq i}$ *in* Ω *and increasing sequences of positive numbers* $(r_n^{(i)})_{n\geq i}$ *such that* $r_n^{(k)} \to \infty$ *as* $n \to \infty$ *for each fixed* k *and the following properties hold:*
   (a) *For each* n, *the balls* $B_{r_n^{(1)}}(x_n^{(1)}), \dots, B_{r_n^{(n)}}(x_n^{(n)})$ *are disjoint.*
   (b) *The same as* (b) *in* (i) *above.*
   (c) *Denote by* $\tilde{q}_n^\ell$ *is the concentration function of* $\mu_{j(n)}|_{\Omega\setminus\bigcup_{i=1}^\ell B_{r_n^{(i)}}(x_n^{(i)})}$ *for* $\ell \geq n$. *Then*

$$\lim_{\ell\to\infty} \lim_{t\to\infty} \limsup_{n\to\infty} \tilde{q}_n^\ell(t) = 0.$$

   (d) *The sequence of measures* $(\mu_{j(n)}|_{\Omega\setminus\bigcup_{i=1}^n B_{r_n^{(i)}}(x_n^{(i)})})_{n\in\mathbb{N}}$ *is a vanishing sequence.*

In [87], this theorem is used to prove a weaker version (from [58]) of the profile decomposition of Solimini.

Common notations

Spaces

$E \hookrightarrow F$	continuous embedding
$C_0^\infty(\Omega)$	space of smooth functions with compact support
$C(\mathbb{R}^N, V(x)), L^\infty(\mathbb{R}^N, V(x))$	space of continuous, resp. measurable, functions with the norm $\sup_{x\in\mathbb{R}^N} \lvert u(x)\rvert V(x)$.
$\dot{C}^{0,\lambda}(\mathbb{R}^N)$	space of Hölder continuous functions modulo constants
$\lVert u\rVert_p$, $\lVert u\rVert_{p,\Omega}$	L^p-norm
$E_{\mathrm{rad}}(\mathbb{R}^N)$	for a space of functions on $\mathbb{R}^N$, its subspace of radially symmetric functions
$\exp L^q$	Orlicz space with the modulus $e^{\lvert u\rvert^q} - 1$
$L^{p,q}$	Lorentz space
$\dot{H}^{s,p}(\mathbb{R}^N)$	homogeneous Sobolev space
$H^{s,p}(\mathbb{R}^N)$	Sobolev space
$H_0^{s,p}(\Omega)$	closure of $C_0^\infty(\Omega)$ in the $H^{s,p}$-norm
$W^{s,p}$	Sobolev–Slobodecki space

Sets

$\mathbb{N}_0$	set $\mathbb{N}$ of natural numbers with added zero
$B(x,r)$	geodesic ball of radius r centered at x on a Riemannian manifold
$B_r(x)$	ball of radius r centered at x on $\mathbb{R}^N$
Ω_r	ball of radius r centered at the origin in $\mathbb{R}^N$

Groups

$\mathrm{Iso}(M)$	group of isometries on a Riemannian manifold
$O(N)$	the orthogonal group on $\mathbb{R}^N$
$\mathrm{SL}(N)$	special linear group on $\mathbb{R}^N$

Miscellaneous

ω_N	area of the unit $N-1$-dimensional sphere
$r(M)$	injectivity radius of a Riemannian manifold
$\mathbb{1}_A$	characteristic (index) function of a set A
id	identity mapping; identity element of a group
p'	conjugate of $p \in [1,\infty]$: $\frac{1}{p} + \frac{1}{p}' = 1$

https://doi.org/10.1515/9783110532432-011

$o_{n\to\infty}(1)$	a sequence of real numbers convergent to zero
$o^{w}_{n\to\infty}(1)$	a sequence in a Banach space weakly convergent to zero
p_s^*	critical Sobolev exponent, $p_s^* = \frac{pN}{N-sp}$ for $1 \le p < N/s$
p^*	same as p_1^*
$\mathcal{F}(u)$	Fourier transform, normalized as a unitary operator in $L^2(\mathbb{R}^N)$.
$u_n \overset{\mathcal{G}}{\rightharpoonup} u$	$\mathcal{G}$-weak convergence
$u_n \overset{\mathcal{G}}{\rightharpoondown} u$	$\mathcal{G}$-Delta convergence

Bibliography

[1] Adams, D. R., A sharp inequality of J. Moser for higher order derivatives, *Ann. Math. (2)* **128** (1988), 385–398.

[2] Adams, R. A.; Fournier, J., Sobolev spaces, 2nd edition, *Pure and applied mathematics 140*, Academic Press, 2003.

[3] Adimurthi, Existence of positive solutions of the semilinear Dirichlet problem with critical growth for the N-Laplacian, *Ann. Sc. Norm. Super. Pisa* **17** (1990), 393–413.

[4] Adimurthi; do Ó, J. M.; Tintarev, K., Cocompactness and minimizers for inequalities of Hardy–Sobolev type involving N-Laplacian, *Nonlinear Differ. Equ. Appl.* **17** (2010), 467–477.

[5] Adimurthi; Prashanth, S., Failure of Palais–Smale condition and blow-up analysis for the critical exponent problem in $\mathbf{R}^2$, *Proc. Indian Acad. Sci. Math. Sci.* **107** (1997), 283–317.

[6] Adimurthi; Sandeep, K., Existence and non-existence of the first eigenvalue of the perturbed Hardy–Sobolev operator, *Proc. R. Soc. Edinb., Sect. A* **132** (2002), 1021–1043.

[7] Adimurthi; Tintarev, C., On compactness in the Trudinger–Moser inequality, *Ann. SNS Pisa Cl. Sci. (5)* **XIII** (2014), 1–18.

[8] Adimurthi; Tintarev, C., On the Brezis–Lieb lemma without pointwise convergence, *Nonlinear Differ. Equ. Appl.* **22** (2015), 1515–1521.

[9] Adimurthi; Tintarev, C., Defect of compactness in spaces of bounded variation, *J. Funct. Anal.* **271** (2016), 37–48.

[10] Ambrosetti, A.; Rabinowitz, P., Dual variational methods in critical point theory and applications, *J. Funct. Anal.* **14** (1973), 349–381.

[11] Ambrosio, L.; Fusco, N.; Pallara, D., Functions of bounded variation and free discontinuity problems, *Oxford mathematical monographs*, Oxford University Press, 2000.

[12] Arioli, G.; Szulkin, A., A semilinear Schrödinger equation in the presence of a magnetic field, *Arch. Ration. Mech. Anal.* **170** (2003) 277–295.

[13] Bahouri, H.; Cohen, A.; Koch, G., A general wavelet-based profile decomposition in the critical embedding of function spaces, *Confluentes Matematicae* **3** (2011), 387–411.

[14] Bahouri, H.; Perelman, G., A Fourier approach to the profile decomposition in Orlicz spaces, *Math. Res. Lett.* **21** (2014), 33–54.

[15] Begout, P.; Vargas, A., Mass concentration phenomena for the L^2-critical nonlinear Schrödinger equation, *Trans. Am. Math. Soc.* **359** (2007), 5257–5282.

[16] Ben-Naoum, A. K.; Troestler, C.; Willem, M., Extrema problems with critical Sobolev exponents on unbounded domains, Nonlinear.

[17] Benci, V; Cerami, G., Positive solutions of some nonlinear elliptic problems in exterior domains, *Arch. Ration. Mech. Anal.* **99** (1987), 283–300.

[18] Benci, V.; Cerami, G., Existence of positive solutions of the equation $-\Delta u + a(x)u = u^{(N+2)(N-2)}$ in $\mathbb{R}^N$, *J. Funct. Anal.* **88** (1990), 90–117.

[19] Bergh, J.; Löfström, J., Interpolation spaces. An introduction, *Grundlehren der mathematischen Wissenschaften, 223*, Springer-Verlag, 1976.

[20] Biagini, S., Positive solutions for a semilinear equation on the Heisenberg group, *Boll. Unione Mat. Ital., B* **9** (1995), 883–900.

[21] Bianchi, G.; Chabrowski, J.; Szulkin, A., On symmetric solutions of an elliptic equation with a nonlinearity involving critical Sobolev exponent, *Nonlinear Anal., Theory Methods Appl.* **25** (1995), 41–59.

[22] Bourgain, J., Refinements of Straichartz inequality and applications to 2d-NLS with critical nonlinearity, *Int. Math. Res. Not.* (1998), 253–283.

[23] Brezis, H., Functional analysis, Sobolev spaces and partial differential equations, Springer-Verlag, 2011.

https://doi.org/10.1515/9783110532432-012

[24] Brézis, H.; Coron, J.-M., Convergence of solutions of H-systems or how to blow bubbles, *Arch. Rat. Mech. Anal.* **89** (1985), 21–56.

[25] Brézis, H.; Lieb, E., A relation between pointwise convergence of functions and convergence of functionals, *Proc. Am. Math. Soc.* **88** (1983), 486–490.

[26] Calderon, A. P., Intermediate spaces and interpolation, the complex method, *Stud. Math.* **24** (1964), 113–190.

[27] Carles, R.; Keraani, S., On the role of quadratic oscillations in nonlinear Schrödinger equation II. The L^2-critical case, *Trans. Am. Math. Soc.* **359** (2007), 33–62.

[28] Carleson, L.; Chang, S.-Y. A., On the existence of an extremal function for an inequality of J. Moser, *Bull. Sci. Math. (2)* **110** (1986), 113–127.

[29] Cerami, G., Some nonlinear elliptic problems in unbounded domains, *Milan J. Math.* **74** (2006), 47–77.

[30] Chabrowski, J., Weak convergence methods for semilinear elliptic equations, World Scientific Publishing Co., 1999.

[31] Clark, C., An embedding theorem for function spaces, *Pac. J. Math.* **19** (1966), 243–251.

[32] Coifman, R.; Lions, P.-L.; Meyer, Y.; Semmes, S., Compensated compactness and Hardy spaces, *J. Math. Pures Appl. (9)* **72** (1993), 247–286.

[33] Costa, D.; Tintarev, C., Concentration profiles for the Trudinger–Moser functional are shaped like toy pyramids, *J. Funct. Anal.* **266** (2014), 676–692.

[34] Cwikel, M., Opial's condition and cocompactness for Besov and Triebel–Lizorkin spaces, in preparation.

[35] Cwikel, M.; Tintarev, C., On interpolation of cocompact imbeddings, *Rev. Mat. Complut.* **26** (2013), 33–55.

[36] del Pino, M.; Felmer, P., Least energy solutions for elliptic equations in unbounded domains, *Proc. R. Soc. Edinb.* **126A** (1996), 195–208.

[37] Devillanova, G., Tintarev, C., Nonlinear Schrödinger equation with bounded magnetic field, ArXiv:1911.01657.

[38] Devillanova, G.; Solimini, S.; Tintarev, C., On weak convergence in metric spaces, in: Nonlinear analysis and optimization (B. S. Mordukhovich; S. Reich; A. J. Zaslavski, eds.), *Contemporary mathematics 659*, Amer. Math. Soc., Providence, RI, 2016, pp. 43–63.

[39] Druet, O; Hebey, E.; Robert, F., Blow-up theory for elliptic PDEs in Riemannian geometry, Princeton University Press, 2004.

[40] Edelstein, M., The construction of an asymptotic center with a fixed-point property, *Bull. Am. Math. Soc.* **78** (1972), 206–208.

[41] Eichhorn, J., The boundedness of connection coefficients and their derivatives, *Math. Nachr.* **152** (1991), 144–158.

[42] Ekeland, I., On the variational principle, *J. Math. Anal. Appl.* **47** (1974), 324–353.

[43] Enflo, P., A counterexample to the approximation property in Banach spaces, *Acta Math.* **130** (1973), 309–317.

[44] Esteban, M. J.; Lions, P.-L., Stationary solutions of nonlinear Schrödinger equations with an external magnetic field, in: Partial differential equations and the calculus of variations, vol. I, *Progr. nonlinear differential equations appl., 1*, Birkhäuser, 1989, pp. 401–449.

[45] Falconer, K. J., Semilinear PDEs on self-similar fractals, *Commun. Math. Phys.* **206** (1999), 235–245.

[46] Falconer, K. J.; Hu, J., Non-linear elliptical equations on the Sierpinski gasket, *J. Math. Anal. Appl.* **240** (1999), 552–573.

[47] Fieseler, K.-H.; Tintarev, K., Semilinear elliptic problems and concentration compactness on non-compact Riemannian manifolds, *J. Geom. Anal.* **13** (2003), 67–75.

[48] Figiel, T., On the moduli of convexity and smoothness, *Stud. Math.* **56** (1976), 121–155.

[49] Folland, G. B., Subelliptic estimates and function spaces on nilpotent Lie groups, *Ark. Math.* **13**, (1975) 161–207.

[50] Folland, G. B.; Stein, E. M., Estimates for the $\bar{\partial}_b$ complex and analysis on the Heisenberg group. *Commun. Pure Appl. Math.* **27** (1974), 429–522.

[51] Frazier, M.; Jawerth, B.; Weiss, G., Littlewood–Paley theory and the study of function spaces, *CBMS regional conference series on mathematics 79*, Providence, RI, 1991.

[52] Furioli, G.; Melzi, C.; Veneruso, A., Littlewood–Paley decompositions and Besov spaces on Lie groups of polynomial growth, *Math. Nachr.* **279** (2006), 1028–1040.

[53] Gallagher, I., Décomposition en profils pour les solutions des équations de Navier–Stokes, Séminaire équations aux dérivées partielles de l'Ecole Polytechnique, mai 2000.

[54] Gallier, J.; Quaintance, J., Notes on differential geometry and Lie groups, Book in progress (2017), http://www.cis.upenn.edu/~jean/gbooks/manif.html.

[55] Gallier, J.; Xu, J.; Siqueira, M., Parametric pseudo-manifolds, *Differ. Geom. Appl.* **30** (2012), 702–736.

[56] Gallot, S.; Hulin, D.; Lafontain, J., Riemannian geometry, Springer-Verlag, Berlin, Heidelberg, 1987.

[57] Garofalo, N.; Vassilev, D., Symmetry properties of positive entire solutions of Yamabe-type equations on groups of Heisenberg type, *Duke Math. J.* **106** (2001), 411–448.

[58] Gérard, P., Description du défaut de compacité de l'injection de Sobolev. *ESAIM Control Optim. Calc. Var.*, **3** (1998), 213–233 (electronic).

[59] Gérard, P.; Meyer, Y.; Oru, F., Inégalités de Sobolev precisées, in: Séminaire EDP 1996–1997, Exposé 4, Ecole Polytechnique, 1996.

[60] Gidas, B.; Ni, W.-M.; Nirenberg, L., Symmetry and related properties via the maximum principle, *Commun. Math. Phys.* **68** (1979), 209–243.

[61] Ginibre, J.; Velo, G., Smoothing properties and retarded estimates for some dispersive evolution equations, *Commun. Math. Phys.* **144** (1992), 163–188.

[62] Goebel, K.; Kirk, W. A., Topics in metric fixed point theory, *Cambridge studies in advanced mathematics*, Cambridge University Press, 1990.

[63] Goebel, K.; Reich, S., Uniform convexity, hyperbolic geometry, and nonexpansive mappings, *Monogrpahs and Textbooks in Pure and Applied Mathematics* **83**, Marcel Dekker, 1984.

[64] Große, N.; Schneider, C., Sobolev spaces on Riemannian manifolds with bounded geometry: general coordinates and traces, *Math. Nachr.* **286** (2013), 1586–1613.

[65] Hebey, E., Sobolev spaces on Riemannian manifolds, *Lecture notes in mathematics 1365*, Springer-Verlag, 1996.

[66] Hebey, E.; Vaugon, M., Sobolev spaces in the presence of symmetry, *J. Math. Pures Appl.* **76** (1997), 859–881.

[67] Huang, Q.-Z.; Li, A.-J., Optimal Sobolev norms in the affine class, *J. Math. Anal. Appl.* **436** (2016), 568–585.

[68] Jaffard, S., Analysis of the lack of compactness in the critical Sobolev embeddings, *J. Funct. Anal.* **161** (1999), 384–396.

[69] Jawerth, B., Some observations on Besov and Lizorkin–Triebel spaces, *Math. Scand.* **40** (1977), 94–104.

[70] Jones, P. W.; Journe, J.-L., On weak convergence in $H^1(\mathbf{R}^d)$, *Proc. Am. Math. Soc.* **120** (1994), 137–138.

[71] Keraani, S., On the defect of compactness for the Strichartz estimates for the Schrödinger equations, *J. Differ. Equ.* **175** (2001), 353–392.

[72] Kigami, Analysis on fractals, *Cambridge tracts in mathematics 143*, Cambridge University Press, Cambridge, 2001, 226 pp.

[73] Killip, R.; Visan, M., Nonlinear Schrödinger equations at critical regularity, in: Evolution equations, *Clay math. Proc. 17*, Amer. Math. Soc., Providence, RI, 2013, pp. 325–437.

[74] Kuczumow, T., An almost convergence and its applications, *Ann. Univ. Mariae Curie-Skłodowska, Sect. A* **32** (1978), 79–88.

[75] Kwong, M. K., Uniqueness of positive solutions of $\Delta u - u + u^p = 0$, *Arch. Ration. Mech. Anal.* **105** (1989), 243–266.

[76] Lieb, E., On the lowest eigenvalue of the Laplacian for the intersection of two domains, *Invent. Math.* **74** (1983), 441–448.

[77] Lieb, E. H.; Loss, M., Analysis, 2nd edition, *Graduate studies in mathematics 14*, American Mathematical Society, 2001.

[78] Lim, T. C., Remarks on some fixed point theorems, *Proc. Am. Math. Soc.* **60** (1976), 179–182.

[79] Lindenstrauss, J.; Tzafriri, L., Classical Banach spaces. I. Sequence spaces, *Ergebnisse der Mathematik und ihrer Grenzgebiete [Results in mathematics and related areas] 92*, Springer-Verlag, 1977.

[80] Lindenstrauss, J.; Tzafriri, L., Classical Banach spaces. II. Sequence spaces, *Ergebnisse der Mathematik und ihrer Grenzgebiete [Results in mathematics and related areas] 97*, Springer-Verlag, 1979.

[81] Lions, P.-L., Symétrie et compacité dans les espaces de Sobolev, *J. Funct. Anal.* **49** (1982), 315–334.

[82] Lions, P.-L., On the concentration-compactness principle, in: Contributions to nonlinear partial differential equations (Madrid, 1981), *Res. notes in math. 89*, Pitman, 1983, pp. 225–238.

[83] Lions, P.-L., The concentration-compactness principle in the calculus of variations. The locally compact case, parts 1 and 2, *Ann. Inst. Henri Poincaré, Anal. Non Linéaire* **1** (1984), 109–145 and 223–283.

[84] Lions, P.-L., The concentration-compactness principle in the calculus of variations. The limit case, parts 1 and 2, *Rev. Mat. Iberoam.* **1.1** (1985), 145–201 and **1.2** (1985), 45–121.

[85] Lions, P.-L., Solutions of Hartree–Fock equations for Coulomb systems, *Commun. Math. Phys.* **109** (1987), 33–97.

[86] Lutwak, E.; Yang, D.; Zhang, G., Sharp affine L_p Sobolev inequalities, *J. Differ. Geom.* **62** (2002), 17–38.

[87] Mariş, M., Profile decomposition for sequences of Borel measures, ArXiv:140.6125.

[88] Martinazzi, L., Fractional Adams–Moser–Trudinger type inequalities, *Nonlinear Anal.* **127** (2015), 263–278.

[89] Matzeu, M., Mountain pass and linking type solutions for semilinear Dirichlet forms, in: Recent trends in nonlinear analysis, *Progr. nonlinear differential equations appl. 40*, Birkhäuser, 2000, pp. 217–231.

[90] Maz'ya, V. G., Certain integral inequalities for functions of several variables. (Russian), in: Problems of mathematical analysis, no. 3: integral and differential operators. Differential equations (Russian), Izdat. Leningrad. Univ., Leningrad, 1972, pp. 33–68.

[91] Merle, R.; Vega, L., Compactness at blow-up time for L^2 solutions of the critical nonlinear Schrödinger equation in 2D, *Int. Math. Res. Not.* **8** (1998), 399–425.

[92] Meyer, Y., Wavelets and operators, Cambridge University Press, 1993.

[93] Mosco, U., Dirichlet forms and self-similarity, in: New directions in Dirichlet forms, *AMS/IP studies in advanced math. 8*, Amer. Math. Soc., Providence, RI, 1998, pp. 117–155.

[94] Moser, J., A sharp form of an inequality by N. Trudinger, *Indiana Univ. Math. J.* **20** (1971), 1077–1092.

[95] Munoz Flores, A., Nardulli, S. Generalized compactness for finite perimeter sets and applications to the isoperimetric problem, ArXiv:1504.05104.

[96] Opial, Z., Weak convergence of the sequence of successive approximations for nonexpansive mappings, *Bull. Am. Math. Soc.* **73** (1967), 591–597.

[97] Ozawa, T., On critical cases of Sobolev's inequalities, *J. Funct. Anal.* **127** (1995), 259–269.

[98] Palatucci, G.; Pisante, A., Improved Sobolev embeddings, profile decomposition, and concentration-compactness for fractional Sobolev spaces, *Calc. Var.* **50** (2014), 799–829.

[99] Petersen, P., Riemannian Geometry (third ed.), *Graduate texts in mathematics 171*, Springer Verlag, 2006.

[100] Sandeep, K; Tintarev, C., A subset of Caffarelli–Kohn–Nirenberg inequalities in the hyperbolic space $\mathbb{H}^N$, *Ann. Mat. Pura Appl.* **196** (2017), 2005–2021.

[101] Sandeep, K.; Tintarev, C., Profile decomposition of Struwe–Solimini for manifolds with bounded geometry, arXiv:1901.10427.

[102] Schindler, I.; Tintarev, C., Compactness properties and ground states for the affine Laplacian, *Calc. Var.* **57** (2018) https://doi.org/10.1007/s00526-018-1338-6 (online).

[103] Schindler, I.; Tintarev, C., On compactness properties in the affine Sobolev inequality, *Pure Appl. Funct. Anal.*, to appear.

[104] Schindler, I.; Tintarev, K., An abstract version of the concentration compactness principle, *Rev. Mat. Complut.* **15** (2002), 1–20.

[105] Schindler, I.; Tintarev, K., Semilinear subelliptic problems without compactness on Lie groups, *Nonlinear Differ. Equ. Appl.* **11** (2004), 299–309.

[106] Schindler, I.; Tintarev, K., Semilinear elliptic equations on fractal blow-ups, *J. Math. Anal. Appl.* **352** (2009), 57–64.

[107] Shubin, M. A., Spectral theory of elliptic operators on noncompact manifolds, in: Méthodes semi-classiques, vol. 1 (Nantes, 1991), *Astérisque 207*, (1992), pp. 35–108.

[108] Skrzypczak, L., Atomic decompositions on manifolds with bounded geometry, *Forum Math.* **10** (1998), 19–38.

[109] Skrzypczak, L.; Tintarev, C., A geometric criterion for compactness of invariant subspaces, *Arch. Math.* **101** (2013), 259–268.

[110] Skrzypczak, L.; Tintarev, C., Defect of compactness for Sobolev spaces on manifolds with bounded geometry, *Ann. Sc. Norm. Super. Pisa*, to appear.

[111] Skrzypczak, L.; Tintarev, C., On compact subsets of Sobolev spaces on manifolds, in preparation.

[112] Solimini, S., A note on compactness-type properties with respect to Lorentz norms of bounded subsets of a Sobolev space, *Ann. Inst. Henri Poincaré, Anal. Non Linéaire* **12**, 319–337.

[113] Solimini, S.; Tintarev, C., Concentration analysis in Banach spaces, *Commun. Contemp. Math.* **18** (2016), 1550038 (33 pages).

[114] Staples, J., Fixed point theorems in uniformly rotund metric spaces, *Bull. Aust. Math. Soc.* **14** (1976), 181–192.

[115] Strauss, W., Existence of solitary waves in higher dimensions, *Commun. Math. Phys.* **55** (1977), 149–162.

[116] Strichartz, R., Fractals in the large, *Can. J. Math.* **50** (1998), 638–657.

[117] Strichartz, R., Function spaces on fractals, *J. Funct. Anal.* **198** (2003), 43–83.

[118] Strichartz, R. S., Restriction of Fourier transform to quadratic surfaces and deay of solutions of wave equations, *Duke Math. J.* **44** (1977), 705–714.

[119] Struwe, M., A global compactness result for elliptic boundary value problems involving limiting nonlinearities, *Math. Z.* **187** (1984), 511–517.

[120] Struwe, M., Variational methods. Applications to nonlinear partial differential equations and Hamiltonian systems, 4th edition, *Ergebnisse der Mathematik und ihrer Grenzgebiete. 3. Folge. A Series of modern surveys in mathematics 34*, Springer-Verlag, 2000.

[121] Talenti, G., Best constant in Sobolev inequality, *Ann. Mat. Pura Appl. (4)* **110** (1976), 353–372.

[122] Tao, T., Nonlinear dispersive equations: local and global analysis, *CBMS regional conference series in mathematics 106*, American Mathematical Society, Providence, RI, 2006.

[123] Tao, T., A pseudoconformal compactification of the nonlinear Schrödinger equation and applications, *N.Y. J. Math.* **15** (2009), 265–282.

[124] Tartar, L., The compensated compactness method applied to systems of conservation laws, in: Systems of nonlinear partial differential equations, *Proc. nato advanced study inst. 111*, Oxford 1982 (J. M. Ball, ed.) Reidel, Dordrecht, 1983, pp. 263–285.

[125] Tintarev, C., Cocompact imbeddings and structure of weakly convergent sequences, in: Sobolev spaces in mathematics, volume 1: Sobolev type inequalities (V. Maz'ya; V. Isakov, eds.), *International mathematical series 8*, Springer-Verlag, 2009, pp. 361–375.

[126] Tintarev, C., Four proofs of cocompactness for Sobolev embeddings, in: Functional analysis, harmonic analysis and image processing: a collection of papers in honor of Björn Jawerth, *Contemporary mathematics 693*, Amer. Math. Soc., Providence, RI, 2017, pp. 321–329.

[127] Tintarev, K.; Fieseler, K.-H., Concentration compactness: functional-analytic grounds and applications, Imperial College Press, 2007.

[128] Triebel, H., Theory of function spaces, Birkhäuser, 1983.

[129] Triebel, H., Theory of function spaces II, Birkhäuser, 1992.

[130] Trudinger, N., On imbeddings into Orlicz spaces and some applications, *J. Math. Mech.* **17** (1967), 473–483.

[131] Van Dulst, D., Equivalent norms and the fixed point property for nonexpansive mappings, *J. Lond. Math. Soc. (2)* **25** (1982), 139–144.

[132] Varopoulos, N. Th., Analysis on Lie groups, *J. Funct. Anal.* **76** (1988), 346–410.

[133] Willem, M., Minimax theorems, Birkhäuser, 1996.

[134] Willem, M., Functional analysis, fundamentals and applications. Springer-Verlag 2013.

[135] Yudovich, V. I., Some estimates connected with integral operators and with solutions of elliptic equations, *Sov. Math. Dokl.* **2** (1961), 746–749.

[136] Zhang, G., *The affine Sobolev inequality*, *J. Differ. Geom.* **53** (1999), 183–202.

[137] Zhu, X.; Cao, D.-M., The concentration-compactness principle in nonlinear elliptic equations, *Acta Math. Sci.* (1989) **9**, 307–328.

Index

De Gruyter Series in Nonlinear Analysis and Applications

Volume 32
Peter I. Kogut, Olha P. Kupenko
Approximation Methods in Optimization of Nonlinear Systems, 2019
ISBN 978-3-11-066843-8, e-ISBN (PDF) 978-3-11-066852-0, e-ISBN (EPUB) 978-3-11-066859-9

Volume 31
Pablo Blanc, Julio Daniel Rossi
Game Theory and Partial Differential Equations, 2019
ISBN 978-3-11-061925-6, e-ISBN (PDF) 978-3-11-062179-2, e-ISBN (EPUB) 978-3-11-061932-4

Volume 30
Lucio Damascelli, Filomena Pacella
Morse Index of Solutions of Nonlinear Elliptic Equations, 2019
ISBN 978-3-11-053732-1, e-ISBN (PDF) 978-3-11-053824-3, e-ISBN (EPUB) 978-3-11-053824-3

Volume 29
Rafael Ortega
Periodic Differential Equations in the Plane. A Topological Perspective, 2019
ISBN 978-3-11-055040-5, e-ISBN (PDF) 978-3-11-055116-7, e-ISBN (EPUB) 978-3-11-055042-9

Volume 28
Dung Le
Strongly Coupled Parabolic and Elliptic Systems.
Existence and Regularity of Strong and Weak Solutions, 2018
ISBN 978-3-11-060715-4, e-ISBN (PDF) 978-3-11-060876-2, e-ISBN (EPUB) 978-3-11-060717-8

Volume 27
Maxim Olegovich Korpusov, Alexey Vital'evich Ovchinnikov, Alexey Georgievich Sveshnikov, Egor Vladislavovich Yushkov
Blow-Up in Nonlinear Equations of Mathematical Physics. Theory and Methods, 2018
ISBN 978-3-11-060108-4, e-ISBN (PDF) 978-3-11-060207-4, e-ISBN (EPUB) 978-3-11-059900-8

Bei Fragen zur Produktsicherheit wenden Sie sich bitte an:
If you have any questions regarding product safety,
please contact:

Walter de Gruyter GmbH
Genthiner Straße 13
10785 Berlin
productsafety@degruyterbrill.com